Probability, Random Processes,
and Ergodic Properties

Robert M. Gray

Probability, Random Processes, and Ergodic Properties

Springer-Verlag
New York Berlin Heidelberg
London Paris Tokyo

Robert M. Gray
Department of Electrical Engineering
Stanford University
Stanford, CA 94305, USA

Library of Congress Cataloging-in-Publication Data
Gray, Robert M., 1943–
Probability, random processes, and ergodic properties.
Bibliography: p.
Includes index.
1. Probabilities. 2. Measure theory. 3. Stochastic processes. I. Title.
QA273.G683 1987 519.2 87–28429

Camera-ready copy prepared by the author using EQN/TRoff.
Printed and bound by Edward Brothers, Inc., Ann Arbor, Michigan.
Printed in the United States of America.

9 8 7 6 5 4 3 2 1

ISBN 0-387-96655-2 Springer-Verlag New York Berlin Heidelberg
ISBN 3-540-96655-2 Springer-Verlag Berlin Heidelberg New York

This book is affectionately dedicated to

Elizabeth Dubois Jordan Gray

and to the memory of

R. Adm. Augustine Heard Gray, U.S.N.
1888-1981

Sara Jean Dubois

and

William "Billy" Gray
1750-1825

CONTENTS

PREFACE

This book has been written for several reasons, not all of which are academic. This material was for many years the first half of a book in progress on information and ergodic theory. The intent was and is to provide a reasonably self-contained advanced treatment of measure theory, probability theory, and the theory of discrete time random processes with an emphasis on general alphabets and on ergodic and stationary properties of random processes that might be neither ergodic nor stationary. The intended audience was mathematically inclined engineering graduate students and visiting scholars who had not had formal courses in measure theoretic probability. Much of the material is familiar stuff for mathematicians, but many of the topics and results have not previously appeared in books.

The original project grew too large and the first part contained much that would likely bore mathematicians and discourage them from the second part. Hence I finally followed the suggestion to separate the material and split the project in two. The original justification for the present manuscript was the pragmatic one that it would be a shame to waste all the effort thus far expended. A more idealistic motivation was that the presentation had merit as filling a unique, albeit small, hole in the literature. Personal experience indicates that the intended audience rarely has the time to take a complete course in measure and probability theory in a mathematics or statistics department, at least not before they need some of the material in their research. In addition, many of the existing mathematical texts on the subject are hard for this audience to follow, and the emphasis is not well matched to engineering applications. A notable exception is Ash's excellent text [1], which was likely influenced by his original training as an electrical engineer. Still, even that text devotes little effort to ergodic theorems, perhaps the most fundamentally important family of results for applying probability theory to real problems. In addition, there are many other special topics that are given little space (or none at all) in most texts on advanced probability and random processes.

Examples of topics developed in more depth here than in most existing texts are the following:

(i) *Random processes with standard alphabets* We develop the theory of standard spaces as a model of quite general process alphabets. Although not as general (or abstract) as often considered by probability theorists, standard spaces have useful structural properties that simplify the proofs of some general results and yield additional results that may not hold in the more general abstract case. Examples of results holding for standard alphabets that have not been proved in the general abstract case are the Kolmogorov extension theorem, the ergodic decomposition, and the existence of regular conditional probabilities. In fact, Blackwell [2] introduced the notion of a Lusin space, a structure closely related to a standard space, in order to avoid known examples of probability spaces where the Kolmogorov extension theorem does not hold and regular conditional probabilities do not exist. Standard spaces include the common models of finite alphabets (digital processes) and real alphabets as well as more general complete separable metric spaces (Polish spaces). Thus they include many function spaces, Euclidean vector spaces, two-dimensional image intensity rasters, etc. The basic theory of standard Borel spaces may be found in the elegant text of Parthasarathy [12], and treatments of standard spaces and the related Lusin and Suslin spaces may be found in Christensen [4], Schwartz [13] , Bourbaki [3], and Cohn [5]. We here provide a different and more coding oriented development of the basic results and attempt to separate clearly the properties of standard spaces, which are useful and easy to manipulate, from the demonstrations that certain spaces are standard, which are more complicated and can be skipped. Thus, unlike in the traditional treatments, we define and study standard spaces first from a purely probability theory point of view and postpone the topological metric space considerations until later.

(ii) *Nonstationary and nonergodic processes* We develop the theory of asymptotically mean stationary processes and the ergodic decomposition in order to model many physical processes better than can traditional stationary and ergodic processes. Both topics are virtually absent in all books on random processes, yet they are fundamental to understanding the limiting behavior of nonergodic and nonstationary processes. Both topics are considered in Krengel's excellent book on ergodic theorems [10], but the treatment here is more detailed and in greater depth. We consider both the common two-sided processes, which are considered to have been producing outputs forever, and the more difficult one-sided processes, which better model processes that are "turned on" at some specific time and which exhibit transient behavior.

(iii) *Ergodic properties and theorems* We develop the notion of time averages along with that of probabilistic averages to emphasize their similarity and to demonstrate many of the implications of the existence of limiting sample averages. We prove the ergodic theorem theorem for the general case of asymptotically mean stationary processes. In fact, it is shown that asymptotic mean stationarity is both sufficient and necessary for the classical pointwise or almost everywhere ergodic theorem to hold. We also prove the subadditive ergodic theorem of Kingman [9], which is useful for studying the limiting behavior of certain measurements on random processes that are not simple arithmetic averages. The proofs are based on recent simple proofs of the ergodic theorem developed by Ornstein and Weiss [11], Katznelson and Weiss [8], Jones [7], and Shields [14]. These proofs use coding arguments reminiscent of information and communication theory rather than the traditional (and somewhat tricky) maximal ergodic theorem. We consider the interrelations of stationary and ergodic properties of processes that are stationary or ergodic with respect to block shifts, that is, processes that produce stationary or ergodic vectors rather than scalars.

(iv) *Process distance measures* We develop measures of the "distance" between random processes. Such results quantify how "close" one process is to another and are useful for considering spaces of random processes. These in turn provide the means of proving the ergodic decomposition of certain functionals of random processes and of characterizing how close or different the long term behavior of distinct random processes can be expected to be.

Having described the topics treated here that are lacking in most texts, we admit to the omission of many topics usually contained in advanced texts on random processes or second books on random processes for engineers. The most obvious omission is that of continuous time random processes. A variety of excuses explain this: The advent of digital systems and sampled-data systems has made discrete time processes at least equally important as continuous time processes in modeling real world phenomena. The shift in emphasis from continuous time to discrete time in texts on electrical engineering systems can be verified by simply perusing modern texts. The theory of continuous time processes is inherently more difficult than that of discrete time processes. It is harder to construct the models precisely and much harder to demonstrate the existence of measurements on the models, e.g., it is harder to prove that limiting integrals exist than limiting sums. One can approach continuous time models via discrete time models by letting the outputs be pieces of waveforms. Thus, in a sense, discrete time systems can be used as a building block for continuous time systems.

Another topic clearly absent is that of spectral theory and its applications to estimation and prediction. This omission is a matter of taste and there are many books on the subject.

A further topic not given the traditional emphasis is the detailed theory of the most popular particular examples of random processes: Gaussian and Poisson processes. The emphasis of this book is on general properties of random processes rather than the specific properties of special cases.

The final noticeably absent topic is martingale theory. Martingales are only briefly discussed in the treatment of conditional expectation. My excuse is again that of personal taste. In addition, this powerful theory is simply not required in the intended sequel to this book on information and ergodic theory.

The book's original goal of providing the needed machinery for a book on information and ergodic theory remains. That book will rest heavily on this book and will only quote the needed material, freeing it to focus on the information measures and their ergodic theorems and on source and channel coding theorems. In hindsight, this manuscript also serves an alternative purpose. I have been approached by engineering students who have taken a master's level course in random processes using my book with Lee Davisson [6] and who are interested in exploring more deeply into the underlying mathematics that is often referred to, but rarely exposed. This manuscript provides such a sequel and fills in many details only hinted at in the lower level text.

As a final, and perhaps less idealistic, goal, I intended in this book to provide a catalogue of many results that I have found need of in my own research together with proofs that I could follow. This is one goal wherein I can judge the success; I often find myself consulting these notes to find the conditions for some convergence result or the reasons for some required assumption or the generality of the existence of some limit.

If the manuscript provides similar service for others, it will have succeeded in a more global sense.

The book is aimed at graduate engineers and hence does not assume even an undergraduate mathematical background in functional analysis or measure theory. Hence topics from these areas are developed from scratch, although the developments and discussions often diverge from traditional treatments in mathematics texts. Some mathematical sophistication is assumed for the frequent manipulation of deltas and epsilons, and hence some background in elementary real analysis or a strong calculus knowledge is required.

Acknowledgments

The research in information theory that yielded many of the results and some of the new proofs for old results in this book was supported by the National Science Foundation. Portions of the research and much of the early writing were supported by a fellowship from the John Simon Guggenheim Memorial Foundation. The book was written using the eqn and troff utilities on several UNIX (an AT&T trademark) systems supported by the Industrial Affiliates Program of the Stanford University Information Systems Laboratory.

The book benefited greatly from comments from numerous students and colleagues through many years: most notably Paul Shields, Lee Davisson, John Kieffer, Dave Neuhoff, Don Ornstein, Bob Fontana, Jim Dunham, Farivar Saadat, Mari Ostendorf, Michael Sabin, Paul Algoet, Wu Chou, Phil Chou, and Tom Lookabaugh. They should not be blamed, however, for any mistakes I have made in implementing their suggestions.

I would also like to acknowledge my debt to Al Drake for introducing me to elementary probability theory and to Tom Pitcher for introducing me to measure theory. Both are extraordinary teachers.

Finally, I would like to apologize to Lolly, Tim, and Lori for all the time I did not spend with them while writing this book.

REFERENCES

1. R. B. Ash, *Real Analysis and Probability*, Academic Press, New York, 1972.

2. D. Blackwell, ''On a class of probability spaces,'' *Proc. 3rd Berkeley Symposium on Math. Sci. and Prob.*, vol. II, pp. 1-6, Univ. California Press, Berkeley, 1956.
3. N. Bourbaki, *Elements de Mathematique, Livre VI, Integration,* Hermann, Paris, 1956-1965.
4. J. P. R. Christensen, *Topology and Borel Structure,* Mathematics Studies 10, North-Holland/American Elsevier, New York, 1974.
5. D. C. Cohn, *Measure Theory,* Birkhauser, New York, 1980.
6. R. M. Gray and L. D. Davisson, *Random Processes: A Mathematical Approach for Engineers,* Prentice-Hall, Englewood Cliffs, New Jersey, 1986.
7. R. Jones, ''New proof for the maximal ergodic theorem and the Hardy-Littlewood maximal inequality,'' *Proc. AMS,* vol. 87, pp. 681-684, 1983.
8. I. Katznelson and B. Weiss, ''A simple proof of some ergodic theorems,'' *Israel Journal of Mathematics*, vol. 42, pp. 291-296, 1982.
9. J. F. C. Kingman, ''The ergodic theory of subadditive stochastic processes,'' *Ann. Probab.*, vol. 1, pp. 883-909, 1973.
10. U. Krengel, *Ergodic Theorems,* De Gruyter Series in Mathematics, De Gruyter, New York, 1985.
11. D. Ornstein and B. Weiss, ''The Shannon-McMillan-Breiman theorem for a class of amenable groups,'' *Israel J. of Math*, vol. 44, pp. 53-60, 1983.
12. K. R. Parthasarathy, *Probability Measures on Metric Spaces,* Academic Press, New York, 1967.
13. L. Schwartz, *Radon Measures on Arbitrary Topological Spaces and Cylindrical Measures,* Oxford University Press, Oxford, 1973.
14. P. C. Shields, ''The ergodic and entropy theorems revisited,'' *IEEE Transactions on Information Theory*, vol. IT-33, pp. 263-266, March 1987.

1

PROBABILITY AND RANDOM PROCESSES

1.1 INTRODUCTION

In this chapter we develop basic mathematical models of discrete time random processes. Such processes are also called discrete time stochastic processes, information sources, and time series. Physically a random process is something that produces a succession of symbols called ''outputs'' in a random or nondeterministic manner. The symbols produced may be real numbers such as produced by voltage measurements from a transducer, binary numbers as in computer data, two-dimensional intensity fields as in a sequence of images, continuous or discontinuous waveforms, and so on. The space containing all of the possible output symbols is called the *alphabet* of the random process, and a random process is essentially an assignment of a probability measure to events consisting of sets of sequences of symbols from the alphabet. It is useful, however, to treat the notion of time explicitly as a transformation of sequences produced by the random process. Thus in addition to the common random process model we shall also consider modeling random processes by dynamical systems as considered in ergodic theory.

1.2 PROBABILITY SPACES AND RANDOM VARIABLES

The basic tool for describing random phenomena is probability theory. The history of probability theory is long, fascinating, and rich (see, for example, Maistrov [16]); its modern origins begin with the axiomatic development of Kolmogorov in the 1930s [13]. Notable landmarks in the subsequent development of the theory (and often still good reading) are the books by Cramér [6], Loève [15], and Halmos [10]. Modern treatments that I have found useful for background and reference are Ash [1], Breiman [3], Chung [5], and the treatment of probability theory in Billingsley [2].

Measurable Space

A measurable space $(\Omega,\boldsymbol{B})$ is a pair consisting of a sample space Ω together with a σ-field $\boldsymbol{B}$ of subsets of Ω (also called the event space). A σ-field or σ-algebra $\boldsymbol{B}$ is a collection of subsets of Ω with the following properties:

$$\Omega \in \boldsymbol{B}. \tag{1.2.1}$$

$$\text{If } F \in \boldsymbol{B}, \text{ then } F^c = \{\omega: \omega \notin F\} \in \boldsymbol{B}. \tag{1.2.2}$$

$$\text{If } F_i \in \boldsymbol{B};\ i = 1,2,\ldots, \text{ then } \bigcup F_i \in \boldsymbol{B}. \tag{1.2.3}$$

From de Morgan's "laws" of elementary set theory it follows that also

$$\bigcap_{i=1}^{\infty} F_i = (\bigcup_{i=1}^{\infty} F_i^c)^c \in \boldsymbol{B}.$$

An event space is a collection of subsets of a sample space (called events by virtue of belonging to the event space) such that any countable sequence of set theoretic operations (union, intersection, complementation) on events produces other events. Note that there are two extremes: the largest possible σ-field of Ω is the collection of all subsets of Ω (sometimes called the power set), and the smallest possible σ-field is $\{\Omega,\emptyset\}$, the entire space together with the null set $\emptyset = \Omega^c$ (called the *trivial space*).

If instead of the closure under countable unions required by (1.2.3), we only require that the collection of subsets be closed under finite unions, then we say that the collection of subsets is a *field*.

Although the concept of a field is simpler to work with, a σ-field possesses the additional important property that it contains all of the limits of sequences of sets in the collection. That is, if F_n, $n = 1,2,...$ is an increasing sequence of sets in a σ-field, that is, if $F_{n-1} \subset F_n$ and if $F = \bigcup_{n=1}^{\infty} F_n$ (in which case we write $F_n \uparrow F$ or $\lim_{n\to\infty} F_n = F$), then also F is contained in the σ-field. This property may not hold true for fields, that is, fields need not contain the limits of sequences of field elements. Note that if a field has the property that it contains all increasing sequences of its members, then it is also a σ-field. In a similar fashion we can define decreasing sets: If F_n decreases to F in the sense that $F_{n+1} \subset F_n$ and $F = \bigcap_{n=1}^{\infty} F_n$, then we write $F_n \downarrow F$. If $F_n \in \boldsymbol{B}$ for all n, then $F \in \boldsymbol{B}$.

Because of the importance of the notion of converging sequences of sets, we note a generalization of the definition of a σ-field that emphasizes such limits: A collection $\boldsymbol{M}$ of subsets of Ω is called a *monotone class* if it has the property that if $F_n \in \boldsymbol{M}$ for $n = 1,2, \cdots$ and either $F_n \uparrow F$ or $F_n \downarrow F$, then also $F \in \boldsymbol{M}$. Clearly a σ-field is a monotone class, but the reverse need not be true. If a field is also a monotone class, however, then it must be a σ-field.

A σ-field is sometimes referred to as a Borel field in the literature and the resulting measurable space called a Borel space. We will reserve this nomenclature for the more common use of these terms as the special case of a σ-field having a certain topological structure that will be developed later.

Probability Spaces

A *probability space* $(\Omega,\boldsymbol{B},P)$ is a triple consisting of a sample space Ω, a σ-field $\boldsymbol{B}$ of subsets of Ω, and a probability measure P defined on the σ-field; that is, $P(F)$ assigns a real number to every member F of $\boldsymbol{B}$ so that the following conditions are satisfied:

Nonnegativity:

$$P(F) \geq 0, \text{ all } F \in \boldsymbol{B}, \tag{1.2.4}$$

Normalization:

$$P(\Omega) = 1. \tag{1.2.5}$$

Countable Additivity:

If $F_i \in \boldsymbol{B}$, $i = 1,2,...$ are disjoint, then

$$P(\bigcup_{i=1}^{\infty} F_i) = \sum_{i=1}^{\infty} P(F_i) . \tag{1.2.6}$$

A set function P satisfying only (1.2.4) and (1.2.6) but not necessarily (1.2.5) is called a *measure*, and the triple $(\Omega,\boldsymbol{B},P)$ is called a *measure space.* Since the probability measure is defined on a σ-field, such countable unions of subsets of Ω in the σ-field are also events in the σ-field. A set function satisfying (1.2.6) only for finite sequences of disjoint events is said to be *additive* or *finitely additive.*

A straightforward exercise provides an alternative characterization of a probability measure involving only finite additivity, but requiring the addition of a continuity requirement: a set function P defined on events in the σ-field of a measurable space $(\Omega,\boldsymbol{B})$ is a probability measure if (1.2.4) and (1.2.5) hold, if the following condition is met:

Finite Additivity:

if $F_i \in \boldsymbol{B}$, $i = 1,2,\ldots, n$ are disjoint, then

$$P(\bigcup_{i=1}^{n} F_i) = \sum_{i=1}^{n} P(F_i) , \tag{1.2.7}$$

and if $G_n \downarrow \varnothing$ (the empty or null set), that is, if $G_{n+1} \subset G_n$, all n, and $\bigcap_{n=1}^{\infty} G_n = \varnothing$, then we have

Continuity at $\varnothing$:

$$\lim_{n\to\infty} P(G_n) = 0. \tag{1.2.8}$$

The equivalence of continuity and countable additivity is easily seen by making the correspondence $F_n = G_n - G_{n-1}$ and observing that countable additivity for the F_n will hold if and only if the continuity relation holds for the G_n . It is also easy to see that condition (1.2.8) is equivalent to two other forms of continuity:

Continuity from Below:

$$\text{If } F_n \uparrow F, \text{ then } \lim_{n\to\infty} P(F_n) = P(F) . \tag{1.2.9}$$

Continuity from Above:

$$\text{If } F_n \downarrow F , \text{ then } \lim_{n\to\infty} P(F_n) = P(F) . \tag{1.2.10}$$

Thus a probability measure is an additive, nonnegative, normalized set function on a σ-field or event space with the additional property that if a sequence of sets converges to a limit set, then the corresponding probabilities must also converge.

If we wish to demonstrate that a set function P is indeed a valid probability measure, then we must show that it satisfies the preceding properties (1.2.4), (1.2.5), and either (1.2.6) or (1.2.7) and one of (1.2.8), (1.2.9), or (1.2.10).

Observe that if a set function satisfies (1.2.4), (1.2.5), and (1.2.7), then for any disjoint sequence of events $\{F_i\}$ and any n

$$P(\bigcup_{i=0}^{\infty} F_i) = P(\bigcup_{i=0}^{n} F_i) + P(\bigcup_{i=n+1}^{\infty} F_i)$$

$$\geq P(\bigcup_{i=0}^{n} F_i) = \sum_{i=0}^{n} P(F_i) .$$

and hence we have taking the limit as $n \rightarrow \infty$ that

$$P(\bigcup_{i=0}^{\infty} F_i) \geq \sum_{i=0}^{\infty} P(F_i). \tag{1.2.11}$$

Thus to prove that P is a probability measure one must show that the preceding inequality is in fact an equality.

Random Variables.

Let $(A, \boldsymbol{B}_A)$ denote another measurable space. A *random variable* or *measurable function* defined on $(\Omega, \boldsymbol{B})$ and taking values in $(A, \boldsymbol{B}_A)$ is a mapping or function $f: \Omega \rightarrow A$ with the property that

$$\text{if } F \in \boldsymbol{B}_A, \text{ then } f^{-1}(F) = \{\omega: f(\omega) \in F\} \in \boldsymbol{B} . \tag{1.2.12}$$

The name "random variable" is commonly associated with the special case where A is the real line and $\boldsymbol{B}$ the Borel field (which we shall later define) and occasionally a more general sounding name such as "random object" is used for a measurable function to include implicitly random variables (A the real line), random vectors (A a Euclidean space), and random processes (A a sequence or waveform space). We will use the terms "random variable" in the more general sense.

A random variable is just a function or mapping with the property that inverse images of "output events" determined by the random variable are events in the original measurable space. This simple property ensures that the output of the random variable will inherit its own probability measure. For example, with the probability measure P_f defined by

$$P_f(B) = P(f^{-1}(B)) = P(\omega: f(\omega) \in B) \; ; B \in \boldsymbol{B}_A ,$$

$(A, \boldsymbol{B}_A, P_f)$ becomes a probability space since measurability of f and elementary set theory ensure that P_f is indeed a probability measure. The induced probability measure P_f is called the *distribution* of the random variable f. The measurable space $(A, \boldsymbol{B}_A)$ or, simply, the sample space A is called the alphabet of the random variable f. We shall occasionally also use the notation Pf^{-1} which is a mnemonic for the relation $Pf^{-1}(F) = P(f^{-1}(F))$ and which is less awkward when f itself is a function with a complicated name, e.g., $\Pi_{\boldsymbol{I} \to \boldsymbol{M}}$.

If the alphabet A of a random variable f is not clear from context, then we shall refer to f as an A-valued random variable. If f is a measurable function from $(\Omega, \boldsymbol{B})$ to $(A, \boldsymbol{B}_A)$, we will say that f is $\boldsymbol{B}/\boldsymbol{B}_A$-measurable if the σ-fields are not clear from context.

Exercises

1. Set theoretic difference is defined by $F - G = F \cap G^c$ and symmetric difference is defined by

$$F \Delta G = (F \cap G^c) \cup (F^c \cap G) .$$

Show that

$$F \Delta G = (F \cup G) - (F \cap G)$$

and

$$F \cup G = F \cup (F - G) .$$

Hint: When proving two sets F and G are equal, the straightforward approach is to show first that if $\omega \in F$, then also $\omega \in G$ and hence $F \subset G$. Reversing the procedure proves the sets equal.

2. Let Ω be the real line and $\boldsymbol{F}$ the collection of all finite unions of intervals (a,b) with rational endpoints and the complements of such sets. Show that $\boldsymbol{F}$ is a field. Is it a σ-field?

3. Let Ω be an arbitrary space. Suppose that $\boldsymbol{F}_i$, $i = 1,2, \cdots$ are all σ-fields of subsets of Ω. Define the collection $\boldsymbol{F} = \bigcap_i \boldsymbol{F}_i$; that is, the collection of all sets that are in all of the $\boldsymbol{F}_i$. Show that $\boldsymbol{F}$ is a σ-field.

4. Given a measurable space $(\Omega,\boldsymbol{F})$, a collection of sets $\boldsymbol{G}$ is called a sub-σ-field of $\boldsymbol{F}$ if it is a σ-field and if all of its elements belong to $\boldsymbol{F}$, in which case we write $\boldsymbol{G} \subset \boldsymbol{F}$. Show that $\boldsymbol{G}$ is the intersection of all σ-fields of subsets of Ω of which it is a sub-σ-field.

5. Prove deMorgan's laws.

6. Prove that if P satisfies (1.2.4), (1.2.5), and (1.2.7), then (1.2.8)-(1.2.10) are equivalent, that is, any one holds if and only if the other two also hold. Prove the following elementary properties of probability (all sets are assumed to be events).

7. $P(F \cup G) = P(F) + P(G) - P(F \cap G)$.

8. (The union bound)

$$P(\bigcup_{i=1}^{\infty} G_i) = \sum_{i=1}^{\infty} P(G_i) \,.$$

9. $P(F^c) = 1 - P(F)$.

10. For all events F $P(F) \leq 1$.

11 If $G \subset F$, then $P(F - G) = P(F) - P(G)$.

12. $P(F \Delta G) = P(F \cup G) - P(F \cap G)$.

13. $P(F \Delta G) = P(F) + P(G) - 2P(F \cap G)$.

14. $| P(F) - P(G) | \leq P(F \Delta G)$.

15. $P(F \Delta G) \leq P(F \Delta H) + P(H \Delta G)$.

16. If $F \in \boldsymbol{B}$, show that the indicator function 1_F defined by $1_F(x) = 1$ if $x \in F$ and 0 otherwise is a random variable. Describe its distribution. Is the product of indicator functions measurable?

17. If F_i, $i = 1,2, \cdots$ is a sequence of events that all have probability 1, show that $\bigcap_i F_i$ also has probability 1.

18. Suppose that P_i, $i = 1,2, \cdots$ is a countable family of probability measures on a space $(\Omega,\boldsymbol{B})$ and that a_i, $i = 1,2, \cdots$ is a sequence of positive real numbers that sums to one. Show that the set function m defined by

$$m(F) = \sum_{i=1}^{\infty} a_i P_i(F)$$

is also a probability measure on $(\Omega,\boldsymbol{B})$. This is an example of a *mixture* probability measure.

19. Show that for two events F and G,

$$| 1_F(x) - 1_G(x) | = 1_{F \Delta G}(x) \, .$$

20. Let $f : \Omega \to A$ be a random variable. Prove the following properties of inverse images:

$$f^{-1}(G^c) = f^{-1}(G)$$

$$f^{-1}\left(\bigcup_{k=1}^{\infty} G_k\right) = \bigcup_{k=1}^{\infty} f^{-1}(G_k)$$

$$\text{If } G \cap F = \emptyset, \ \text{then } f^{-1}(G) \cap f^{-1}(F) = \emptyset \, .$$

1.3 RANDOM PROCESSES AND DYNAMICAL SYSTEMS

We now consider two mathematical models for a random process. The first is the familiar one in elementary courses: a random process is just a sequence of random variables. The second model is likely less familiar: a random process can also be constructed from an abstract dynamical system consisting of a probability space together with a transformation on the space. The two models are connected by considering a time shift to be a transformation, but an example from communication theory shows that other transformations can be useful. The formulation and proof of ergodic theorems are more natural in the dynamical system context.

Random Processes

A *discrete time random process*, or for our purposes simply a *random process*, is a sequence of random variables $\{X_n\}_{n \in \boldsymbol{I}}$ or $\{X_n \, ; \, n \in \boldsymbol{I}\}$, where $\boldsymbol{I}$ is an index set, defined on a common probability space $(\Omega, \boldsymbol{B}, P)$. We usually assume that all of the random variables share a common alphabet, say A. The two most common index sets of interest are the set of all integers $\mathbf{Z} = \{\cdots, -2, -1, 0, 1, 2, \cdots\}$, in which case the random process is referred to as a two-sided random process, and the set of all nonnegative integers $\mathbf{Z}_+ = \{0, 1, 2, \cdots\}$, in which case the random process is said to be one-sided. One-sided random processes will often prove to be far more difficult in theory, but they provide better models for physical random processes that must be "turned on" at some time or that have transient behavior.

Observe that since the alphabet A is general, we could also model continuous time random processes in the preceding fashion by letting A consist of a family of waveforms defined on an interval, e.g., the random variable X_n could be in fact a continuous time waveform $X(t)$ for $t \in [nT,(n+1)T)$, where T is some fixed positive real number.

The preceding definition does not specify any structural properties of the index set $\boldsymbol{I}$. In particular, it does not exclude the possibility that $\boldsymbol{I}$ be a finite set, in which case "random vector" would be a better name than "random process." In fact, the two cases of $\boldsymbol{I} = \mathbf{Z}$ and $\boldsymbol{I} = \mathbf{Z}_+$ will be the only really important examples for our purposes. The general notation of $\boldsymbol{I}$ will be retained, however, in order to avoid having to state separate results for these two cases. Most of the theory to be considered in this chapter, however, will remain valid if we simply require that $\boldsymbol{I}$ be closed under addition, that is, if n and k are in $\boldsymbol{I}$, then so is $n + k$ (where the "+" denotes a suitably defined addition in the index set). For this reason we henceforth will assume that if $\boldsymbol{I}$ is the index set for a random process, then $\boldsymbol{I}$ is closed in this sense.

Dynamical Systems

An abstract dynamical system consists of a probability space $(\Omega,\boldsymbol{B},P)$ together with a measurable transformation $T:\Omega \to \Omega$ of Ω into itself. Measurability means that if $F \in \boldsymbol{B}$, then also $T^{-1}F = \{\omega: T\omega \in F\} \in \boldsymbol{B}$. The quadruple $(\Omega,\boldsymbol{B},P,T)$ is called a *dynamical system* in ergodic theory. The interested reader can find excellent introductions to classical ergodic theory and dynamical system theory in the books of Halmos [11] and Sinai [20]. More complete treatments may be found in [2, 19, 18, 7, 21, 17, 8, 14]. The name "dynamical systems" comes from the focus of the theory on the long term "dynamics" or "dynamical behavior" of repeated applications of the transformation T on the underlying measure space.

An alternative to modeling a random process as a sequence or family of random variables defined on a common probability space is to consider a single random variable together with a transformation defined on the underlying probability space. The outputs of the random process will then be values of the random variable taken on transformed points in the original space. The transformation will usually be related to shifting in time, and hence this viewpoint will focus on the action of time itself. Suppose now that T is a measurable mapping of points of the sample space Ω into itself. It is easy to see that the cascade or composition of measurable functions is also measurable. Hence the transformation T^n defined as $T^2\omega = T(T\omega)$ and so on ($T^n\omega = T(T^{n-1}\omega)$) is a measurable function for all positive integers n. If f is an A-valued random variable defined on $(\Omega, \boldsymbol{B})$, then the functions $fT^n: \Omega \to A$ defined by $fT^n(\omega) = f(T^n\omega)$ for $\omega \in \Omega$ will also be random variables for all n in $\mathbf{Z}_+$. Thus a dynamical system together with a random variable or measurable function f defines a single-sided random process $\{X_n\}_{n \in \mathbf{Z}_+}$ by $X_n(\omega) = f(T^n\omega)$. If it should be true that T is invertible, that is, T is one-to-one and its inverse T^{-1} is measurable, then one can define a double-sided random process by $X_n(\omega) = f(T^n\omega)$, all n in $\mathbf{Z}$.

The most common dynamical system for modeling random processes is that consisting of a sequence space Ω containing all one- or two-sided A-valued sequences together with the shift transformation T, that is, the transformation that maps a sequence $\{x_n\}$ into the sequence $\{x_{n+1}\}$ wherein each coordinate has been shifted to the left by one time unit. Thus, for example, let $\Omega = A^{\mathbf{Z}_+} = \{\text{all } x = (x_0, x_1, \cdots) \text{ with } x_i \in A \text{ for all } i\}$ and define $T: \Omega \to \Omega$ by $T(x_0, x_1, x_2, \cdots) = (x_1, x_2, x_3, \cdots)$. T is called the *shift* or *left shift* transformation on the one-sided sequence space. The shift for two-sided spaces is defined similarly.

Some interesting dynamical systems in communications applications do not, however, have this structure. As an example, consider the mathematical model of a device called a sigma-delta modulator, that is used for analog-to-digital conversion, that is, encoding a sequence of real numbers into a binary sequence (analog-to-digital conversion), which is then decoded into a reproduction sequence approximating the original sequence (digital-to-analog conversion) [12, 4, 9]. Given an input sequence $\{x_n\}$ and an initial "state" u_0, the operation of the encoder is described by the difference equations

$$e_n = x_n - q(u_n),$$

$$u_n = e_{n-1} + u_{n-1},$$

where $q(u)$ is $+b$ if its argument is nonnegative and $-b$ otherwise (q is called a binary quantizer). The decoder is described by the equation

$$\hat{x}_n = \frac{1}{N} \sum_{i=1}^{N} q(u_{n-i}).$$

The basic idea of the code's operation is this: An incoming continuous time, continuous amplitude waveform is sampled at a rate that is so high that the incoming waveform stays fairly constant over N sample times (in engineering parlance the original waveform is *oversampled* or sampled at many times the Nyquist rate). The binary quantizer then produces outputs for which the average over N samples is very near the input so that the decoder output $\hat{x}_{kN}$ is a good approximation to the input at the corresponding times. Since $\hat{x}_n$ has only a discrete number of possible values (N+1 to be exact), one has thereby accomplished analog-to-digital conversion. Because the system involves only a binary quantizer used repeatedly, it is a popular one for microcircuit implementation.

As an approximation to a very slowly changing input sequence x_n, it is of interest to analyze the response of the system to the special case of a constant input $x_n = x \in [-b,b)$ for all n (called a *quiet* input). This can be accomplished by recasting the system as a dynamical system as follows: Given a fixed input x, define the transformation T by

$$Tu = \begin{cases} u + x - b \; ; \text{ if } u \geq 0 \\ u + x + b \; ; \text{ if } u < 0 \; . \end{cases}$$

Given a constant input $x_n = x$, $n = 1,2, \cdots, N$, and an initial condition u_0 (which may be fixed or random), the resulting U_n sequence is given by

$$u_n = T^n u_0 \, .$$

If the initial condition u_0 is selected at random, then the preceding formula defines a dynamical system which can be analyzed.

The example is provided simply to emphasize the fact that time shifts are not the only interesting transformation when modeling communication systems.

The different models provide equivalent models for a given process: one emphasizing the sequence of outputs and the other emphasising the action of a transformation on the underlying space in producing these outputs. In order to demonstrate in what sense the models are equivalent for given random processes, we next turn to the notion of the distribution of a random process.

Exercises

1. Consider the sigma-delta example with a constant input in the case $b = 1/2$, $u_0 = 0$, and $x = 1/\pi$. Find u_n for $n = 1,2,3,4$.

2. Show by induction in the constant input sigma-delta example that if $u_0 = 0$ and $x \in [-b,b)$, then $u_n \in [-b,b)$ for all $n = 1,2, \cdots$.

3. Let $\Omega = [0,1)$ and $F = [0,1/2)$ and fix an $\alpha \in (0,1)$. Define the transformation $Tx = \lfloor \alpha x \rfloor$, where $\lfloor r \rfloor \in [0,1)$ denotes the fractional part of r; that is, every real number r has a unique representation as $r = K + \lfloor r \rfloor$ for some integer K. Show that if α is rational, then $T^n x$ is a periodic sequence in n.

1.4 DISTRIBUTIONS

Although in principle all probabilistic quantities of a random process can be determined from the underlying probability space, it is often more convenient to deal with the induced probability measures or distributions on the space of possible outputs of the random process. In particular, this allows us to compare different random processes without regard to the underlying probability spaces and thereby permits us to reasonably equate two random processes if their outputs have the same probabilistic structure, even if the underlying probability spaces are quite different.

We have already seen that each random variable X_n of the random process $\{X_n\}$ inherits a distribution because it is measurable. To describe a process, however, we need more than simply probability measures on output values of separate single random variables: we require probability measures on collections of random variables, that is, on sequences of outputs. In order to place probability measures on sequences of outputs of a random process, we first must construct the appropriate measurable spaces. A convenient technique for accomplishing this is to consider product spaces, spaces for sequences formed by concatenating spaces for individual outputs.

Let $\boldsymbol{I}$ denote any finite or infinite set of integers. In particular, $\boldsymbol{I} = \mathbf{Z}(n) = \{0,1,2,\ldots, n-1\}$, $\boldsymbol{I} = \mathbf{Z}$, or $\boldsymbol{I} = \mathbf{Z}_+$. Define $x^{\boldsymbol{I}} = \{x_i\}_{i \in \boldsymbol{I}}$. For example, $x^{\mathbf{Z}} = (\ldots, x_{-1}, x_0, x_1, \ldots)$ is a two-sided infinite sequence. When $\boldsymbol{I} = \mathbf{Z}(n)$ we abbreviate $x^{\mathbf{Z}(n)}$ to simply x^n. Given alphabets A_i, $i \in \boldsymbol{I}$, define the cartesian product spaces

$$\mathop{\times}_{i \in \boldsymbol{I}} A_i = \{ \text{ all } x^{\boldsymbol{I}}: x_i \in A_i \text{ all } i \text{ in } \boldsymbol{I}\}.$$

In most cases all of the A_i will be replicas of a single alphabet A and the preceding product will be denoted simply by $A^{\boldsymbol{I}}$. We shall abbreviate the space $A^{\mathbf{Z}(n)}$, the space of all n dimensional vectors with coordinates in A, by A^n. Thus, for example, $A^{\{m,m+1,\ldots,n\}}$ is the space of all possible outputs of the process from time m to time n; $A^{\mathbf{Z}}$ is the sequence space of all possible outputs of a two-sided process.

To obtain useful σ-fields of the preceding product spaces, we introduce the idea of a rectangle in a product space. A *rectangle* in $A^{\boldsymbol{I}}$ taking values in the coordinate σ-fields $\boldsymbol{B}_i$, $i \in \boldsymbol{J}$, is defined as any set of the form

$$B = \{x^{\boldsymbol{I}} \in A^{\boldsymbol{I}}: x_i \in B_i; \text{ all } i \text{ in } \boldsymbol{J}\}, \tag{1.4.1}$$

where $\boldsymbol{J}$ is a finite subset of the index set $\boldsymbol{I}$ and $B_i \in \boldsymbol{B}_i$ for all $i \in \boldsymbol{J}$. (Hence rectangles are sometimes referred to as finite dimensional rectangles.) A rectangle as in (1.4.1) can be written as a finite intersection of one-dimensional rectangles as

$$B = \bigcap_{i \in \boldsymbol{J}} \{x^{\boldsymbol{I}} \in A^{\boldsymbol{I}}: x_i \in B_i\} = \bigcap_{i \in \boldsymbol{J}} X_i^{-1}(B_i) \tag{1.4.2}$$

where here we consider X_i as the coordinate functions $X_i: A^{\boldsymbol{I}} \to A$ defined by $X_i(x^{\boldsymbol{I}}) = x_i$.

As rectangles in $A^{\boldsymbol{I}}$ are clearly fundamental events, they should be members of any useful σ-field of subsets of $A^{\boldsymbol{I}}$. One approach is simply to *define* the product σ-field $\boldsymbol{B}_A{}^{\boldsymbol{I}}$ as the smallest σ-field containing all of the rectangles, that is, the collection of sets that contains the clearly important class of rectangles and the minimum amount of other stuff required to make the collection a σ-field. In general, given any collection $\boldsymbol{G}$ of subsets of a space Ω, then $\sigma(\boldsymbol{G})$ will denote the smallest σ-field of subsets of Ω that contains $\boldsymbol{G}$ and it will be called the σ-field *generated* by $\boldsymbol{G}$. By *smallest* we mean that any σ-field containing $\boldsymbol{G}$ must also contain $\sigma(\boldsymbol{G})$. The σ-field is well defined since there must exist at least one σ-field containing $\boldsymbol{G}$, the collection of all subsets of Ω. Then the intersection of all σ-fields that contain $\boldsymbol{G}$ must be a σ-field, it must contain $\boldsymbol{G}$, and it must in turn be contained by all σ-fields that contain $\boldsymbol{G}$.

Given an index set $\boldsymbol{I}$ of integers, let $RECT(\boldsymbol{B}_i, i \in \boldsymbol{I})$ denote the set of all rectangles in $A^{\boldsymbol{I}}$ taking coordinate values in sets in $\boldsymbol{B}_i$, $i \in \boldsymbol{I}$. We then define the product σ-field of $A^{\boldsymbol{I}}$ by

$$\boldsymbol{B}_A{}^{\boldsymbol{I}} = \sigma(RECT(\boldsymbol{B}_i, i \in \boldsymbol{I})) .$$

At first glance it would appear that given an index set $\boldsymbol{I}$ and an A-valued random process $\{X_n\}_{n \in \boldsymbol{I}}$ defined on an underlying probability space $(\Omega, \boldsymbol{B}, P)$, then given any index set $\boldsymbol{J} \subset \boldsymbol{I}$, the measurable space $(A^{\boldsymbol{J}}, \boldsymbol{B}_A{}^{\boldsymbol{J}})$ should inherit a probability measure from the underlying space through the random variables $X^{\boldsymbol{J}} = \{X_n;\ n \in \boldsymbol{J}\}$. The only hitch is that so far we only know that individual random variables X_n are measurable (and hence inherit a probability measure). To make sense here we must first show that collections of random variables such as the random sequence $X^{\mathbf{Z}}$ or the random vector $X^n = \{X_0, \ldots, X_{n-1}\}$ are also measurable and hence themselves random variables.

Observe that for any index set $\boldsymbol{I}$ of integers it is easy to show that inverse images of the mapping $X^{\boldsymbol{I}}$ from Ω to $A^{\boldsymbol{I}}$ will yield events in $\boldsymbol{B}$ if we confine attention to rectangles. To see this we simply use the measurability of each individual X_n and observe that since $(X^{\boldsymbol{I}})^{-1}(B) = \bigcap_{i \in \boldsymbol{I}} X_i^{-1}(B_i)$ and since finite and countable unions of events are events, then we have for rectangles that

$$(X^{\boldsymbol{I}})^{-1}(B) \in \boldsymbol{B}. \tag{1.4.3}$$

We will have the desired measurability if we can show that if (1.4.3) is satisfied for all rectangles, then it is also satisfied for all events in the σ-field generated by the rectangles. This result is an application of an approach named the "good sets principle" by Ash [1], p. 5. We shall occasionally wish to prove that all events possess some particular desirable property that is easy to prove for generating events. The good sets principle consists of the following argument: Let $\boldsymbol{S}$ be the collection of "good sets" or of all events $F \in \sigma(\boldsymbol{G})$ possessing the desired property. If (1) $\boldsymbol{G} \subset \boldsymbol{S}$ and hence all the generating events are good, and (2) $\boldsymbol{S}$ is a σ-field, then $\sigma(\boldsymbol{G}) \subset \boldsymbol{S}$ and hence *all* of the events $F \in \sigma(\boldsymbol{G})$ are good.

Lemma 1.4.1. Given measurable spaces $(\Omega_1, \boldsymbol{B})$ and $(\Omega_2, \sigma(\boldsymbol{G}))$, then a function $f: \Omega_1 \to \Omega_2$ is $\boldsymbol{B}$-measurable if and only if $f^{-1}(F) \in \boldsymbol{B}$ for all $F \in \boldsymbol{G}$; that is, measurability can be verified by showing that inverse images of generating events are events.

Proof. If f is $\boldsymbol{B}$-measurable, then $f^{-1}(F) \in \boldsymbol{B}$ for all F and hence for all $F \in \boldsymbol{G}$. Conversely, if $f^{-1}(F) \in \boldsymbol{B}$ for all generating events $F \in \boldsymbol{G}$, then define the class of sets

$$\boldsymbol{S} = \{G\colon G \in \sigma(\boldsymbol{G}),\ f^{-1}(G) \in \boldsymbol{B}\}.$$

It is straightforward to verify that $\boldsymbol{S}$ is a σ-field, clearly $\Omega_1 \in \boldsymbol{S}$ since it is the inverse image of Ω_2. The fact that $\boldsymbol{S}$ contains countable unions of its elements follows from the fact that $\sigma(\boldsymbol{G})$ is closed to countable unions and inverse images preserve set theoretic operations, that is,

$$f^{-1}(\bigcup_i G_i) = \bigcup_i f^{-1}(G_i) .$$

Furthermore, $\boldsymbol{S}$ contains every member of $\boldsymbol{G}$ by assumption. Since $\boldsymbol{S}$ contains $\boldsymbol{G}$ and is a σ-field, $\sigma(\boldsymbol{G}) \subset \boldsymbol{S}$ by the good sets principle. □

We have shown that the mappings $X^{\boldsymbol{I}}: \Omega \to A^{\boldsymbol{I}}$ are measurable and hence the output measurable space $(A^{\boldsymbol{I}}, \boldsymbol{B}_A{}^{\boldsymbol{I}})$ will inherit a probability measure from the underlying probability space and thereby determine a new probability space $(A^{\boldsymbol{I}}, \boldsymbol{B}_A{}^{\boldsymbol{I}}, P_{X^{\boldsymbol{I}}})$, where the induced probability measure is defined by

$$P_{X^{\boldsymbol{I}}}(F) = P((X^{\boldsymbol{I}})^{-1}(F)) = P(\omega : X^{\boldsymbol{I}}(\omega) \in F), \quad F \in \boldsymbol{B}_A{}^{\boldsymbol{I}}. \quad (1.4.4)$$

Such probability measures induced on the outputs of random variables are referred to as *distributions* for the random variables, exactly as in the simpler case first treated. When $\boldsymbol{I} = \{m, m+1, \ldots, m+n-1\}$, e.g., when we are treating X^n taking values in A^n, the distribution is referred to as an n-dimensional or nth order distribution and it describes the behavior of an n-dimensional random variable. If $\boldsymbol{I}$ is the entire process index set, e.g., if $\boldsymbol{I} = \mathbf{Z}$ for a two-sided process or $\boldsymbol{I} = \mathbf{Z}_+$ for a one-sided process, then the induced probability measure is defined to be the distribution of the process. Thus, for example, a probability space $(\Omega, \boldsymbol{B}, P)$ together with a doubly infinite sequence of random variables $\{X_n\}_{n \in \mathbf{Z}}$ induces a new probability space $(A^{\mathbf{Z}}, \boldsymbol{B}_A{}^{\mathbf{Z}}, P_{X^{\mathbf{Z}}})$ and $P_{X^{\mathbf{Z}}}$ is the distribution of the process. For simplicity, let us now denote the process distribution simply by m. We shall call the probability space $(A^{\boldsymbol{I}}, \boldsymbol{B}_A{}^{\boldsymbol{I}}, m)$ induced in this way by a random process $\{X_n\}_{n \in \mathbf{Z}}$ the output space or sequence space of the random process.

Equivalence

Since the sequence space $(A^{\boldsymbol{I}}, \boldsymbol{B}_A{}^{\boldsymbol{I}}, m)$ of a random process $\{X_n\}_{n \in \mathbf{Z}}$ is a probability space, we can define random variables and hence also random processes on this space. One simple and useful such definition is that of a sampling or coordinate or projection function defined as follows: Given a product space $A^{\boldsymbol{I}}$, define the sampling functions $\Pi_n: A^{\boldsymbol{I}} \to A$ by

$$\Pi_n(x^{\boldsymbol{I}}) = x_n, \; x^{\boldsymbol{I}} \in A^{\boldsymbol{I}}, \; n \in \boldsymbol{I}. \quad (1.4.5)$$

The sampling function is named Π since it is also a projection. Observe that the distribution of the random process $\{\Pi_n\}_{n \in \mathbf{I}}$ defined on the probability space $(A^{\mathbf{I}}, \mathbf{B}_A{}^{\mathbf{I}}, m)$ is exactly the same as the distribution of the random process $\{X_n\}_{n \in \mathbf{I}}$ defined on the probability space $(\Omega, \mathbf{B}, P)$. In fact, so far they are the same process since the $\{\Pi_n\}$ simply read off the values of the $\{X_n\}$.

What happens, however, if we no longer build the Π_n on the X_n, that is, we no longer first select ω from Ω according to P, then form the sequence $x^{\mathbf{I}} = X^{\mathbf{I}}(\omega) = \{X_n(\omega)\}_{n \in \mathbf{I}}$, and then define $\Pi_n(x^{\mathbf{I}}) = X_n(\omega)$? Instead we directly choose an x in $A^{\mathbf{I}}$ using the probability measure m and then view the sequence of coordinate values. In other words, we are considering two completely separate experiments, one described by the probability space $(\Omega, \mathbf{B}, P)$ and the random variables $\{X_n\}$ and the other described by the probability space $(A^{\mathbf{I}}, \mathbf{B}_A{}^{\mathbf{I}}, m)$ and the random variables $\{\Pi_n\}$. In these two separate experiments, the actual sequences selected may be completely different. Yet intuitively the processes should be the "same" in the sense that their statistical structures are identical, that is, they have the same distribution. We make this intuition formal by defining two processes to be *equivalent* if their process distributions are identical, that is, if the probability measures on the output sequence spaces are the same, regardless of the functional form of the random variables of the underlying probability spaces. In the same way, we consider two random variables to be equivalent if their distributions are identical.

We have described two equivalent processes or two equivalent models for the same random process, one defined as a sequence of perhaps very complicated random variables on an underlying probability space, the other defined as a probability measure directly on the measurable space of possible output sequences. The second model will be referred to as a *directly given* random process.

Which model is "better" depends on the application. For example, a directly given model for a random process may focus on the random process itself and not its origin and hence may be simpler to deal with. If the random process is then coded or measurements are taken on the random process, then it may be better to model the encoded random process in terms of random variables defined on the original random process and not as a directly given random process. This model will then focus on the input process and the coding operation. We shall let convenience determine the most appropriate model.

We can now describe yet another model for the random process described previously, that is, another means of describing a random process with the same distribution. This time the model is in terms of a dynamical system. Given the probability space $(A^{\boldsymbol{I}},\boldsymbol{B}_A{}^{\boldsymbol{I}},m)$, define the (left) shift transformation T: $A^{\boldsymbol{I}}\to A^{\boldsymbol{I}}$ by

$$T(x^{\boldsymbol{I}}) = T(\{x_n\}_{n\in \boldsymbol{I}}) = y^{\boldsymbol{I}} = \{y_n\}_{n\in \boldsymbol{I}},$$

where

$$y_n = x_{n+1} \ , \ n \in \boldsymbol{I} \ .$$

Thus the nth coordinate of $y^{\boldsymbol{I}}$ is simply the $(n + 1)$ coordinate of $x^{\boldsymbol{I}}$. (Recall that we assume for random processes that $\boldsymbol{I}$ is closed under addition and hence if n and 1 are in $\boldsymbol{I}$, then so is $(n + 1)$.) If the alphabet of such a shift is not clear from context, we will occasionally denote it by T_A or $T_{A^{\boldsymbol{I}}}$. It can easily be shown that the shift is indeed measurable by showing it for rectangles and then invoking Lemma 1.4.1.

Consider next the dynamical system $(A^{\boldsymbol{I}},\boldsymbol{B}_A{}^{\boldsymbol{I}},P, T)$ and the random process formed by combining the dynamical system with the zero time sampling function Π_0 (we assume that 0 is a member of $\boldsymbol{I}$). If we define $Y_n(x) = \Pi_0(T^n x)$ for $x = x^{\boldsymbol{I}} \in A^{\boldsymbol{I}}$, or, in abbreviated form, $Y_n = \Pi_0 T^n$, then the random process $\{Y_n\}_{n\in \boldsymbol{I}}$ is equivalent to the processes developed previously. Thus we have developed three different, but equivalent, means of producing the same random process. Each will be seen to have its uses.

The preceding development shows that a dynamical system is a more fundamental entity than a random process since we can always construct an equivalent model for a random process in terms of a dynamical system: use the directly given representation, shift transformation, and zero time sampling function.

The shift transformation introduced previously on a sequence space is the most important transformation that we shall encounter. It is not, however, the only important transformation. Hence when dealing with transformations we will usually use the notation T to reflect the fact that it is often related to the action of a simple left shift of a sequence, yet we should keep in mind that occasionally other operators will be considered and the theory to be developed will remain valid,; that is, T is not required to be a simple time shift. For example, we will also consider block shifts of vectors instead of samples and variable length shifts.

Most texts on ergodic theory deal with the case of an invertible transformation, that is, where T is a one-to-one transformation and the inverse mapping T^{-1} is measurable. This is the case for the shift on $A^{\mathbf{Z}}$, the so-called two-sided shift. It is not the case, however, for the one-sided shift defined on $A^{\mathbf{Z}_+}$ and hence we will avoid use of this assumption. We will, however, often point out in the discussion and exercises what simplifications or special properties arise for invertible transformations.

Since random processes are considered equivalent if their distributions are the same, we shall adopt the notation $[A,m,X]$ for a random process $\{X_n;\ n \in \boldsymbol{I}\}$ with alphabet A and process distribution m, the index set $\boldsymbol{I}$ usually being clear from context. We will occasionally abbreviate this to the more common notation $[A,m]$, but it is often convenient to note the name of the output random variables as there may be several; e.g., a random process may have an input X and output Y. By "the associated probability space" of a random process $[A,m,X]$ we shall mean the sequence probability space $(A^{\boldsymbol{I}},\boldsymbol{B}_A{}^{\boldsymbol{I}},m)$. It will often be convenient to consider the random process as a directly given random process, that is, to view X_n as the coordinate functions Π_n on the sequence space $A^{\boldsymbol{I}}$ rather than as being defined on some other abstract space. This will not always be the case, however, as often processes will be formed by coding or communicating other random processes. Context should render such bookkeeping details clear.

Monotone Classes

Unfortunately there is no constructive means of describing the σ-field generated by a class of sets. That is, we cannot give a prescription of adding all countable unions, then all complements, and so on, and be ensured of thereby giving an algorithm for obtaining all σ-field members as sequences of set theoretic operations on members of the original collection. We can, however, provide some insight into the structure of such generated σ-fields when the original collection is a field. This structure will prove useful when considering extensions of measures.

Recall that a collection $\boldsymbol{M}$ of subsets of Ω is a monotone class if whenever $F_n \in \boldsymbol{M}$, $n = 1,2,\cdots$, and $F_n \uparrow F$ or $F_n \downarrow F$, then also $F \in \boldsymbol{M}$.

Lemma 1.4.2. Given a field $\boldsymbol{F}$, then $\sigma(\boldsymbol{F})$ is the smallest monotone class containing $\boldsymbol{F}$.

Proof. Let $\boldsymbol{M}$ be the smallest monotone class containing $\boldsymbol{F}$ and let $F \in \boldsymbol{M}$. Define $\boldsymbol{M}_F$ as the collection of all sets $G \in \boldsymbol{M}$ for which $F \cap G$, $F \cap G^c$, and $F^c \cap G$ are all in $\boldsymbol{M}$. Then $\boldsymbol{M}_F$ is a monotone class. If $F \in \boldsymbol{F}$, then all members of $\boldsymbol{F}$ must also be in $\boldsymbol{M}_F$ since they are in $\boldsymbol{M}$ and since $\boldsymbol{F}$ is a field. Since both classes contain $\boldsymbol{F}$ and $\boldsymbol{M}$ is the minimal monotone class, $\boldsymbol{M} \subset \boldsymbol{M}_F$. Since the members of $\boldsymbol{M}_F$ are all chosen from $\boldsymbol{M}$, $\boldsymbol{M} = \boldsymbol{M}_F$. This implies in turn that for any $G \in \boldsymbol{M}$, then all sets of the form $G \cap F$, $G \cap F^c$, and $G^c \cap F$ for any $F \in \boldsymbol{F}$ are in $\boldsymbol{M}$. Thus for this G, all $F \in \boldsymbol{F}$ are members of $\boldsymbol{M}_G$ or $\boldsymbol{F} \subset \boldsymbol{M}_G$ for any $G \in \boldsymbol{M}$. By the minimality of $\boldsymbol{M}$, this means that $\boldsymbol{M}_G = \boldsymbol{M}$. We have now shown that for $F, G \in \boldsymbol{M} = \boldsymbol{M}_F$, then $F \cap G$, $F \cap G^c$, and $F^c \cap G$ are also in $\boldsymbol{M}$. Thus $\boldsymbol{M}$ is a field. Since it also contains increasing limits of its members, it must be a σ-field and hence it must contain $\sigma(\boldsymbol{F})$ since it contains $\boldsymbol{F}$. Since $\sigma(\boldsymbol{F})$ is a monotone class containing $\boldsymbol{F}$, it must contain $\boldsymbol{M}$; hence the two classes are identical. □

Exercises

1. Given a random process $\{X_n\}$ with alphabet A, show that the class $\boldsymbol{F}_0 = RECT(\boldsymbol{B}_i; i \in \boldsymbol{I})$ of all rectangles is a field.

2. Let *field*$(\boldsymbol{G})$ denote the field generated by a class of sets $\boldsymbol{G}$, that is, *field*$(\boldsymbol{G})$ contains the given class and is in turn contained by all other fields containing $\boldsymbol{G}$. Show that $\sigma(\boldsymbol{G}) = \sigma(\mathit{field}(\boldsymbol{G}))$.

1.5 EXTENSION

We have seen one example where a σ-field is formed by generating it from a class of sets. Just as we construct event spaces by generating them from important collections of sets, we will often develop probability measures by specifying their values on an important class of sets and then extending the measure to the full σ-field generated by the class. The goal of this section is to develop the fundamental result for extending probability measures from fields to σ-fields, the Carathéodory extension theorem. The theorem states that if we have a probability measure on a field, then there exists a unique probability measure on the σ-field that agrees with the given probability measure on events in the field. We shall develop the result in a series of steps. The development is patterned on that of Halmos [10].

Suppose that $\boldsymbol{F}$ is a field of subsets of a space Ω and that P is a probability measure on a field $\boldsymbol{F}$; that is, P is a nonnegative, normalized, countably additive set function when confined to sets in $\boldsymbol{F}$. We wish to obtain a probability measure, say λ, on $\sigma(\boldsymbol{F})$ with the property that for all $F \in \boldsymbol{F}$, $\lambda(F) = P(F)$. Eventually we will also wish to demonstrate that there is only one such λ. Toward this end define the set function

$$\lambda(F) = \inf_{\{F_i\}:\, F_i \in \boldsymbol{F},\, F \subset \bigcup_i F_i} \sum_i P(F_i) \,. \tag{1.5.1}$$

The infimum is over all countable collections of field elements whose unions contain the set F. We will call such a collection of field members whose union contains F a *cover* of F. Note that we could confine interest to covers whose members are all disjoint since if $\{F_i\}$ is an arbitrary cover of F, then the collection $\{G_i\}$ with $G_1 = F_1$, $G_i = F_i - F_{i-1}$, $i = 1,2, \cdots$ is a disjoint cover for F. Observe that this set function is defined for *all* subsets of Ω. Note that from the definition, given any set F and any $\varepsilon > 0$, there exists a cover $\{F_i\}$ such that

$$\sum_i P(F_i) - \varepsilon \le \lambda(F) \le \sum_i P(F_i) \,. \tag{1.5.2}$$

A cover satisfying (1.5.2) will be called an ε-cover for F.

The goal is to show that λ is in fact a probability measure on $\sigma(\boldsymbol{F})$. Obviously λ is nonnegative, so we need to show that it is normalized and countably additive. This we will do in a series of steps, beginning with the simplest:

Lemma 1.5.1. The set function λ of (1.5.1) satisfies

(a) $\lambda(\emptyset) = 0$.

(b) *Monotonicity*: If $F \subset G$, then $\lambda(F) \le \lambda(G)$.

(c) *Subadditivity*: For any two sets F, G,

$$\lambda(F \cup G) \le \lambda(F) + \lambda(G) \,. \tag{1.5.3}$$

(d) *Countable Subadditivity*: Given sets $\{F_i\}$,

$$\lambda(\bigcup_i F_i) \le \sum_{i=1}^{\infty} \lambda(F_i) \,. \tag{1.5.4}$$

Proof. Property (a) follows immediately from the definition since $\emptyset \in \boldsymbol{F}$ and contains itself, hence

$$\lambda(\emptyset) \leq P(\emptyset) = 0 .$$

From (1.5.2) given G and ε we can choose a cover $\{G_n\}$ for G such that

$$\sum_i P(G_i) \leq \lambda(G) + \varepsilon .$$

Since $F \subset G$, a cover for G is also a cover for F and hence

$$\lambda(F) \leq \sum_i P(G_i) \leq \lambda(G) + \varepsilon .$$

Since ε is arbitrary, (b) is proved. To prove (c) let $\{F_i\}$ and $\{G_i\}$ be $\varepsilon/2$ covers for F and G. Then $\{F_i \cup G_i\}$ is a cover for $F \cup G$ and hence

$$\lambda(F \cup G) \leq \sum_i P(F_i \cup G_i) \leq \sum_i P(F_i) + \sum_i P(G_i) \leq \lambda(F) + \lambda(G) + \varepsilon .$$

Since ε is arbitrary, this proves (c).

To prove (d), for each F_i let $\{F_{ik}; k=1,2, \cdots\}$ be an $\varepsilon 2^{-i}$ cover for F_i. Then $\{F_{ik} ; i = 1,2, \cdots ; j = 1,2, \cdots\}$ is a cover for $\bigcup_i F_i$ and hence

$$\lambda(\bigcup_i F_i) \leq \sum_{i=1}^{\infty} \sum_{k=1}^{\infty} P(F_{ik}) \leq \sum_{i=1}^{\infty} (\lambda(F_i) + \varepsilon 2^{-i}) = \sum_{i=1}^{\infty} \lambda(F_i) + \varepsilon ,$$

which completes the proof since ε is arbitrary. □

We note in passing that a set function λ on a collection of sets having properties (a)-(d) of the lemma is called an *outer measure* on the collection of sets.

The simple properties have an immediate corollary: The set function λ agrees with P on field events.

Corollary 1.5.1. If $F \in \boldsymbol{F}$, then $\lambda(F) = P(F)$. Thus, for example, $\lambda(\Omega) = 1$.

Proof. Since a set covers itself, we have immediately that $\lambda(F) \leq P(F)$ for all field events F. Suppose that $\{F_i\}$ is an ε cover for F. Then

$$\lambda(F) \geq \sum_i P(F_i) - \varepsilon \geq \sum_i P(F \cap F_i) - \varepsilon .$$

Since $F \cap F_i \in \mathbf{F}$ and $F \subset \bigcup_i F_i$, $\bigcup_i (F \cap F_i) = F \cap \bigcup_i F_i = F \in \mathbf{F}$ and hence, invoking the countable additivity of P on $\mathbf{F}$,

$$\sum_i P(F \cap F_i) = P(\bigcup_i F \cap F_i) = P(F)$$

and hence

$$\lambda(F) \geq P(F) - \varepsilon . \square$$

Thus far the properties of λ have depended primarily on manipulating the definition. In order to show that λ is countably additive (or even only finitely additive) over $\sigma(\mathbf{F})$, we need to introduce a new concept and a new collection of sets that we will later see contains $\sigma(\mathbf{F})$. The definitions seem a bit artificial, but some similar form of tool seems to be necessary to get to the desired goal. By way of motivation, we are trying to show that a set function is finitely additive on some class of sets. Perhaps the simplest form of finite additivity looks like

$$\lambda(F) = \lambda(F \cap R) + \lambda(F \cap R^c) .$$

Hence it should not seem too strange to build at least this form into the class of sets considered. To do this, *define* a set $R \in \sigma(\mathbf{F})$ to be λ-*measurable* if

$$\lambda(F) = \lambda(F \cap R) + \lambda(F \cap R^c) \text{ , all } F \in \sigma(\mathbf{F}) .$$

In words, a set R is λ-measurable if it splits all events in $\sigma(\mathbf{F})$ in an additive fashion. Let $\mathbf{H}$ denote the collection of all λ-measurable sets. We shall see that indeed λ is countably additive on the collection $\mathbf{H}$ and that $\mathbf{H}$ contains $\sigma(\mathbf{F})$. Observe for later use that since λ is subadditive, to prove that $R \in \mathbf{H}$ requires only that we prove

$$\lambda(F) \geq \lambda(F \cap R) + \lambda(F \cap R^c) \text{ , all } F \in \sigma(\mathbf{F}) .$$

Lemma 1.5.2. $\mathbf{H}$ is a field.

Proof. Clearly $\Omega \in \mathbf{H}$ since $\lambda(\emptyset) = 0$ and $\Omega \cap F = F$. Equally clearly, $F \in \mathbf{H}$ implies that $F^c \in \mathbf{H}$. The only work required is show that if F, $G \in \mathbf{H}$, then also $F \cup G \in \mathbf{H}$. To accomplish this begin by recalling that for F, $G \in \mathbf{H}$ and for any $H \in \sigma(\mathbf{F})$ we have that

$$\lambda(H) = \lambda(H \cap F) + \lambda(H \cap F^c)$$

and hence since both arguments on the right are members of $\sigma(\boldsymbol{F})$

$$\lambda(H) = \lambda(H \cap F \cap G) + \lambda(H \cap F \cap G^c)$$
$$+ \lambda(H \cap F^c \cap G) + \lambda(H \cap F^c \cap G^c) . \tag{1.5.5}$$

As (1.5.5) is valid for any event $H \in \sigma(\boldsymbol{F})$, we can replace H by $H \cap (F \cup G)$ to obtain

$$\lambda(H \cap (F \cup G)) = \lambda(H \cap (F \cap G))$$
$$+ \lambda(H \cap (F \cap G^c)) + \lambda(H \cap (F^c \cap G)) , \tag{1.5.6}$$

where the fourth term has disappeared since the argument is null. Plugging (1.5.6) into (1.5.5) yields

$$\lambda(H) = \lambda(H \cap (F \cup G)) + \lambda(H \cap (F^c \cap G^c))$$

which implies that $F \cup G \in \boldsymbol{H}$ since $F^c \cap G^c = (F \cup G)^c$. □

Lemma 1.5.3. $\boldsymbol{H}$ is a σ-field.

Proof. Suppose that $F_n \in \boldsymbol{H}$, $n = 1,2, \cdots$ and

$$F = \bigcup_{i=1}^{\infty} F_i .$$

It will be convenient to work with disjoint sets, so we formally introduce the sets

$$G_1 = F_1$$
$$G_k = F_k - \bigcup_{i=1}^{k-1} F_i \tag{1.5.7}$$

with the property that the G_k are disjoint and

$$\bigcup_{i=1}^{n} G_i = \bigcup_{i=1}^{n} F_i$$

for all integers n and for $n = \infty$. Since $\boldsymbol{H}$ is a field, clearly the G_i and all finite unions of F_i or G_i are also in $\boldsymbol{H}$. First apply (1.5.6) with G_1 and G_2 replacing F and G. Using the fact that G_1 and G_2 are disjoint,

$$\lambda(H \cap (G_1 \cap G_2)) = \lambda(H \cap G_1) + \lambda(H \cap G_2) .$$

Using this equation and mathematical induction yields

$$\lambda(H \cap \bigcup_{i=1}^{n} G_i) = \sum_{i=1}^{n} \lambda(H \cap G_i). \tag{1.5.8}$$

For convenience define the set

$$F(n) = \bigcup_{i=1}^{n} F_i = \bigcup_{i=1}^{n} G_i .$$

Since $F(n) \in \boldsymbol{H}$ and since $\boldsymbol{H}$ is a field, $F(n)^c \in \boldsymbol{H}$ and hence for any event $H \in \boldsymbol{H}$

$$\lambda(H) = \lambda(H \cap F(n)) + \lambda(H \cap F(n)^c).$$

Using the preceding formula, (1.5.8), the fact that $F(n) \subset F$ and hence $F^c \subset F(n)^c$, the monotonicity of λ, and the countable subadditivity of λ,

$$\begin{aligned} \lambda(H) &\geq \sum_{i=1}^{\infty} \lambda(H \cap G_i) + \lambda(H \cap F^c) \\ &\geq \lambda(H \cap F) + \lambda(H \cap F^c). \end{aligned} \tag{1.5.9}$$

Since H is an arbitrary event in $\boldsymbol{H}$, this proves that the countable union F is in $\boldsymbol{H}$ and hence that $\boldsymbol{H}$ is a σ-field. □

In fact much more was proved than the stated lemma; the implicit conclusion is made explicit in the following corollary.

Corollary 1.5.2. λ is countably additive on $\boldsymbol{H}$

Proof. Take G_n to be an arbitrary sequence of disjoint events in $\boldsymbol{H}$ in the preceding proof (instead of obtaining them as differences of another arbitrary sequence) and let F denote the union of the G_n. Then from the lemma $F \in \boldsymbol{H}$ and (1.5.9) with $H = F$ implies that

$$\lambda(F) \geq \sum_{i=1} \lambda(G_i).$$

Since λ is countably subadditive, the inequality must be an equality, proving the corollary. □

We now demonstrate that the strange class $\boldsymbol{H}$ is exactly the σ-field generated by $\boldsymbol{F}$.

Lemma 1.5.4. $\boldsymbol{H} = \sigma(\boldsymbol{F})$.

Proof. Since the members of $\boldsymbol{H}$ were all chosen from $\sigma(\boldsymbol{F})$, $\boldsymbol{H} \subset \sigma(\boldsymbol{F})$. Let $F \in \boldsymbol{F}$ and $G \in \sigma(\boldsymbol{F})$ and let $\{G_n\}$ be an ε-cover for G. Then

$$\lambda(G) + \varepsilon \geq \sum_{i=1}^{\infty} P(G_n) = \sum_{i=1}^{\infty} (P(G_n \cap F) + P(G_n \cap F^c))$$

$$\geq \lambda(G \cap F) + \lambda(G \cap F^c)$$

which implies that for $F \in \boldsymbol{F}$

$$\lambda(G) \geq \lambda(G \cap F) + \lambda(G \cap F^c)$$

for all events $G \in \sigma(\boldsymbol{F})$ and hence that $F \in \boldsymbol{H}$. This implies that $\boldsymbol{H}$ contains $\boldsymbol{F}$. Since $\boldsymbol{H}$ is a σ-field, however, it therefore must contain $\sigma(\boldsymbol{F})$. □

We have now proved that λ is a probability measure on $\sigma(\boldsymbol{F})$ which agrees with P on $\boldsymbol{F}$. The only remaining question is whether it is unique. The following lemma resolves this issue.

Lemma 1.5.5. If two probability measures λ and μ on $\sigma(\boldsymbol{F})$ satisfy $\lambda(F) = \mu(F)$ for all F in the field $\boldsymbol{F}$, then the measures are identical; that is, $\lambda(F) = \mu(F)$ for all $F \in \sigma(\boldsymbol{F})$.

Proof. Let $\boldsymbol{M}$ denote the collection of all sets F such that $\lambda(F) = \mu(F)$. From the continuity of probability, $\boldsymbol{M}$ is a monotone class. Since it contains $\boldsymbol{F}$, it must contain the smallest monotone class containing $\boldsymbol{F}$. That class is $\sigma(\boldsymbol{F})$ from Lemma 1.4.2 and hence $\boldsymbol{M}$ contains the entire σ-field.

Combining all of the pieces of this section yields the principal result:

Theorem 1.5.1. *The Carathéodory Extension Theorem.* If a set function P satisfies the properties of a probability measure (nonnegativity, normalization, and countable additivity or finite additivity plus continuity) for all sets in a field $\boldsymbol{F}$ of subsets of a space Ω, then there exists a unique measure λ defined by (1.5.1) on the measurable space $(\Omega, \sigma(\boldsymbol{F}))$ that agrees with P on $\boldsymbol{F}$.

When no confusion is possible, we will usually denote the extension of a measure P by P rather than by a different symbol.

We will this section with an important corollary to the extension theorem that shows that an arbitrary event can be approximated closely by a field event in a probabilistic way. First, however, we derive a useful property of the symmetric difference operation defined by

$$F \Delta G = (F \cap G^c) \cup (F^c \cap G = F \cup G - F \cap G . \quad (1.5.10)$$

Lemma 1.5.6. For any events G, F, H,

$$P(F \Delta G) \leq P(F \Delta H) + P(H \Delta G) ;$$

that is, probabilities of symmetric differences satisfy a triangle inequality.

Proof. A little set theory shows that

$$F \Delta G \subset (F \Delta H) \cup (H \Delta G)$$

and hence the subadditivity and monotonicity of probability imply the lemma. □

Corollary 1.5.3. (Approximation Theorem) Given a probability space $(\Omega,\boldsymbol{B},P)$ and a generating field $\boldsymbol{F}$, that is, $\boldsymbol{B} = \sigma(\boldsymbol{F})$, then given $F \in \boldsymbol{B}$ and $\varepsilon > 0$, there exists an $F_0 \in \boldsymbol{F}$ such that $P(F \Delta F_0) \leq \varepsilon$, where Δ denotes the symmetric difference (all points in the union that are not in the intersection).

Proof. From the definition of λ in (1.5.1) (which yielded the extension of the original measure P) and the ensuing discussion one can find a countable collection of disjoint field events $\{F_n\}$ such that $F \subset \bigcup_n F_n$ and

$$\sum_{i=1}^{\infty} P(F_i) - \varepsilon/2 \leq \lambda(F) \leq \sum_{i=1}^{\infty} P(F_i)$$

Renaming λ as P

$$P(F \Delta \bigcup_{i=1}^{\infty} F_n) = P(\bigcup_{i=1}^{\infty} F_n - F) = P(\bigcup_{i=1}^{\infty} F_n) - P(F) \leq \frac{\varepsilon}{2} .$$

Choose a finite union $\tilde{F} = \bigcup_{i=1}^{n} F_i$ with n chosen large enough to ensure that

$$P(\bigcup_{i=1}^{\infty} F_i \,\Delta\, \bar{F}) = P(\bigcup_{i=1}^{\infty} F_i - \bar{F}) = \sum_{i=n+1}^{\infty} P(F_i) \le \frac{\varepsilon}{2} .$$

From the previous lemma we see that

$$P(F \,\Delta\, \bar{F}) \le P(F \,\Delta\, \bigcup_{i=1}^{\infty} F_i) + P(\bigcup_{i=1}^{\infty} F_i \,\Delta\, \bar{F}) \le \varepsilon . \square$$

Exercises

1. Show that if m and p are two probability measures on $(\Omega,\sigma(\boldsymbol{F}))$, where $\boldsymbol{F}$ is a field, then given an arbitrary event F and $\varepsilon > 0$ there is a field event $F_0 \in \boldsymbol{F}$ such that $m(F \Delta F_0) \le \varepsilon$ and $p(F \Delta F_0) \le \varepsilon$.

1.6 ISOMORPHISM

We have defined random variables or random processes to be equivalent if they have the same output probability spaces, that is, if they have the same probability measure or distribution on their output value measurable space. This is not the only possible notion of two random processes being the same. For example, suppose we have a random process with a binary output alphabet and hence an output space made up of binary sequences. We form a new directly given random process by taking each successive pair of the binary process and considering the outputs to be a sequence of quaternary symbols, that is, the original random process is "coded" into a new random process via the mapping

$$00 \to 0$$
$$01 \to 1$$
$$10 \to 2$$
$$11 \to 3$$

applied to successive pairs of binary symbols.

The two random processes are not equivalent since their output sequence measurable spaces are different; yet clearly they are the same since each can be obtained by a simple relabeling or coding of the other. This leads to a more fundamental definition of "sameness": isomorphism. The definition of isomorphism is for dynamical systems rather than for random processes since, as previously noted, this is the more fundamental notion and hence the definition applies to random processes.

There are, in fact, several notions of isomorphism: isomorphic measurable spaces, isomorphic probability spaces, and isomorphic dynamical systems. We present these definitions together as they are intimately connected.

Two measurable spaces $(\Omega,\boldsymbol{B})$ and $(\Lambda,\boldsymbol{S})$ are *isomorphic* if there exists a measurable function $f: \Omega \to \Lambda$ that is one-to-one and has a measurable inverse f^{-1}. In other words, the inverse image $f^{-1}(\lambda)$ of a point $\lambda \in \Lambda$ consists of exactly one point in Ω and the inverse mapping so defined, say $g: \Lambda \to \Omega$, $g(\lambda) = f^{-1}(\lambda)$, is itself a measurable mapping. The function f (or its inverse g) with these properties is called an *isomorphism*. An isomorphism between two measurable spaces is thus an invertible mapping between the two sample spaces that is measurable in both directions.

Two probability spaces $(\Omega,\boldsymbol{B},P)$ and $(\Lambda,\boldsymbol{S},Q)$ are *isomorphic* if there is an isomorphism $f: \Omega \to \Lambda$ between the two measurable spaces $(\Omega,\boldsymbol{B})$ and $(\Lambda,\boldsymbol{S})$ with the added property that

$$Q = P_f \text{ and } P = Q_{f^{-1}}$$

that is,

$$Q(F) = P(f^{-1}(F)) \; ; F \in \boldsymbol{S}$$

$$P(G) = Q(f(G)) \; ; G \in \boldsymbol{B} \, .$$

Two probability spaces are isomorphic (1) if one can find for each space a random variable defined on that space that has the other as its output space, and (2) the random variables can be chosen to be inverses of each other; that is, if the two random variables are f and g, then $f(g(\lambda)) = \lambda$ and $g(f(\omega)) = \omega$.

Note that if the two probability spaces $(\Omega,\boldsymbol{B},P)$ and $(\Lambda,\boldsymbol{S},Q)$ are isomorphic and $f: \Omega \to \Lambda$ is an isomorphism with inverse g, then the random variable fg defined by $fg(\lambda) = f(g(\lambda))$ is equivalent to the identity random variable $i: \Lambda \to \Lambda$ defined by $i(\lambda) = \lambda$.

With the ideas of isomorphic measurable spaces and isomorphic probability spaces in hand, we now can define isomorphic dynamical systems. Roughly speaking, two dynamical systems are isomorphic if one can be coded onto the other in an invertible way so that the coding carries one transformation into the other , that is, one can code from one system into the other and back again and coding and transformations commute.

Two dynamical systems $(\Omega,\boldsymbol{B},P,S)$ and $(\Lambda,\boldsymbol{S},m,T)$ are *isomorphic* if there exists an isomorphism $f:\Omega\to\Lambda$ such that

$$Tf(\omega)=f(S\omega)\ ;\ \omega\in\Omega.$$

If the probability space $(\Lambda,\boldsymbol{S},m)$ is the sequence space of a directly given random process, T is the shift on this space, and Π_0 the sampling function on this space, then the random process $\Pi_0 T^n=\Pi_n$ defined on $(\Lambda,\boldsymbol{S},m)$ is equivalent to the random process $\Pi_0(f\,S^n)$ defined on the probability space $(\Omega,\boldsymbol{B},P)$. More generally, any random process of the form gT^n defined on $(\Lambda,\boldsymbol{S},m)$ is equivalent to the random process $g(f\,S^n)$ defined on the probability space $(\Omega,\boldsymbol{B},P)$. A similar conclusion holds in the opposite direction. Thus, any random process that can be defined on one dynamical system as a function of transformed points possesses an equivalent model in terms of the other dynamical system and its transformation. In addition, not only can one code from one system into the other, one can recover the original sample point by inverting the code.

The binary example introduced at the beginning of this section is easily seen to meet this definition of sameness. Let $\boldsymbol{B}(\mathbf{Z}(n))$ denote the σ-field of subsets of $\mathbf{Z}(n)$ comprising all possible subsets of $\mathbf{Z}(n)$ (the *power set* of $\mathbf{Z}(n)$). The described mapping of binary pairs or members of $\mathbf{Z}(1)^2$ into $\mathbf{Z}(3)$ induces a mapping $f:\mathbf{Z}(1)^{\mathbf{Z}_+}\to\mathbf{Z}(3)^{\mathbf{Z}_+}$ mapping binary sequences into quaternary sequences. This mapping is easily seen to be invertible (by construction) and measurable (use the good sets principle and focus on rectangles). Let T be the shift on the binary sequence space and S be the shift on the quaternary sequence space; then the dynamical systems $(\mathbf{Z}(1)^{\mathbf{Z}_+},\boldsymbol{B}(\mathbf{Z}(1))^{\mathbf{Z}_+},m,T^2)$ and $(\mathbf{Z}(3)^{\mathbf{Z}_+},\boldsymbol{B}(\mathbf{Z}(3))^{\mathbf{Z}_+},m_f,T)$ are isomorphic; that is, the quaternary model with an ordinary shift is isomorphic to the binary model with the two-shift that shifts symbols a pair at a time.

Isomorphism will often provide a variety of equivalent models for random processes. Unlike the previous notion of equivalence, however, isomorphism will often yield dynamical systems that are decidedly different, but that are isomorphic and hence produce equivalent random processes by coding.

REFERENCES

1. R. B. Ash, *Real Analysis and Probability,* Academic Press, New York, 1972.
2. P. Billingsley, *Ergodic Theory and Information,* Wiley, New York, 1965.
3. L. Breiman, *Probability,* Addison-Wesley, Menlo Park, Calif., 1968.
4. J. C. Candy, "A use of limit cycle oscillations to obtain robust analog-to-digital converters," *IEEE Transactions on Communications*, vol. COM-22, pp. 298-305, March 1974.
5. K. L. Chung, *A Course in Probability Theory,* Academic Press, New York, 1974.
6. H. Cramér, *Mathematical Methods of Statistics,* Princeton University Press, Princeton, NJ, 1946.
7. M. Denker, C. Grillenberger, and K. Sigmund, *Ergodic Theory on Compact Spaces,* 57, Lecture Notes in Mathematics, Springer-Verlag, New York, 1970.
8. N. A. Friedman, *Introduction to Ergodic Theory,* Van Nostrand Reinhold Company, New York, 1970.
9. R. M. Gray, "Oversampled Sigma-Delta Modulation," *IEEE Transactions on Communications*, vol. COM-35, pp. 481-489, April 1987.
10. P. R. Halmos, *Measure Theory,* Van Nostrand Reinhold, New York, 1950.
11. P. R. Halmos, *Lectures on Ergodic Theory,* Chelsea, New York, 1956.
12. H. Inose and Y. Yasuda, "A unity bit coding method by negative feedback," *Proceedings of the IEEE*, vol. 51, pp. 1524-1535, November 1963.
13. A. N. Kolmogorov, *Foundations of the Theory of Probability,* Chelsea, New York, 1950.
14. U. Krengel, *Ergodic Theorems,* De Gruyter Series in Mathematics, De Gruyter, New York, 1985.
15. M. Loève, *Probability Theory,* D. Van Nostrand, Princeton, New Jersey, 1963. Third Edition.
16. L. E. Maistrov, *Probability Theory: A Historical Sketch,* Academic Press, New York, 1974. Translated by S. Kotz.

17. D. Ornstein, *Ergodic Theory, Randomness, and Dynamical Systems,* Yale University Press, New Haven, 1975.

18. K. Petersen, *Ergodic Theory,* Cambridge University Press, Cambridge, 1983.

19. P. C. Shields, *The Theory of Bernoulli Shifts,* The University of Chicago Press, Chicago, Ill., 1973.

20. Ya. G. Sinai, *Introduction to Ergodic Theory,* Mathematical Notes, Princeton University Press, Princeton, 1976.

21. P. Walters, *Ergodic Theory-Introductory Lectures,* Lecture Notes in Mathematics No. 458, Springer-Verlag, New York, 1975.

2

STANDARD ALPHABETS

It is desirable to develop a theory under the most general possible assumptions. Random process models with very general alphabets are useful because they include all conceivable cases of practical importance. On the other hand, considering only the abstract spaces of the previous chapter can result in both weaker properties and more complicated proofs. Restricting the alphabets to possess some structure is necessary for some results and convenient for others. Ideally, however, we can focus on a class of alphabets that both possesses useful structure and still is sufficiently general to well model all examples likely to be encountered in the real world. Standard spaces are a candidate for this goal and are the topic of this chapter and the next. In this chapter we focus on the definitions and properties of standard spaces, leaving the more complicated demonstration that specific spaces are standard to the next chapter. The reader in a hurry can skip the next chapter. The theory of standard spaces is usually somewhat hidden in theories of topological measure spaces. Standard spaces are related to or include as special cases standard Borel spaces, analytic spaces, Lusin spaces, Suslin spaces, and Radon spaces. Such spaces are usually defined by their relation via a mapping to a complete separable metric space, a topic to be introduced in Chapter 3. Good supporting texts are Parthasarathy [7], Christensen [3], Schwartz [9], Bourbaki [2], and Cohn [4], and the papers by Mackey [6] and Bjornsson [1].

The presentation here differs from the traditional one in that we focus on the fundamental properties of standard spaces purely from a probability theory viewpoint and postpone introduction of the topological and metric space ideas until later. As a result, we can define standard spaces by the properties that we will need and not by an isomorphism to a particular kind of probability space with a metric space alphabet. This provides a shorter route to a description of such a space and to its basic properties. The next chapter will show that indeed such topological probability spaces satisfy the required properties, but we will also see in this chapter that certain simple spaces also meet the criteria.

2.1 EXTENSION OF PROBABILITY MEASURES

We shall often wish to construct probability measures and sources by specifying the values of the probability of certain simple events and then by finding a probability measure on the σ-field containing these events. It was shown in Section 1.5 that if one has a set function meeting the required conditions of a probability measure on a field, then there is a unique extension of that set function to a consistent probability measure on the σ-field generated by the field. Unfortunately, the extension theorem is often difficult to apply since countable additivity or continuity of a set function can be difficult to prove, even on a field. Nonnegativity, normalization, and finite additivity are, however, usually quite simple to test. Because several of the results to be developed will require such extensions of finitely additive measures to countably additive measures, we shall develop in this chapter a class of alphabets for which such extension will always be possible.

To apply the Carathéodory extension theorem, we will first have to pin down a candidate probability measure on a generating field. In most such constructions we will be able at best to force a set function to be nice on a countable collection of sets. Hence we will focus on σ-fields that are *countably generated,* that is, for which there is a countable collection of sets $\boldsymbol{G}$ that generates the σ-field. Say that we have a σ-field $\boldsymbol{B} = \sigma(\boldsymbol{G})$ for some countable class $\boldsymbol{G}$. Let *field*$(\boldsymbol{G})$ denote the field generated by $\boldsymbol{G}$, that is, the smallest field containing $\boldsymbol{G}$. Unlike σ-fields, there is a simple constructive definition of the field generated by a class: *field*$(\boldsymbol{G})$ consists exactly of all elements of $\boldsymbol{G}$ together with all sets obtainable from finite sequences of set theoretic operations on $\boldsymbol{G}$. Thus if

$\boldsymbol{G}$ is countable, so is *field*($\boldsymbol{G}$). It is easy to show using the good sets principle that if $\boldsymbol{B} = \sigma(\boldsymbol{G})$, then also $\boldsymbol{B} = \sigma(field(\boldsymbol{G}))$ and hence if $\boldsymbol{B}$ is countably generated, then it is generated by a countable field.

Our goal is to find countable generating fields which have the property that every nonnegative, normalized, finitely additive set function on the field is also countably additive on the field (and hence will have a unique extension to a probability measure on the σ-field generated by the field). We formalize this property in a definition:

A field $\boldsymbol{F}$ is said to have the *countable extension property* if it has a countable number of elements and if *every* set function P satisfying (1.2.4), (1.2.5), and (1.2.7) on $\boldsymbol{F}$ also satisfies (1.2.8) on $\boldsymbol{F}$. A measurable space $(\Omega,\boldsymbol{B})$ is said to have the *countable extension property* if $\boldsymbol{B} = \sigma(\boldsymbol{F})$ for some field $\boldsymbol{F}$ with the countable extension property.

Thus a measurable space has the countable extension property if we can find a countable generating field such that all normalized, nonnegative, finitely additive set functions on the field extend to a probability measure on the full σ-field. This chapter is devoted to characterizing those fields and measurable spaces possessing the countable extension property and to prove one of the most important results for such spaces–the Kolmogorov extension theorem. We also develop some simple properties of standard spaces in preparation for the next chapter, where we develop the most important and most general known class of such spaces.

2.2 STANDARD SPACES

As a first step towards characterizing fields with the countable extension property, we consider the special case of fields having only a finite number of elements. Such finite fields trivially possess the countable extension property. We shall then proceed to construct a countable generating field from a sequence of finite fields and we will determine conditions under which the limiting field will inherit the extendibility properties of the finite fields.

Let $\boldsymbol{F} = \{F_i,\ i = 0,1,2, \cdots, n-1\}$ be a finite field of a sample space Ω, that is, $\boldsymbol{F}$ is a finite collection of sets in Ω that is closed under finite set theoretic operations. Note that $\boldsymbol{F}$ is trivially also a σ-field. $\boldsymbol{F}$ itself can be represented as the field generated by a more fundamental class of sets. A set F in $\boldsymbol{F}$ will be called an *atom* if its only subsets that are also field members are itself and the empty set, that is, it cannot be broken up into smaller pieces that are also in the field. Let $\boldsymbol{A}$ denote the collection of atoms of $\boldsymbol{F}$. Clearly there are fewer than n atoms. It is easy to show that $\boldsymbol{A}$ consists exactly of all nonempty sets of the form

$$\bigcap_{i=0}^{n-1} F_i^*$$

where F_i^* is either F_i or F_i^c. In fact, let us call such sets *intersection sets* and observe that any two intersection sets must be disjoint since for at least one i one intersection set must lie inside F_i and the other within F_i^c. Thus all intersection sets must be disjoint. Next observe that any field element can be written as a finite union of intersection sets–just take the union of all intersection sets contained in the given field element. Let G be an atom of $\boldsymbol{F}$. Since it is an element of $\boldsymbol{F}$, it is the union of disjoint intersection sets. There can be only one nonempty intersection set in the union, however, or G would not be an atom. Hence every atom is an intersection set. Conversely, if G is an intersection set, then it must also be an atom since otherwise it would contain more than one atom and hence contain more than one intersection set, contradicting the disjointness of the intersection sets.

In summary, given any finite field $\boldsymbol{F}$ of a space Ω we can find a unique collection of atoms $\boldsymbol{A}$ of the field such that the sets in $\boldsymbol{A}$ are disjoint, nonempty, and have the entire space Ω as their union (since Ω is a field element and hence can be written as a union of atoms). Thus $\boldsymbol{A}$ is a *partition* of Ω. Furthermore, since every field element can be written as a union of atoms, $\boldsymbol{F}$ is generated by $\boldsymbol{A}$ in the sense that it is the smallest field containing $\boldsymbol{A}$. Hence we write $\boldsymbol{F} = field(\boldsymbol{A})$. Observe that if we assign nonnegative numbers p_i to the atoms G_i in $\boldsymbol{A}$ such that their sum is 1, then this immediately gives a finitely additive, nonnegative, normalized set function on $\boldsymbol{F}$ by the formula

$$P(F) = \sum_{i:\, G_i \subset F} p_i \, .$$

Furthermore, this set function is trivially countably additive since there are only a finite number of elements in $\boldsymbol{F}$.

The next step is to consider an increasing sequence of finite fields that in the limit will give a countable generating field. A sequence of finite fields $\boldsymbol{F}_n$, $n = 1,2, \cdots$ is said to be *increasing* if the elements of each field are also elements of all of the fields with higher indices, that is, if $\boldsymbol{F}_n \subset \boldsymbol{F}_{n+1}$, all n. This implies that if $\boldsymbol{A}_n$ are the corresponding collections of atoms, then the atoms in $\boldsymbol{A}_{n+1}$ are formed by splitting up the atoms in $\boldsymbol{A}_n$. Given an increasing sequence of fields $\boldsymbol{F}_n$, define the limit $\boldsymbol{F}$ as the union of all of the elements in all of the $\boldsymbol{F}_n$, that is,

$$\boldsymbol{F} = \bigcup_{i=1}^{\infty} \boldsymbol{F}_i \,.$$

$\boldsymbol{F}$ is easily seen to be itself a field. For example, any $F \in \boldsymbol{F}$ must be in $\boldsymbol{F}_n$ for some n, hence the complement F^c must also be in $\boldsymbol{F}_n$ and hence also in $\boldsymbol{F}$. Similarly, any finite collection of sets F_n, $n = 1,2, \cdots, m$ must all be contained in some $\boldsymbol{F}_k$ for sufficiently large k. The latter field must hence contain the union and hence so must $\boldsymbol{F}$. Thus we can think of the increasing sequence $\boldsymbol{F}_n$ as increasing to a limit field $\boldsymbol{F}$ and we write

$$\boldsymbol{F}_n \uparrow \boldsymbol{F}$$

if $\boldsymbol{F}_n \subset \boldsymbol{F}_{n+1}$, all n, and $\boldsymbol{F}$ is the union of all of the elements of the $\boldsymbol{F}_n$. Note that $\boldsymbol{F}$ has by construction a countable number of elements. When $\boldsymbol{F}_n \uparrow \boldsymbol{F}$, we shall say that the sequence $\boldsymbol{F}_n$ *asymptotically generates* $\boldsymbol{F}$.

Lemma 2.2.1. Given any countable field $\boldsymbol{F}$ of subsets of a space Ω, then there is a sequence of finite fields $\{\boldsymbol{F}_n;\, n = 1,2, \cdots\}$ that asymptotically generates $\boldsymbol{F}$. In addition, the sequence can be constructed so that the corresponding sets of atoms $\boldsymbol{A}_n$ of the field $\boldsymbol{F}_n$ can be indexed by a subset of the set of all binary n-tuples, that is, $\boldsymbol{A}_n = \{G_{u^n},\, u^n \in \mathrm{B}\}$, where B is a subset of $\{0,1\}^n$, with the property that $G_{u^n} \subset G_{u^{n-1}}$. Thus if u^n is a prefix of u^m for $m>n$, then G_{u^m} is contained in G_{u^n}. (We shall refer to such a sequence of atoms of finite fields as a *binary indexed* sequence.)

Proof. Let $\boldsymbol{F} = \{F_i,\, i = 0,1, \cdots\}$. Consider the sequence of finite fields defined by $\boldsymbol{F}_n = field(F_i,\, i = 0,1, \cdots, n-1)$, that is, the field generated by the first n elements in $\boldsymbol{F}$. The sequence is increasing since the generating classes are increasing. Any given element in $\boldsymbol{F}$ is eventually in one of the $\boldsymbol{F}_n$ and hence the union of the $\boldsymbol{F}_n$ contains all of the elements in $\boldsymbol{F}$ and is itself a field, hence it must contain $\boldsymbol{F}$. Any element in the union of the $\boldsymbol{F}_n$, however, must be in an $\boldsymbol{F}_n$ for some n and hence must be an element of $\boldsymbol{F}$. Thus the two fields are the same. A similar argument to that used above to construct the atoms of an arbitrary finite field will demonstrate that the atoms in $field(F_0, \cdots, F_{n-1})$ are simply all nonempty sets of the form $\bigcap_{k=0}^{n-1} F_k^*$, where F_k^* is either F_k or F_k^c. For each such intersection set let u^n denote the binary n-tuple having a one in each coordinate i for which $F_i^* = F_i$ and zeros in the remaining coordinates and define G_{u^n} as the corresponding intersection set. Then each G_{u^n} is either an atom or empty and all atoms are obtained as u varies through the set $\{0,1\}^n$ of all binary n-tuples. By construction,

$$\omega \in G_{u^n} \text{ if and only if } u_i = 1_{F_i}(\omega),\ i = 0,1, \cdots, n-1, \quad (2.2.1)$$

where for any set F $1_F(\omega)$ is the indicator function of the set F, that is,

$$1_F(\omega) = \begin{cases} 1 \text{ if } \omega \in F, \\ 0 \text{ if } \omega \notin F. \end{cases}$$

Eq.(2.2.1) implies that for any n and any binary n-tuple u^n that $G_{u^n} \subset G_{u^{n-1}}$, completing the proof. □

Before proceeding to further study sequences of finite fields, we note that the construction of the binary indexed sequence of atoms for a countable field presented above provides an easy means of determining which atoms contain a given sample point. For later use we formalize this notion in a definition.

Given an enumeration $\{F_n;\ n = 0,1, \cdots\}$ of a countable field $\boldsymbol{F}$ of subsets of a sample space Ω and the single-sided binary sequence space

$$M = \{0,1\}^{\mathbf{Z}_+} = \mathop{\times}_{i=0}^{\infty} \{0,1\},$$

define the *canonical binary sequence function* $f : \Omega \to M$ by

$$f(\omega) = \{1_{F_i}(\omega);\ i = 0,1, \cdots\}. \quad (2.2.2)$$

Given an enumeration of a countable field and the corresponding binary indexed set of atoms $\boldsymbol{A}_n = \{G_{u^n}\}$ as above, then for any point $\omega \in \Omega$ we have that

$$\omega \in G_{f(\omega)^n},\ n = 1,2, \cdots, \quad (2.2.3)$$

where $f(\omega)^n$ denotes the binary n-tuple comprising the first n symbols in $f(\omega)$.

Thus the sequence of decreasing atoms containing a given point ω can be found as a prefix of the canonical binary sequence function. Observe, however, that f is only an into mapping, that is, some binary sequences may not correspond to points in Ω. In addition, the function may be many to one, that is, different points in Ω may yield the same binary sequences.

Unfortunately, the sequence of finite fields converging upward to a countable field developed above does not have sufficient structure to guarantee that probability measures on the finite fields will imply a probability measure on the limit field. The missing item that will prove to be the key is specified in the following definitions:

A sequence of finite fields $\boldsymbol{F}_n$, $n = 0,1, \cdots$, is said to be a *basis* for a field $\boldsymbol{F}$ if it has the following properties:

It asymptotically generates $\boldsymbol{F}$, that is,

$$\boldsymbol{F}_n \uparrow \boldsymbol{F} . \tag{2.2.4}$$

If G_n is a sequence of atoms of $\boldsymbol{F}_n$, then

$$\text{if } G_n \in \boldsymbol{A}_n \text{ and } G_{n+1} \subset G_n,\ n = 0,1,2, \cdots, \text{ then}$$

$$\bigcap_{n=1}^{\infty} G_n \neq \emptyset . \tag{2.2.5}$$

A basis for a field is an asymptotically generating sequence of finite fields with the property that a decreasing sequence of atoms cannot converge to the empty set. The property (2.2.4) of a sequence of fields is called the *finite intersection property*. The name comes from the similar use in topology and reflects the fact that if all finite intersections of the sequence of sets G_n are not empty (which is true since the G_n are a decreasing sequence of nonempty sets), then the intersection of all of the G_n cannot be empty.

A sequence $\boldsymbol{F}_n$, $n = 1, \cdots$ is said to be a *basis* for a measurable space $(\Omega, \boldsymbol{B})$ if the $\{\boldsymbol{F}_n\}$ form a basis for a field that generates $\boldsymbol{B}$.

If the sequence $\boldsymbol{F}_n \uparrow \boldsymbol{F}$ and $\boldsymbol{F}$ generates $\boldsymbol{B}$, that is, if

$$\boldsymbol{B} = \sigma(\boldsymbol{F}) = \sigma(\bigcup_{n=1}^{\infty} \boldsymbol{F}_n),$$

then the $\boldsymbol{F}_n$ are said to *asymptotically generate* the σ-field $\boldsymbol{B}$.

A field $\boldsymbol{F}$ is said to be *standard* if it possesses a basis. A measurable space $(\Omega, \boldsymbol{B})$ is said to be *standard* if $\boldsymbol{B}$ can be generated by a standard field, that is, if $\boldsymbol{B}$ possesses a basis.

The requirement that a σ-field be generated by the limit of a sequence of simple finite fields is a reasonably intuitive one if we hope for the σ-field to inherit the extendibility properties of the finite fields. The second condition–that a decreasing sequence of atoms has a nonempty limit–is less intuitive, however, and will prove harder to demonstrate. Although nonintuitive at this point, we shall see that the existence of a basis is a sufficient *and necessary* condition for extending arbitrary finitely additive measures on countable fields to countably additive measures.

The proof that the standard property is sufficient to ensure that any finitely additive set function is also countably additive requires additional machinery that will be developed later. The proof of necessity, however, can be presented now and will perhaps provide some insight into the standard property by showing what can go wrong in its absence.

Lemma 2.2.2. Let $\boldsymbol{F}$ be a field of subsets of a space Ω. A necessary condition for $\boldsymbol{F}$ to have the countable extension property is that it be standard, that is, that it possess a basis.

Proof. We assume that $\boldsymbol{F}$ does not possess a basis and we construct a finitely additive set function that is not continuous at $\emptyset$ and hence not countably additive. To have the countable extension property, $\boldsymbol{F}$ must be countable. From Lemma 2.2.1 we can construct a sequence of finite fields $\boldsymbol{F}_n$ such that $\boldsymbol{F}_n \uparrow \boldsymbol{F}$. Since $\boldsymbol{F}$ does not possess a basis, we know that for any such sequence $\boldsymbol{F}_n$ there must exist a decreasing sequence of atoms G_n of $\boldsymbol{F}_n$ such that $G_n \downarrow \emptyset$. Define set functions P_n on $\boldsymbol{F}_n$ as follows: If $G_n \subset F$, then $P_n(F) = 1$, if $F \cap G_n = \emptyset$, then $P_n(F) = 0$. Since $F \in \boldsymbol{F}_n$, F either wholly contains the atom G_n or F and G_n are disjoint, hence the P_n are well defined. Next define the set function P on the limit field $\boldsymbol{F}$ in the natural way: If $F \in \boldsymbol{F}$, then $F \in \boldsymbol{F}_n$ for some smallest value of n (e.g., if the $\boldsymbol{F}_n$ are constructed as before as the field generated by the first n elements of $\boldsymbol{F}$, then eventually every element of the countable field $\boldsymbol{F}$ must appear in one of the $\boldsymbol{F}_n$). Thus we can set $P(F) = P_n(F)$. By construction, if $m \geq n$ then also $P_m(F) = P_n(F)$ and hence $P(F) = P_n(F)$ for any n such that $F \in \boldsymbol{F}_n$. P is obviously nonnegative, $P(\Omega) = 1$ (since all of the atoms G_n in the given sequence are in the sample space), and P is finitely additive. To see the latter fact, let F_i, $i = 1, \cdots, m$ be a finite collection of disjoint sets in $\boldsymbol{F}$. By construction, all must lie in some field $\boldsymbol{F}_n$ for sufficiently large n. If G_n lies in one of the sets F_i (it can lie in at most one since the sets are disjoint), then (1.2.7) holds with both sides equal to one. If none of the sets contains G_n, then (1.2.7) holds with both sides equal to zero. Thus P satisfies (1.2.4), (1.2.5), and (1.2.7). To prove P to be countably additive, then, we must verify (1.2.8). By construction $P(G_n) = P_n(G_n) = 1$ for all

n and hence

$$\lim_{n\to\infty} P(G_n) = 1 \neq 0,$$

and therefore (1.2.8) is violated since by assumption the G_n decrease to the empty set $\emptyset$ which has zero probability. Thus P is not continuous at $\emptyset$ and hence not countably additive. □

If a field is not standard, then finitely additive probability measures that are not countably additive can always be constructed by putting probability on a sequence of atoms that collapses down to nothing. Thus there can always be probability on ever smaller sets, but the limit cannot support the probability since it is empty. Thus the necessity of the standard property for the extension of arbitrary additive probability measures justifies its candidacy for a general, useful alphabet.

Corollary 2.2.1. A necessary condition for a countably generated measurable space $(\Omega, \boldsymbol{B})$ to have the countable extension property is that it be a standard space.

Proof. To have the countable extension property, a measurable space must have a countable generating field. If the measurable space is not standard, then no such field can possess a basis and hence no such field will possess the countable extension property. In particular, one can always find as in the proof of the lemma a generating field and an additive set function on that field which is not countably additive and hence does not extend. □

Exercises

1. A class of subsets $\boldsymbol{V}$ of A is said to be *separating* if given any two points $x,y \in A$, there is a $V \in \boldsymbol{V}$ that contains only one of the two points and not the other. Suppose that a separable σ-field $\boldsymbol{B}$ has a countable generating class $\boldsymbol{V} = \{V_i;\ i = 1,2, \cdots\}$ that is also separating. Describe the intersection sets

$$\bigcap_{n=1}^{\infty} V_n^*,$$

where V_n^* is either V_n or V_n^c.

2.3 SOME PROPERTIES OF STANDARD SPACES

The following results provide some useful properties of standard spaces. In particular, they show how certain combinations of or mappings on standard spaces yield other standard spaces. These results will prove useful for demonstrating that certain spaces are indeed standard. The first result shows that if we form a product space from a countable number of standard spaces as in Section 1.4, then the product space is also standard. Thus if the alphabet of a source or random process for one sample time is standard, then the space of all sequences produced by the source is also standard.

Lemma 2.3.1. Let $\mathbf{F}_i$, $i \in \mathbf{I}$, be a family of standard fields for some countable index set $\mathbf{I}$. Let $\mathbf{F}$ be the product field generated by all rectangles of the form $F = \{x^{\mathbf{I}}: x_i \in F_i, i \in \mathbf{M}\}$, where $F_i \in \mathbf{F}_i$ all i and $\mathbf{M}$ is any finite subset of $\mathbf{I}$. That is,

$$\mathbf{F} = field(RECT(\mathbf{F}_i, i \in \mathbf{I})),$$

then $\mathbf{F}$ is also standard.

Proof. Since $\mathbf{I}$ is countable, we may assume that $\mathbf{I} = \{1,2, \ldots\}$. For each $i \in \mathbf{I}$, $\mathbf{F}_i$ is standard and hence possesses a basis, say $\{\mathbf{F}_i(n), n = 1,2, \cdots\}$. Consider the sequence

$$\mathbf{G}_n = field(RECT(\mathbf{F}_i(n), i = 1,2, \cdots, n)), \tag{2.3.1}$$

that is, $\mathbf{G}_n$ is the field of subsets formed by taking all rectangles formed from the nth order basis fields $\mathbf{F}_i(n)$, $i = 1, \cdots, n$ in the first n coordinates. The lemma will be proved by showing that $\mathbf{G}_n$ forms a basis for the field $\mathbf{F}$ and hence the field is standard. The fields $\mathbf{G}_n$ are clearly finite and increasing since the coordinate fields are. The field generated by the union of all the fields $\mathbf{G}_n$ will contain all of the rectangles in $\mathbf{F}$ since for each i the union in that coordinate contains the full coordinate field $\mathbf{F}_i$. Thus $\mathbf{G}_n \uparrow \mathbf{F}$. Say we have a sequence G_n of atoms of $\mathbf{G}_n$ ($G_n \in \mathbf{G}_n$ for all n) decreasing to the null set. Each such atom must have the form

$$G_n = \{x^{\mathbf{I}}: x_i \in G_n(i); i = 1,2, \cdots, n\}.$$

where $G_n(i)$ is an atom of the coordinate field $\mathbf{F}_i(n)$. For $G_n \downarrow \emptyset$, however, this requires that $G_n(i) \downarrow \emptyset$ at least for one i, violating the definition of a basis for the ith coordinate. Thus $\mathbf{F}_n$ must be a basis. □

Corollary 2.3.1. Let $(A_i,\boldsymbol{B}_i)$, $i \in \boldsymbol{I}$, be a family of standard spaces for some countable index set $\boldsymbol{I}$. Let $(A,\boldsymbol{B})$ be the product measurable space, that is,

$$(A,\boldsymbol{B}) = \mathop{\times}_{i \in \boldsymbol{I}} (A_i,\boldsymbol{B}_i) = (\mathop{\times}_{i \in \boldsymbol{I}} A_i, \mathop{\times}_{i \in \boldsymbol{I}} \boldsymbol{B}_i),$$

where

$$\mathop{\times}_{i \in \boldsymbol{I}} A_i = \{ \text{ all } x^{\boldsymbol{I}} = (x_i,\ i \in \boldsymbol{I}) \colon x_i \in A_i;\ i \in \boldsymbol{I}\}$$

is the cartesian product of the alphabets A_i, and

$$\mathop{\times}_{i \in \boldsymbol{I}} \boldsymbol{B}_i = \sigma(RECT(\boldsymbol{B}_i, i \in \boldsymbol{I})).$$

Then $(A,\boldsymbol{B})$ is standard.

Proof. Since each $(A_i,\boldsymbol{B}_i)$, $i \in \boldsymbol{I}$, is standard, each possesses a basis, say $\{\boldsymbol{F}_i(n);\ n = 1,2, \cdots\}$, which asymptotically generates a coordinate field $\boldsymbol{F}_i$ which in turn generates $\boldsymbol{B}_i$. (Note that these are not the same as the $\boldsymbol{F}_i(n)$ of (2.3.1).) From Lemma 2.3.1 the product field of the nth order basis fields given by (2.3.1) is a basis for $\boldsymbol{F}$. Thus we will be done if we can show that $\boldsymbol{F}$ generates $\boldsymbol{B}$. It is an easy exercise, however, to show that if $\boldsymbol{F}_i$ generates $\boldsymbol{B}_i$, all $i \in \boldsymbol{I}$, then

$$\boldsymbol{B} = \sigma(RECT(\boldsymbol{B}_i,\ i \in \boldsymbol{I})) =$$

$$\sigma(RECT(\boldsymbol{F}_i,\ i \in \boldsymbol{I})) = \sigma(field(RECT(\boldsymbol{F}_i,\ i \in \boldsymbol{I}))) = \sigma(\boldsymbol{F}) \ . \ \square \quad (2.3.2)$$

We now turn from cartesian products of different standard spaces to certain subspaces of standard spaces. First, however, we need to define subspaces of measurable spaces.

Let $(A,\boldsymbol{B})$ be a measurable space (not necessarily standard). For any $F \in \boldsymbol{B}$ and any class of sets $\boldsymbol{G}$ define the new class $\boldsymbol{G} \cap F = \{$all sets of the form $G \cap F$ for some $G \in \boldsymbol{G}\}$. Given a set $F \in \boldsymbol{B}$ we can define a measurable space $(F,\boldsymbol{B} \cap F)$ called a *subspace* of $(A,\boldsymbol{B})$. It is easy to verify that $\boldsymbol{B} \cap F$ is a σ-field of subsets of F. Intuitively, it is the full σ-field $\boldsymbol{B}$ "cut down" to the set F.

We cannot show in general that arbitrary subspaces of standard spaces are standard. One case where they are is described next.

Lemma 2.3.2. Suppose that $(A,\boldsymbol{B})$ is standard with a basis $\boldsymbol{F}_n$, $n = 1,2,\cdots$. If F is any countable union of members of the $\boldsymbol{F}_n$, that is, if

$$F = \bigcup_{i=1}^{\infty} F_i \,;\, F_i \in \bigcup_{n=1}^{\infty} \boldsymbol{F}_n \,,$$

then the space $(A - F, \boldsymbol{B} \cap (A - F))$ is also standard.

Proof. The basic idea is that if you take a standard space with a basis and remove some atoms, the original basis with these atoms removed still works since decreasing sequences of atoms must still be empty for some finite n or have a nonempty intersection. Given a basis $\boldsymbol{F}_n$ for $(A,\boldsymbol{B})$, form a candidate basis $\boldsymbol{G}_n = \boldsymbol{F}_n \cap F^c$. This is essentially the original basis with all of the atoms in the sets in F removed. Let us call the countable collection of basis field atoms in F "bad atoms." From the good sets principle, the $\boldsymbol{G}_n$ asymptotically generate $B \cap F^c$. Suppose that $G_n \in \boldsymbol{G}_n$ is a decreasing sequence of nonempty atoms. By construction $G_n = A_n \cap F^c$ for some sequence of atoms A_n in $\boldsymbol{F}_n$. Since the original space is standard, the intersection $\bigcap_{n=1}^{\infty} A_n$ is nonempty, say some set $\underline{A}$. Suppose that $\underline{A} \cap F^c$ is empty, then $\underline{A} \subset F$ and hence $\underline{A}$ must be contained in one of the bad atoms, say $\underline{A} \subset B_m \in \boldsymbol{F}_m$. But the atoms of $\boldsymbol{F}_m$ are disjoint and $\underline{A} \subset A_m$ with $A_m \cap B_m = \emptyset$, a contradiction unless A_m is empty in violation of the assumptions. □

Next we consider spaces isomorphic to standard spaces. Recall that given two measurable spaces $(A,\boldsymbol{B})$ and $(B,\boldsymbol{S})$, an isomorphism f is a measurable function $f: A \rightarrow B$ that is one-to-one and has a measurable inverse f^{-1}. Recall also that if such an isomorphism exists, the two measurable spaces are said to be isomorphic. We first provide a useful technical lemma and then show that if two spaces are isomorphic, then one is standard if and only if the other one is standard.

Lemma 2.3.3. If $(A,\boldsymbol{B})$ and $(B,\boldsymbol{S})$ are isomorphic measurable spaces, let $f: A \rightarrow B$ denote an isomorphism. Then $\boldsymbol{B} = f^{-1}(\boldsymbol{S}) = \{$all sets of the form $f^{-1}(G)$ for G in $\boldsymbol{S}\}$.

Proof. Since f is measurable, the class of sets $\boldsymbol{F} = f^{-1}(\boldsymbol{S})$ has all its elements in $\boldsymbol{B}$. Conversely, since f is an isomorphism, its inverse function, say g, is well defined and measurable. Thus if $F \in \boldsymbol{B}$, then $g^{-1}(F) = f(F) = D \in \boldsymbol{S}$ and hence $F = f^{-1}(D)$. Thus every element of $\boldsymbol{B}$ is given by $f^{-1}(D)$ for some D in $\boldsymbol{S}$. In addition, $\boldsymbol{F}$ is a σ-field since the inverse image operation preserves all countable set theoretic operations.

Suppose that there is an element D in $\boldsymbol{B}$ that is not in $\boldsymbol{F}$. Let $g: B \to A$ denote the inverse operation of f. Since g is measurable and one-to-one, $g^{-1}(D) = f(D)$ is in $\boldsymbol{G}$ and therefore its inverse image under f, $f^{-1}(f(D)) = D$, must be in $\boldsymbol{F} = f^{-1}(\boldsymbol{G})$ since f is measurable. But this is a contradiction since D was assumed to not be in $\boldsymbol{F}$ and hence $\boldsymbol{F} = \boldsymbol{B}$. □

Corollary 2.3.2. If $(A,\boldsymbol{B})$ and $(B,\boldsymbol{S})$ are isomorphic, then $(A,\boldsymbol{B})$ is standard if and only if $(B,\boldsymbol{S})$ is standard.

Proof. Say that $(B,\boldsymbol{S})$ is standard and hence has a basis $\boldsymbol{G}_n$, $n = 1,2,\cdots$ and that $\boldsymbol{G}$ is the generating field formed by the union of all of the $\boldsymbol{G}_n$. Let $f: A \to B$ denote an isomorphism and define $\boldsymbol{F}_n = f^{-1}(\boldsymbol{G}_n)$, $n = 1,2,\cdots$ Since inverse images preserve set theoretic operations, the $\boldsymbol{F}_n$ are increasing finite fields with $\bigcup \boldsymbol{F}_n = f^{-1}(\boldsymbol{G})$. A straightforward application of the good sets principle show that if $\boldsymbol{G}$ generates $\boldsymbol{S}$ and $\boldsymbol{B} = f^{-1}(\boldsymbol{S})$, then $f^{-1}(\boldsymbol{G}) = \boldsymbol{F}$ also generates $\boldsymbol{B}$ and hence the $\boldsymbol{F}_n$ asymptotically generate $\boldsymbol{B}$. Given any decreasing set of atoms, say D_n, $n = 1,2,\cdots$, of $\boldsymbol{F}_n$, then $f(D_n) = g^{-1}(D_n)$ must also be a decreasing sequence of atoms of $\boldsymbol{G}_n$ (again, since inverse images preserve set theoretic operations). The latter sequence cannot be empty since the $\boldsymbol{G}_n$ form a basis, hence neither can the former sequence since the mappings are one-to-one. Reversing the argument shows that if either space is standard, then so is the other, completing the proof. □

As a final result of this section we show that if a field is standard, then we can always find a basis that is a binary indexed sequence of fields in the sense of Lemma 2.2.1. Indexing the atoms by binary vectors will prove a simple and useful representation.

Lemma 2.3.4. If a field $\boldsymbol{F}$ is standard, then there exists a binary indexed basis and a corresponding canonical binary sequence function.

Proof. The proof consists simply of relabeling of a basis in binary vectors. Since $\boldsymbol{F}$ is standard, then it possesses a basis, say $\boldsymbol{F}_n$, $n = 0,1,\cdots$. Use this basis to enumerate all of the elements of $\boldsymbol{F}$ in order, that is, first list the elements of $\boldsymbol{F}_0$, then the elements of $\boldsymbol{F}_1$, and so on. This produces a sequence $\{F_n, n = 0,1,\cdots\} = \boldsymbol{F}$. Now construct a corresponding binary indexed asymptotically generating sequence of fields, say $\boldsymbol{G}_n$, $n = 0,1,\cdots$, as in Lemma 2.2.1. To prove $\boldsymbol{G}_n$ is a basis we need only show that it has the finite intersection property of (2.2.4). First observe that for every n there is an $m \geq n$ such that $\boldsymbol{F}_n = \boldsymbol{G}_m$. That is, eventually the binary indexed fields $\boldsymbol{G}_m$ match the original fields, they just take longer to get all of the elements since they include them one new one at a time. If G_n is a decreasing sequence of atoms in $\boldsymbol{G}_n$, then there is a sequence of integers $n(k) \geq k$, $k = 1,2, \cdots$, such that $G_{n(k)} = B_k$, where B_k is an atom of $\boldsymbol{F}_k$. If G_n collapses to the empty set, then so must the subsequence $G_{n(k)}$ and hence the sequence of atoms B_k of $\boldsymbol{F}_k$. But this is impossible since the $\boldsymbol{F}_k$ are a basis. Thus the $\boldsymbol{G}_n$ must also possess the finite intersection property and be a basis. The remaining conclusion follows immediately from Lemma 2.2.1 and the definition that follows it. □

2.4 SIMPLE STANDARD SPACES

First consider a measurable space $(A,\boldsymbol{B})$ with A a finite set and $\boldsymbol{B}$ the class of all subsets of A. This space is trivially standard–simply let all fields in the basis be $\boldsymbol{B}$. Alternatively one can observe that since $\boldsymbol{B}$ is finite, the space trivially possesses the countable extension property and hence must be standard.

Next assume that $(A,\boldsymbol{B})$ has a countable alphabet and that $\boldsymbol{B}$ is again the class of all subsets of A. A natural first guess for a basis is to first enumerate the points of A as $\{a_0,a_1, \cdots\}$ and to define the increasing sequence of finite fields $\boldsymbol{F}_n = \mathit{field}(\{a_0\}, \cdots, \{a_{n-1}\})$. The sequence is indeed asymptotically generating, but it does not have the finite intersection property. The problem is that each field has as an atom the leftover or garbage set $G_n = \{a_n,a_{n+1}, \cdots\}$ containing all of the points of A not yet included in the generating class. This sequence of atoms is nonempty and decreasing, yet it converges to the empty set and hence $\boldsymbol{F}_n$ is not a basis.

In this case there is a simple fix. Simply adjoin the undesirable atoms G_n containing the garbage to a good atom, say a_0, to form a new sequence of finite fields $\boldsymbol{G}_n = \mathit{field}(b_0(n),b_1(n), \cdots, b_{n-1}(n))$ with $b_0(n) = a_0 \bigcup G_n$, $b_i(n) = a_i$, $i = 1, \cdots, n-1$. This sequence of fields also generates the full σ-field (note for example that $a_0 = \bigcap_{n=0}^{\infty} b_0(n)$). It also possesses the finite intersection property and hence is a basis. Thus any countable measurable space is standard.

The above example points out that it is easy to find sequences of asymptotically generating finite fields for a standard space that are not bases and hence which cannot be used to extend arbitrary finitely additive set functions. In this case, however, we were able to rig the field so that it did provide a basis.

From Corollary 2.3.1 any countable product of standard spaces must be standard. Thus products of finite or countable alphabets give standard spaces, a conclusion summarized in the following lemma.

Lemma 2.4.1. Given a countable index set $\boldsymbol{I}$, if A_i are finite or countably infinite alphabets and $\boldsymbol{B}_i$ the σ-fields of all subsets of A_i for $i \in \boldsymbol{I}$, then

$$\mathop{\times}_{i \in \boldsymbol{I}} (A_i, \boldsymbol{B}_i)$$

is standard. Thus, for example, the following spaces are standard:

$$(\boldsymbol{N}, \boldsymbol{B}_{\boldsymbol{N}}) = (\mathbf{Z}_+^{\mathbf{Z}_+}, \boldsymbol{B}_{\mathbf{Z}_+}^{\mathbf{Z}_+}), \tag{2.4.1}$$

where $\mathbf{Z}_+ = \{0,1, \cdots \}$ and $\boldsymbol{B}_{\mathbf{Z}_+}$ is the σ-field of all subsets of $\mathbf{Z}_+$ (called the *power set* of $\mathbf{Z}_+$), and

$$(\boldsymbol{M}, \boldsymbol{B}_{\boldsymbol{M}}) = (\{0,1\}^{\mathbf{Z}_+}, \boldsymbol{B}_{\{0,1\}}^{\mathbf{Z}_+}), \tag{2.4.2}$$

where $\boldsymbol{B}_{\{0,1\}}$ is the σ-field of all subsets of $\{0,1\}$.

The proof of Lemma 2.3.1 coupled with the discussion for the finite and countable cases provide a specific construction for the bases of the above two processes. For the case of the binary sequence space $\boldsymbol{M}$, the fields $\boldsymbol{F}_n$ consist of n dimensional rectangles of the form $\{x: x_i \in B_i, i = 0,1, \cdots, n-1\}$, $B_i \subset \{0,1\}$ all i. These rectangles are themselves finite unions of simpler rectangles of the form

$$c(b^n) = \{x : x_i = b_i \,;\, i = 0,1, \cdots, n-1\} \tag{2.4.3}$$

for $b^n \in \{0,1\}^n$. These sets are called *thin cylinders* and are the atoms of $\boldsymbol{F}_n$. The thin cylinders generate the full σ-field and any decreasing sequence of such rectangles converges down to a single binary sequence. Hence it is easy to show that this space is standard. Observe that this also implies that individual binary sequences $x = \{x_i \,;\, i = 0,1, \cdots \}$ are themselves events since they are a countable intersection of thin cylinders. For the case of the integer sequence space $\boldsymbol{N}$, the field $\boldsymbol{F}_n$ contains all rectangles of the form $\{x: x_i \in F_i, i = 0,1, \cdots, n\}$, where $F_i \in \boldsymbol{F}_i(n)$ for all i, and $F_i(n) = \{b_0(n), b_1(n), \cdots, b_{n-1}(n)\}$ for all i with $b_0(n) = \{0, n, n+1, n+2, \cdots \}$ and $b_j = \{j\}$ for $j = 1,2, \cdots, n-1$.

As previously mentioned, it is not true in general that an arbitrary subset of a standard space is standard, but the previous construction of a basis combined with Lemma 2.3.2 does provide a means of showing that certain subspaces of $\boldsymbol{M}$ and $\boldsymbol{N}$ are standard.

Lemma 2.4.2. Given the binary sequence space $\boldsymbol{M}$ or the integer sequence space $\boldsymbol{N}$, if we remove from these spaces any countable collection of rectangles, then the remaining space is standard.

Proof. The rectangles are eventually in the fields of the basis and the result follows from Lemma 2.3.2. □

A natural next step would be to look at the unit interval or the real line as alphabets. We first need, however, to construct appropriate σ-fields for these cases. Little effort is saved by focusing on these special cases instead of the general case of metric spaces, which will be introduced in the next section and considered in depth in the next chapter.

Exercises

1. Consider the space of (2.4.2) of all binary sequences and the corresponding event space. Is the class of all finite unions of thin cylinders a field? Show that a decreasing sequence of nonempty atoms of $\boldsymbol{F}_n$ must converge to a point, i.e., a single binary sequence.

2.5 METRIC SPACES

In this section we collect together several definitions, examples, and elementary convergence properties of metric spaces that are required to prove the fundamental extension theorem for standard spaces.

A set A with elements called *points* is called a *metric space* if for every pair of points a,b in A there is an associated nonnegative real number $d(a,b)$ such that

$$d(a,b) > 0 \text{ if and only if } a \neq b, \tag{2.5.1}$$

$$d(a,b) = d(b,a) \quad \text{(symmetry)}, \tag{2.5.2}$$

$$d(a,b) \leq d(a,c) + d(c,b) \text{ all } c \in A \text{ (triangle inequality)}. \tag{2.5.3}$$

The function $d: A \times A \to [0,\infty)$ is called a *metric* or *distance*. If $d(a,b) = 0$ does not necessarily imply that $a=b$, then d is called a *pseudo-metric* and

A is a pseudo-metric space.

Metric spaces provide mathematical models of many interesting alphabets. Several common examples are listed below.

Example 2.5.1: A a discrete (finite or countable) set with metric

$$d(a,b) = d_H(a,b) = 1-\delta_{a,b},$$

where $\delta_{a,b}$ is the Kronecker delta function which is 1 if $a = b$ and 0 otherwise. This metric is called the *Hamming distance* and assigns a distance of 0 between a letter and itself and 1 between any two distinct letters.

Example 2.5.2: $A = \mathbf{Z}(M) = \{0,1, \cdots, M-1\}$ with metric $d(a,b) = d_L(a,b) = |a-b| \bmod M$, where $k \bmod M = r$ in the unique representation $k = aM + r$, a an integer, $0 \le r \le M-1$. d_L is called the *Lee metric* or *circular distance.*

Example 2.5.3: A_0 a discrete (finite or countable) set and $A = A_0^n$ (the space of all n-dimensional vectors with coordinates in A_0) and metric

$$d(x^n,y^n) = n^{-1} \sum_{i=0}^{n-1} d_H(x_i,y_i) ,$$

the arithmetic average of the Hamming distances in the coordinates. This distance is also called the Hamming distance, but the name is ambiguous since the Hamming distance in the sense of Example 2.5.1 applied to A can take only values of 0 and 1 while the distance of this example can take on values of i/n; $i = 0,1, \cdots, n-1$. Better names for this example are the average Hamming distance or mean Hamming distance. This distance is useful in error correcting and error detecting coding examples as a measure of the change in digital codewords caused by noisy communication media.

Example 2.5.4: A as in the preceding example. Given two elements x^n and y^n of A, an ordered set of integers $\boldsymbol{K} = \{(k_i,j_i) ; i = 1, \cdots, r\}$ is called a *match* of x^n and y^n if $x_{k_i} = y_{j_i}$ for $i = 1,2, \cdots, r$. The number $r = r(\boldsymbol{K})$ is called the *size* of the match $\boldsymbol{K}$. Let $\boldsymbol{K}(x^n,y^n)$ denote the collection of all matches of x^n and y^n. The *best match size* $m_n(x^n,y^n)$ is the largest r such that there is a match of size r, that is,

$$m_n(x^n,y^n) = \max_{\boldsymbol{K} \in \boldsymbol{K}(x^n,y^n)} r(\boldsymbol{K}) .$$

The *Levenshtein distance* λ_n on A is defined by

$$\lambda_n(x^n,y^n) = n - m_n(x^n,y^n) = \min_{\boldsymbol{K} \in \boldsymbol{K}(x^n,y^n)} (n - r(\boldsymbol{K})) ,$$

that is, it is the minimum number of insertions, deletions, and substitutions required to map one of the n-tuples into the other [5, 10]. The Levenshtein distance is useful in correcting and detecting errors in communication systems where the communication medium may not only cause symbol errors: it can also drop symbols or insert erroneous symbols. Note that unlike the previous example, the distance between vectors is not the sum of the component distances, that is, it is not additive.

Example 2.5.5: $A = \mathbf{R} = (-\infty,\infty)$, the real line, and $d(a,b) = | a - b |$, the absolute magnitude of the difference.

Example 2.5.6: $A = \mathbf{R}^n$, n-dimensional Euclidean space, with

$$d(a^n,b^n) = \left\{ \sum_{i=0}^{n-1} | a_i - b_i |^2 \right\}^{1/2} ,$$

the Euclidean distance.

Example 2.5.7: $A = \mathbf{R}^n$ with

$$d(a^n,b^n) = \max_{i=0,\cdots,n-1} | a_i - b_i |$$

Example 2.5.8: A normed linear vector space A with norm $| . |$ and distance $d(a,b) = | a - b |$. A vector space or linear space A is a space consisting of points called vectors and two operations–one called *addition* which associates with each pair of vectors a,b a new vector $a + b \in A$ and one called *multiplication* which associates with each vector a and each number $r \in \mathbf{R}$ (called a scalar) a vector $ra \in A$–for which the usual rules of arithmetic hold, that is, if $a,b,c \in A$ and $r,s \in \mathbf{R}$, then

$$a + b = b + a \text{ (commutative law)}$$
$$(a + b) + c = a + (b + c) \text{ (associative law)}$$
$$r(a + b) = ra + rb \text{ ; } (r + s)a = ra + sa \text{ (distributive laws)}$$
$$(rs)a = r(sa) \text{ (associative law for multiplication)}$$
$$1a = a.$$

In addition, it is assumed that there is a zero vector, also denoted 0, for which

$$a + 0 = a$$

$$0a = 0.$$

We also define the vector $-a = (-1)a$. A normed linear space is a linear vector space A together with a function $\|a\|$ called a norm defined for each a such that for all $a,b \in A$ and $r \in \mathbf{R}$, $\| a \|$ is nonnegative and

$$\| a \| = 0 \text{ if and only if } a = 0, \tag{2.5.4}$$

$$\| a + b \| \leq \| a \| + \| b \| \text{ (triangle inequality)}, \tag{2.5.5}$$

$$\| ra \| = | r | \, \| a \| . \tag{2.5.6}$$

If $\| a \| = 0$ does not imply that $a = 0$, then $\| \, . \, \|$ is called a seminorm. For example, the Euclidean space example above is a normed linear space with norm

$$\| a \| = \left[\sum_{i=0}^{n-1} a_i^2 \right]^{1/2} .$$

Example 2.5.9: For some dimension k, let $\mathbf{P}$ denote the space of all stochastic matrices, that is, nonnegative matrices whose rows sum to one. (These matrices arise in the study of Markov chains.) Let $\| P \|$, $P \in \mathbf{P}$ be a matrix norm such as

$$\| P \| = (\sum_{i,j} P_{ij}^2)^{1/2} ,$$

the Euclidean norm. Then $\mathbf{P}$ is a normed linear space. (See, e.g., Paz [8] for discussion and applications.)

Example 2.5.10: An inner product space A with inner product $(\, . \, , \, . \,)$ and distance $d(a,b) = \| a - b \|$, where $\| a \| = (a,a)^{1/2}$ is a norm. An inner product space (or pre-Hilbert space) is a linear vector space A such that for each pair of vectors $a,b \in A$ there is a real number (a,b) called an inner product such that for $a,b,c \in A$, $r \in \mathbf{R}$

$$(a,b) = (b,a),$$

$$(a + b,c) = (a,c) + (b,c),$$

$$(ra,b) = r(a,b),$$

$$(a,a) \geq 0 \text{ and } (a,a) = 0 \text{ if and only if } a = 0.$$

Inner product spaces and normed linear vector spaces include a variety of general alphabets and function spaces, many of which will be encountered in this book.

Example 2.5.11: Product spaces. Given a metric space A with metric d and an index set $\boldsymbol{I}$ that is some subset of the integers (e.g., $\mathbf{Z}_+ = \{0,1,2, \cdots\}$), the cartesian product

$$A^{\boldsymbol{I}} = \times_{i \in \boldsymbol{I}} A_i,$$

where the A_i are all replicas of A, is a metric space with metric

$$d^{\boldsymbol{I}}(a,b) = \sum_{i \in \boldsymbol{I}} 2^{-|i|} \frac{d(a_i,b_i)}{1 + d(a_i,b_i)} .$$

The somewhat strange form above ensures that the metric on the sequences in $A^{\boldsymbol{I}}$ is finite–simply adding up the component metrics would likely yield a sum that blows up. Dividing by 1 plus the metric ensures that each term is bounded and the $2^{-|i|}$ ensure convergence of the infinite sum. In the special case where the index set $\boldsymbol{I}$ is finite, then the product space is a metric space with the simpler metric

$$d(a,b) = \sum_{i \in \boldsymbol{I}} d(a_i,b_i) .$$

Example 2.5.12: And now for something completely different: Suppose that $(\Omega,\boldsymbol{B},m)$ is a probability space. Then we can form a psuedo metric space $A = \boldsymbol{B}$ with

$$d(F,G) = m(F \Delta G) .$$

This is only a pseudo-metric because events which differ on a set of probability 0 yield a distance of 0. It can be considered as a metric space if we consider events to be identical if they differ on a set of probability zero.

Example 2.5.13: Suppose that $(\Omega,\boldsymbol{B},m)$ is a probability space and let P denote the class of all measurable partitions of Ω with K atoms, that is, a member of **P** has the form $P = \{P_k, i = 1,2, \cdots, K\}$, where the $P_i \in \boldsymbol{B}$ all i are disjoint. A metric on the space is

$$d(P,Q) = \sum_{i=1}^{K} m(P_i \Delta Q_i) .$$

This metric is called the *partition distance.*

Convergence

A sequence a_n, $n = 1,2, \cdots$ of points in a metric space A is said to *converge* to a point a if

$$\lim_{n\to\infty} d(a_n,a) = 0,$$

that is, if given $\varepsilon > 0$ there is an N such that $n \geq N$ implies that $d(a_n,a) \leq \varepsilon$. If a_n converges to a, then we write $a = \lim_{n\to\infty} a_n$ or $a_n \to_{n\to\infty} a$. A metric space A is said to be *sequentially compact* if every sequence has a convergent subsequence, that is, if $a_n \in A$, $n = 1,2, \cdots$, then there is a subsequence $n(k)$ of increasing integers and a point a in A such that $\lim_{k\to\infty} a_{n(k)} = a$. The notion of sequential compactness is considered somewhat old-fashioned in modern analysis texts, but it is exactly what we need.

We will need the following two easy results on convergence in product spaces.

Lemma 2.5.1. Given a metric space A with metric d and a countable index set $\boldsymbol{I}$, let $A^{\boldsymbol{I}}$ denote the cartesian product with the metric of Example 2.5.11. A sequence $x_n^{\boldsymbol{I}} = \{x_{n,i};\ i \in \boldsymbol{I}\}$ converges to a sequence $x^{\boldsymbol{I}} = \{x_i,\ i \in \boldsymbol{I}\}$ if and only if $\lim_{n\to\infty} x_{n,i} = x_i$ for all $i \in \boldsymbol{I}$.

Proof. The metric on the product space of Example 2.5.11 goes to zero if and only if all of the individual terms of the sum go to zero. □

Corollary 2.5.1. Let A and $A^{\boldsymbol{I}}$ be as in the lemma. If A is sequentially compact, then so is $A^{\boldsymbol{I}}$.

Proof. Let x_n, $n = 1,2, \cdots$ be a sequence in the product space A. Since the first coordinate alphabet is sequentially compact, we can choose a subsequence, say x_n^1, that converges in the first coordinate. Since the second coordinate alphabet is sequentially compact, we can choose a further subsequence, say x_n^2, of x_n^1 that converges in the second coordinate. We continue in this manner, each time taking a subsequence of the previous subsequence that converges in one additional coordinate. The so-called diagonal sequence x_n^n will then converge in all coordinates as $n\to\infty$ and hence by the lemma will converge with respect to the given product space metric. This method of taking subsequences of subsequences with desirable properties is referred to as a *standard diagonalization.* □

The convergence notions developed in this section yield an alternative characterization of standard spaces that provides some additional insight into their structure and plays a key role in the extension to come.

Theorem 2.5.1. A field $\boldsymbol{F}$ of subsets of Ω is standard if and only if (a) $\boldsymbol{F}$ is countable, and (b) if F_n; $n = 1,2,\cdots$, is a sequence of nonempty disjoint elements in $\boldsymbol{F}$, then $\bigcup_n F_n \notin \boldsymbol{F}$. Given (a), condition (b) is equivalent to the condition (b') that if G_n; $n = 1,2,\cdots$, is a sequence of strictly decreasing nonempty elements in $\boldsymbol{F}$, then $\bigcap_n G_n \neq \emptyset$.

Proof. We begin by showing that (b) and (b') are equivalent. We first show that (b) implies (b'). Suppose that G_n satisfy the hypothesis of (b'). Then

$$H = (\bigcap_{n=1}^{\infty} G_n)^c = \bigcup_{n=1}^{\infty} G_n^c = \bigcup_{n=1}^{\infty} (G_n^c - G_{n-1}^c),$$

where $G_0 = \Omega$. Defining $F_n = G_n^c - G_{n-1}^c$ yields a sequence satisfying the conditions of (a) and hence H and therefore H^c are not in $\boldsymbol{F}$, showing that $\bigcap_n G_n = H^c$ cannot be empty (or it would be in $\boldsymbol{F}$ since a field must contain the empty set). To prove that (b') implies (b), suppose that the sequence F_n satisfies the hypothesis of (b) and define $F = \bigcup_n F_n$ and $G_n = F - \bigcup_{j=1}^n F_j$. Then G_n is a strictly decreasing sequence of nonempty sets with empty intersection. From (b'), the sequence G_n cannot all be in $\boldsymbol{F}$, which implies that $F \notin \boldsymbol{F}$.

We now show that the equivalent conditions (b) and (b') hold if and only if the field is standard. First suppose that (b) and (b') hold and that $\boldsymbol{F} = \{F_n, n = 1,2,\cdots\}$. Define $\boldsymbol{F}_n = \mathit{field}(F_1,F_2,\cdots,F_n)$ for $n = 1,2,\cdots$. Clearly $\boldsymbol{F} = \bigcup_{n=1}^{\infty}\boldsymbol{F}_n$ since $\boldsymbol{F}$ contains all of the elements of all of the $\boldsymbol{F}_n$ and each element in $\boldsymbol{F}$ is in some $\boldsymbol{F}_n$. Let A_n be a nonincreasing sequence of nonempty atoms in $\boldsymbol{F}_n$ as in the hypothesis for a basis. If the A_n are all equal for large n, then clearly $\bigcap_{n=1}^{\infty} A_n \neq \emptyset$. If this is not true, then there is a strictly decreasing subsequence A_{n_k} and hence from (b')

$$\bigcap_{n=1}^{\infty} A_n = \bigcap_{k=1}^{\infty} A_{n_k} \neq \emptyset .$$

This means that the $\boldsymbol{F}_n$ form a basis for $\boldsymbol{F}$ and hence $\boldsymbol{F}$ is standard.

Lastly, suppose that $\boldsymbol{F}$ is standard with basis $\boldsymbol{F}_n$, $n = 1,2, \cdots$, and suppose that G_n is a sequence of strictly decreasing elements in $\boldsymbol{F}$ as in (b'). Define the nonincreasing sequence of nonempty sets $H_n \in \boldsymbol{F}_n$ as follows: If $G_1 \in \boldsymbol{F}_1$, then set $H_1 = G_1$ and set $n_1 = 1$. If this is not the case, then G_1 is too "fine" for the field $\boldsymbol{F}_1$ so we set $H_1 = \Omega$. Continue looking for an n_1 such that $G_1 \in \boldsymbol{F}_{n_1}$. When it is found (as it must be since G_1 is a field element), set $H_i = \Omega$ for $i < n_1$ and $H_{n_1} = G_1$. Next consider G_2. If $G_2 \in \boldsymbol{F}_{n_1+1}$, then set $H_{n_1+1} = G_2$. Otherwise set $H_{n_1+1} = G_1$ and find an n_2 such that $G_2 \in \boldsymbol{F}_{n_2}$. Then set $H_i = G_1$ for $n_1 \leq i < n_2$ and $H_{n_2} = G_2$. Continue in this way "filling in" the sequence of G_n with enough repeats to meet the condition. The sequence $H_n \in \boldsymbol{F}_n$ is thus a nonempty nonincreasing sequence of field elements. Clearly by construction

$$\bigcap_{n=1}^{\infty} H_n = \bigcap_{n=1}^{\infty} G_n$$

and hence we will be done if we can prove this intersection nonempty. We accomplish this by assuming that the intersection is empty and show that this leads to a contradiction with the assumption of a standard space. Roughly speaking, a decreasing set of nonempty field elements collapsing to the empty set must contain a decreasing sequence of nonempty basis atoms collapsing to the empty set and that violates the properties of a basis.

Pick for each n an $\omega_n \in H_n$ and consider the sequence $f(\omega_n)$ in the binary sequence space $\boldsymbol{M}$. The binary alphabet $\{0,1\}$ is trivially sequentially compact and hence so is $\boldsymbol{M}$ from Corollary 2.5.1. Thus there is a subsequence, say $f(\omega_{n(k)})$, which converges to some $u \in M$ as $k \to \infty$. From Lemma 2.5.1 this means that $f(\omega_{n(k)})$ must converge to u in each coordinate as $k \to \infty$. Since there are only two possible values for each coordinate in u, this means that given any positive integer n, there is an N such that for $n(k) \geq N$, $f(\omega_{n(k)})_i = u_i$ for $i = 0,1, \cdots, n-1$, and hence from (2.2.1) that $\omega_{n(k)} \in A_{u^n}$, the atom of $\boldsymbol{F}_n$ indexed by u^n. Assuming that we also choose k large enough to ensure that $n(k) \geq n$, then the fact that the fields are decreasing implies that $\omega_{n(k)} \in H_{n(k)} \subset H_n$. Thus we must have that G_{u^n} is contained in H_n. That is, the point $\omega_{n(k)}$ is contained in an atom of $\boldsymbol{F}_{n(k)}$ which is itself contained in an atom of $\boldsymbol{F}_n$. Since the point is also in the set H_n of $\boldsymbol{F}_n$, the set must contain the entire atom.

Thus we have constructed a sequence of atoms contained in a sequence of sets that decreases to the empty set, which violates the standard assumption and hence completes the proof. □

Exercises

1. Show that in inner product space with norm $\| . \|$ satisfies the parallelogram law

$$\| a+b \|^2 + \| a-b \|^2 = 2\| a \|^2 + 2\| b \|^2$$

2. Show that d' and d of Example 2.5.11 are metrics.

3. Show that the partition distance in the above example satisfies

$$d(P,Q) = 2\sum_{i \neq j} m(P_i \cap Q_j) = 2(1 - \sum_i m(P_i \cap Q_i)) .$$

4. Show that if d_i; $i = 0,1, \cdots, n-1$ are all metrics on A, then so is $d(x,y) = \max_i d_i(x,y)$.

5. Suppose that d is a metric on A and r is a fixed positive real number. Define $\rho(x,y)$ to be 0 if $d(x,y) \leq r$ and K if $d(x,y) > r$. Is ρ a metric? a pseudo-metric?

6. Suppose that $X:\Omega \to A$ and $Y:\Omega \to A$ are two random variables defined on a common probability space $(\Omega, \boldsymbol{B}, m)$ and that A is a finite set $\{a_1, \cdots, a_K\}$. Define the sets $P_i = X^{-1}(a_i)$ and $Q_i = Y^{-1}(a_i)$, $i = 1,2, \cdots, K$. Show that the collections of sets $\boldsymbol{P} = \{P_i\}$ and $\boldsymbol{Q} = \{Q_i\}$ are partitions of Ω. Let d denote the partition distance of Example 2.5.12. Show that

$$d(\boldsymbol{P},\boldsymbol{Q}) = 2m(X \neq Y) .$$

2.6 EXTENSION IN STANDARD SPACES

In this section we combine the ideas of the previous two sections to complete the proofs of the following results.

Theorem 2.6.1. A field has the countable extension property if and only if it is standard.

Corollary 2.6.1. A measurable space has the countable extension property if and only if it is standard.

These results will complete the characterization of those spaces having the desirable extension property discussed at the beginning of the section. The "only if" portions of the results are contained in Section 2.2. The corollary follows immediately from the theorem and the definition of standard. Hence we need only now show that a standard field has the countable extension property.

Proof of the theorem. Since $\boldsymbol{F}$ is standard, construct as in Lemma 2.2.1 a basis with a binary indexed sequence of atoms and let $f : \Omega \to M$ denote the corresponding canonical binary sequence function of (2.2.2). Let P be a nonnegative, normalized, finitely additive set function of $\boldsymbol{F}$. We wish to show that (1.2.8) is satisfied. Let F_n be a sequence of field elements decreasing to the null set. If for some finite $n = N$, F_N is empty, then we are done since then

$$\lim_{n \to \infty} P(F_n) = P(F_N) = P(\emptyset) = 0 . \tag{2.6.1}$$

Thus we need only consider sequences F_n of nonempty decreasing field elements that converge to the empty set. From Theorem 2.5.1, however, there cannot be such sequences since the field is standard. Thus (2.6.1) trivially implies (1.2.8) and the theorem is proved by the Carathéodory extension theorem. □

The key to proving sufficiency above is the characterization of standard fields in Theorem 2.5.1. This has the interpretation using part (b) that a field is standard if and only if truly countable unions of nonempty disjoint field elements are not themselves in the field. Thus finitely additive set functions are trivially countably additive on the field because all countable disjoint unions of field elements can be written as a finite union of disjoint nonempty field elements.

2.7 THE KOLMOGOROV EXTENSION THEOREM

In Chapter 1 in the section on distributions we saw that given a random process, we can determine the distributions of all finite dimensional random vectors produced by the process. We now possess the machinery to prove the reverse result–the Kolmogorov extension theorem–which states that given a family of distributions of finite dimensional random vectors that is consistent, that is, higher dimensional distributions imply the lower dimensional ones, then there is a process distribution that agrees with the given family. Thus a consistent family of finite dimensional distributions is sufficient to completely specify a random process possessing the distributions. This result is one of the most important in probability theory since it provides realistic requirements for the construction of mathematical models of physical phenomena.

Theorem 2.7.1. Let $\boldsymbol{I}$ be a countable index set, let $(A_i,\boldsymbol{B}_i)$, $i \in \boldsymbol{I}$, be a family of standard measurable spaces, and let

$$(A^{\boldsymbol{I}},\boldsymbol{B}^{\boldsymbol{I}}) = \times_{i \in \boldsymbol{I}} (A_i,\boldsymbol{B}_i)$$

be the product space. Similarly define for any index set $\boldsymbol{M} \subset \boldsymbol{I}$ the product space $(A^{\boldsymbol{M}},\boldsymbol{B}^{\boldsymbol{M}})$. Given index subsets $\boldsymbol{K} \subset \boldsymbol{M} \subset \boldsymbol{I}$, define the projections $\Pi_{\boldsymbol{M}\to\boldsymbol{K}}$: $A^{\boldsymbol{M}} \to A^{\boldsymbol{K}}$ by

$$\Pi_{\boldsymbol{M}\to\boldsymbol{K}}(x^{\boldsymbol{M}}) = x^{\boldsymbol{K}} ,$$

that is, the projection $\Pi_{\boldsymbol{M}\to\boldsymbol{K}}$ simply looks at the vector or sequence $x^{\boldsymbol{M}} = \{x_i, i \in \boldsymbol{M}\}$ and produces the subsequence $x^{\boldsymbol{K}}$. A family of probability measures $P_{\boldsymbol{M}}$ on the measurable spaces $(A^{\boldsymbol{M}},\boldsymbol{B}^{\boldsymbol{M}})$ for all finite index subsets $\boldsymbol{M} \subset \boldsymbol{I}$ is said to be *consistent* if whenever $\boldsymbol{K} \subset \boldsymbol{M}$, then

$$P_{\boldsymbol{K}}(F) = P_{\boldsymbol{M}}(\Pi^{-1}_{\boldsymbol{M}\to\boldsymbol{K}}(F)), \text{ all } F \in \boldsymbol{B}^{\boldsymbol{K}}, \tag{2.7.1}$$

that is, probability measures on index subsets agree with those on further subsets. Given any consistent family of probability measures, then there exists a unique process probability measure P on $(A^{\boldsymbol{I}},\boldsymbol{B}^{\boldsymbol{I}})$ which agrees with the given family, that is, such that for all $\boldsymbol{M} \subset \boldsymbol{I}$

$$P_{\boldsymbol{M}}(F) = P(\Pi^{-1}_{\boldsymbol{I}\to\boldsymbol{M}}(F)), \text{ all } F \in B^{\boldsymbol{M}}, \text{all finite } \boldsymbol{M} \subset \boldsymbol{I}. \tag{2.7.2}$$

Proof. Abbreviate by Π_i: $A^{\boldsymbol{I}} \to A_i$ the one dimensional sampling function defined by $\Pi_i(x^{\boldsymbol{I}}) = x_i$, $i \in \boldsymbol{I}$. The consistent families of probability measures induces a set function P on all rectangles of the form

$$B = \bigcap_{i \in \boldsymbol{M}} \Pi_i^{-1}(B_i)$$

for any $B_i \in \boldsymbol{B}_i$ via the relation

$$P(B) = P_{\boldsymbol{M}}(\times_{i \in \boldsymbol{M}} B_i) , \tag{2.7.3}$$

that is, we can take (2.7.2) as a definition of P on rectangles. Since the spaces $(A_i,\boldsymbol{B}_i)$ are standard, each has a basis $\{\boldsymbol{F}_i(n),\ n = 1,2, \cdots\}$ and the corresponding generating field $\boldsymbol{F}_i$. From Lemma 2.3.1 and Corollary 2.3.1 the product fields given by (2.3.1) form a basis for $\boldsymbol{F} = field(RECT(\boldsymbol{F}_i,\ i \in \boldsymbol{I}))$, which in turn generates $\boldsymbol{B}^{\boldsymbol{I}}$; hence P is also countably additive on $\boldsymbol{F}$. Hence from the Carathéodory extension theorem, P extends to a unique probability measure on $(A^{\boldsymbol{I}},\boldsymbol{B}^{\boldsymbol{I}})$. Lastly observe that for any finite index set $\boldsymbol{M} \subset \boldsymbol{I}$, both $P_{\boldsymbol{M}}$ and $P(\Pi_{\boldsymbol{I}\to\boldsymbol{M}})^{-1}$ (recall that $Pf^{-1}(F)$ is defined as $P(f^{-1}(F))$ for any function f: it is the distribution of the random variable f) are probability measures on $(A^{\boldsymbol{M}},\boldsymbol{B}^{\boldsymbol{M}})$ that agree on a generating field $field(RECT(\boldsymbol{F}_i, i \in \boldsymbol{M}))$ and hence they must be identical. Thus the process distribution indeed agrees with the finite dimensional distributions for all possible index subsets, completing the proof. □

By simply observing in the above proof that we only required the probability measure on the coordinate generating fields and that these fields are themselves standard, we can weaken the hypotheses of the theorem somewhat to obtain the following corollary. The details of the proof are left as an exercise.

Corollary 2.7.1. Given standard measurable spaces $(A_i,\boldsymbol{B}_i)$, $i \in \boldsymbol{I}$, for a countable index set $\boldsymbol{I}$, let $\boldsymbol{F}_i$ be corresponding generating fields possessing a basis. If we are given a family of normalized, nonnegative, finitely additive set functions $P_{\boldsymbol{M}}$ that are defined on all sets F in $field(RECT(\boldsymbol{F}_i,\ i \in \boldsymbol{I}))$ and are consistent in the sense of satisfying (2.7.2) for all such sets, then the conclusions of the previous theorem hold.

In summary, if we have a sequence space composed of standard measurable spaces and a family of finitely additive candidate probability measures that are consistent, then there exists a random unique process or a process distribution that agrees with the given probabilities on the generating events.

2.8 EXTENSION WITHOUT A BASIS

The principal advantage of a standard space is the countable extension property ensuring that *any* finitely additive candidate probability measure on a standard field is also countably additive. We have not yet, however, actually demonstrated an example of the construction of a probability measure by extension. To accomplish this in the manner indicated we must explicitly construct a basis for a sample space and then provide a finitely additive set function on the associated finite fields.

We have demonstrated an explicit basis for the special case of a discrete space and for the relatively complicated example of a product space with discrete coordinate spaces. Hence in principle we can construct measures on these spaces by providing a finitely additive set function for this basis. As a specific example, consider the binary sequence space $(\boldsymbol{M},\boldsymbol{B_M})$ of (2.4.2). Define the set function m on thin cylinders $c(b^n) = \{x{:}x^n = b^n\}$ by

$$m(c(b^n)) = 2^{-n} \text{ ; all } b^n \in \{0,1\}^n \text{ , } n = 1,2,3, \cdots$$

and extend this to $\boldsymbol{F}_n$ and hence to $\boldsymbol{F} = \bigcup_{n=1}^{\infty}\boldsymbol{F}_n$ in a finitely additive way, that is, any $G \in \boldsymbol{F}$ is in $\boldsymbol{F}_n$ for some n and hence can be written as a disjoint union of thin cylinders, say $G = \bigcup_{i=1}^{N} c(b_i^n)$. In this case define $m(G) = \sum_{i=1}^{N} m(c(b_i^N))$. This defines m for all events in $\boldsymbol{F}$ and the resulting m is finitely additive by construction. It was shown in Section 2.4.2 that the $\boldsymbol{F}_n$, $n = 1,2, \cdots$ form a basis for $\boldsymbol{F}$, however, and hence m extends uniquely to a measure on $\boldsymbol{B_M}$. Thus with a basis we can easily construct probability measures by defining them on simple sets and demonstrating only finite and not countable additivity.

Observe that in this case of equal probability on all thin cylinders of a given length, the probability of individual sequences must be 0 from the continuity of probability. For example, given a sequence $x = (x_0,x_1, \cdots)$, we have that

$$\bigcap_{i=1}^{n} c(x^n) \downarrow x$$

and hence

$$m(\{x\}) = \lim_{n\to\infty} m(c(x^n)) = \lim_{n\to\infty} 2^{-n} = 0 \text{ .}$$

One goal of this section is to show that it is not easy to construct such a basis in general, even in an apparently simple case such as the unit interval [0,1). Thus although a standard space may be useful in theory, it may be less useful in practice if one only knows that a basis exists but is unable to construct one.

A second goal of this section is to show that the inability to construct a basis may not be a problem when trying to explicitly construct probability measures. A basis is nice and important for theory because *all* finitely additive candidate probability measures on a basis are also countably additive and hence have an extension. If we are trying to prove a *specific* candidate probability measure is countably additive, we can sometimes use the structure of a related standard space without having to explicitly describe a basis.

To demonstrate the above considerations we focus on a particular example: The Lebesgue measure on the unit interval. We shall show the difficulties in trying to find a basis and how to circumvent them by instead using the structure of binary sequence space.

Suppose that we wish to construct a probability measure m on the unit interval $[0,1) = \{r: 0 \leq r < 1\}$. We would like the probability measure to have the property that the probability of an interval is proportional to its length, regardless of whether or not the end points are included. Thus, if $0 \leq a < b < 1$, then

$$m([a,b)) = m((a,b)) = b - a . \tag{2.8.1}$$

The question is, is this enough to completely describe a measure on [0,1)? That is, does this set function extend to a normalized, nonnegative, countably additive set function on a suitably defined σ-field?

We first need a useful σ-field for this alphabet. Deferring a general development to Chapter 3, we here content ourselves to observe that the σ-field should at least contain all intervals of the form (a,b) for $0 < a < b < 1$. Hence define $\boldsymbol{B}_{[0,1)} = \sigma(\text{all intervals})$. As we shall later see, this is the *Borel field* of [0,1). This σ-field is countably generated, e.g., by the countable sequence of sets $[0,k/n)$ for all positive integers k and n (since any interval can be formed as an increasing or decreasing limit of simple combinations of such sets). Thus a natural field to consider is $\boldsymbol{F} = \{$all finite set theoretic combinations of sets of the form $[0,k/n)\}$. This field is countable, generates $\boldsymbol{B}_{[0,1)}$, and contains all the intervals with rational endpoints, e.g., all sets of the form $[a,b)$ for a and b rational. It should seem reasonable that this is the correct field to consider.

It is easy to show that all elements of $\boldsymbol{F}$ can be written as finite unions of disjoint sets of the form $[a,b)$ for rational a and b. Clearly m can thus be defined on $\boldsymbol{F}$ in a reasonable way, e.g., if

$$F = \bigcup_{i=1}^{n} [a_i,b_i)$$

and the intervals are disjoint, then

$$m(F) = \sum_{i=1}^{n} m([a_i,b_i)) = \sum_{i=1}^{n} b_i - a_i \ . \tag{2.8.2}$$

Again, the question is whether m is countably additive on $\boldsymbol{F}$.

The discussion of standard spaces suggests that we should construct a basis $\boldsymbol{F}_n \uparrow \boldsymbol{F}$. Knowing that m is finitely additive on $\boldsymbol{F}_n$ will then suffice to prove that it is countably additive on $\boldsymbol{F}$ and hence uniquely extends to $\boldsymbol{B}_{[0,1)}$.

So far so good. Unfortunately, however, the author is unable to find a simple, easily describable basis for $\boldsymbol{F}$. A natural guess is to have ever larger and finer $\boldsymbol{F}_n$ generated by intervals with rational endpoints. For example, let $\boldsymbol{F}_n$ be all finite unions of intervals of the form

$$G_{n,k} = [\ k2^{-n},\ (k+1)2^{-n}\ )\ ,\ k = 0,1, \cdots, 2^n \ . \tag{2.8.3}$$

Analogous to the binary sequence example, the measure of the atoms $G_{n,k}$ of $\boldsymbol{F}_n$ is 2^{-n} and the set function extends to a finitely additive set function on $\boldsymbol{F}$. Clearly $\boldsymbol{F}_n \uparrow \boldsymbol{F}$, but unfortunately this simple sequence does not form a basis because atoms can collapse to empty sets! As an example, consider the atoms $G_n \in \boldsymbol{F}_n$ defined by

$$G_n = [k_n 2^{-n},\ (k_n + 1)2^{-n})\ ,$$

where we choose $k_n = 2^{n-1} - 1$ so that $G_n = [2^{-1} - 2^{-n}, 2^{-1})$. The atoms G_n are nonempty and have nonempty finite intersections, but

$$\bigcap_{n=1}^{\infty} G_n = \emptyset\ ,$$

which violates the conditions required for $\boldsymbol{F}_n$ to form a basis.

Unlike the discrete or product space case, there does not appear to be a simple fix and no simple variation on the above $\boldsymbol{F}_n$ appears to provide a basis. Thus we cannot immediately infer that m is countably additive on $\boldsymbol{F}$.

A second approach is to try to use an isomorphism as in Lemma 2.3.3 to extend m by actually extending a related measure in a known standard space. Again there is a natural first guess. Let $(\boldsymbol{M},\boldsymbol{B}_{\boldsymbol{M}})$ again be the binary sequence space of (2.4.2), where $\boldsymbol{B}_{\boldsymbol{M}}$ is the σ-field generated by the thin cylinders. The space is standard and there is a natural mapping $f : \boldsymbol{M} \rightarrow [0,1)$ defined by

$$f(u) = \sum_{k=0}^{\infty} u_k 2^{-k-1} , \tag{2.8.4}$$

where $u = (u_0, u_1, u_2, \cdots)$. The mapping $f(u)$ is simply the real number whose binary expansion is $.u_0u_1u_2 \cdots$. We shall show shortly that f is measurable and hence one might hope that $[0,1)$ would indeed inherit a basis from $\boldsymbol{M}$ through the mapping f. Sadly, this is not immediately the case–the problem being that f is a many-to-one mapping and not a one-to-one mapping. Hence it cannot be an isomorphism. For example, both $011111 \cdots$ and $10000 \cdots$ yield the value $f(u) = 1/2$. We could make f a one-to-one mapping by removing from $\boldsymbol{M}$ all sequences with only a finite number of 0's, that is, form a new space $\boldsymbol{H} \subset \boldsymbol{M}$ by removing the countable collection of sequences $11111 \cdots$, $01111 \cdots$, $00111 \cdots$, $10111 \cdots$, and so on (remove any binary vector followed by all ones) and consider f as a mapping from $\boldsymbol{H}$ to $[0,1)$. Here again we are stymied–none of the results proved so far permit us to conclude that the reduced space $\boldsymbol{H}$ is standard. In fact, since we have removed individual sequences (rather than cylinders as in Lemma 2.4.2), there will be atoms collapsing to the empty set and the previously constructed basis for $\boldsymbol{M}$ does not work for the smaller space $\boldsymbol{H}$.

We have reached an apparent impasse in that we can not (yet) either show that $\boldsymbol{F}$ has a basis or that m on $\boldsymbol{F}$ is countably additive. We resolve this dilemma by giving up trying to find a basis for $\boldsymbol{F}$ and instead use the standard space $\boldsymbol{M}$ to show directly that m is countably additive.

We begin by showing that $f: \boldsymbol{M} \to [0,1)$ is measurable. From Lemma 1.4.1 this will follow if we can show that for any $G \in \boldsymbol{F}$, $f^{-1}(G) \in \boldsymbol{B_M}$. Since every $G \in \boldsymbol{F}$ can be written as a finite union of disjoint sets of the form (2.8.3), it suffices to show that $f^{-1}(G) \in \boldsymbol{B_M}$ for all sets $G_{n,k}$ of the form (2.8.3). To find

$$f^{-1}(G_{n,k}) = \{u: k2^{-n} \le f(u) < (k+1)2^{-n}\} ,$$

let $b^n = (b_0, b_1, \cdots, b_{n-1})$ satisfy

$$\sum_{j=0}^{n-1} b_j 2^{n-j-1} = \sum_{j=0}^{n-1} b_{n-j-1} 2^j = k , \tag{2.8.5}$$

that is, $(b_{n-1}, \cdots, b_0)$ is the unique binary representation of the integer k. The thin cylinder $c(b^n)$ will then have the property that if $u \in c(b^n)$, then

$$f(u) = \sum_{j=0}^{\infty} b_j 2^{-j-1} \ge \sum_{j=0}^{n-1} b_j 2^{-j-1}$$

$$= 2^{-n} \sum_{j=0}^{n-1} b_j 2^{n-j-1} = k2^{-n}$$

and

$$f(u) \leq \sum_{j=0}^{n-1} b_j 2^{-j-1} + \sum_{j=n}^{\infty} 2^{-j-1}$$
$$= k2^{-n} + 2^{-n}$$

with equality on the upper bound if and only if $b_j = 1$ for all $j \geq n$. Thus if $u \in c(b^n) - \{b^n 111 \cdots\}$, then $u \in f^{-1}(G_{n,k})$ and hence $c(b^n) - \{b^n 111 \cdots\} \subset f^{-1}(G_{n,k})$. Conversely, if $u \in f^{-1}(G_{n,k})$, then $f(u) \in G_{n,k}$ and hence

$$k2^{-n} \leq \sum_{j=0}^{\infty} u_j 2^{-j-1} < (k+1)2^{-n} . \tag{2.8.6}$$

The left-hand inequality of (2.8.6) implies

$$k \leq \sum_{j=0}^{n-1} u_j 2^{n-j-1} + 2^n \sum_{j=n}^{\infty} u_j 2^{-j-1}$$
$$\leq \sum_{j=0}^{n-1} u_j 2^{n-j-1} + 1 , \tag{2.8.7}$$

with equality on the right if and only if $u_j = 1$ for all $j > n$. The right-hand inequality of (2.8.6) implies that

$$\sum_{j=0}^{n-1} u_j 2^{n-j-1} \leq \sum_{j=0}^{n-1} u_j 2^{n-j-1} + 2^n \sum_{j=n}^{\infty} u_j 2^{-j-1}$$
$$2^n \sum_{j=0}^{\infty} u_j 2^{n-j-1} < k + 1 . \tag{2.8.8}$$

Combining (2.8.7) and (2.8.8) yields for $u \in f^{-1}(G_{n,k})$

$$k - 1 \leq \sum_{j=0}^{n-1} u_j 2^{n-j-1} < k + 1 , \tag{2.8.9}$$

with equality on the left if and only if $u_j = 1$ for all $j \geq n$. Since the sum in (2.8.9) must be an integer, it can only be k or $k - 1$. It can only be $k - 1$, however, if $u_j = 1$ for all $j \geq n$. Thus either

$$\sum_{j=0}^{n-1} u_j 2^{n-j-1} = k ,$$

in which case $u \in c(b^n)$ with b_n defined as in (2.8.5), or

$$\sum_{j=0}^{n-1} u_j 2^{n-j-1} = k - 1 \text{ and } u_j = 1, \text{ all } j \geq n .$$

The second relation can only occur if for b^n as in (2.8.5) we have that $b_{n-1} = 1$ and $u_j = b_j$, $j = 0, \cdots, n-2$, $u_{n-1} = 0$, and $u_j = 1$, $j \geq n$. Letting $w = (b^{n-1}0111 \cdots)$ denote this sequence, we therefore have that

$$f^{-1}(G_{n,k}) \subset c(b^n) \cup w .$$

In summary, every sequence in $c(b^n)$ is contained in $f^{-1}(G_{n,k})$ except the single sequence $(b^n 111 \cdots)$ and every sequence in $f^{-1}(G_{n,k})$ is contained in $c(b^n)$ except the sequence $(b^{n-1} 0111 \cdots)$ if $b^n = 1$ in (2.8.5). Thus given $G_{n,k}$, if $b_n = 0$

$$f^{-1}(G_{n,k}) = c(b^n) - (b^n 111 \cdots) \tag{2.8.9a}$$

and if $b_n = 1$

$$\left[c(b^n) - (b^n 111 \cdots)\right] \cup (b^{n-1}0111 \cdots) . \tag{2.8.9b}$$

In both cases, $f^{-1}(G_{n,k}) \in \boldsymbol{B_M}$, proving that f is measurable.

Measurability of f almost immediately provides the desired conclusions: Simply specify a measure P on the standard space $(\boldsymbol{M},\boldsymbol{B_M})$ via

$$P(c(b^n)) = 2^{-n} \text{ ; all } b^n \in \{0,1\}^n$$

and extend this to the field generated by the cylinders in an additive fashion as before. Since this field is standard, this provides a measure on $(\boldsymbol{M},\boldsymbol{B_M})$. Now define a measure m on $([0,1),\boldsymbol{B}_{[0,1)})$ by

$$m(G) = P(f^{-1}(G)) \text{ , all } G \in \boldsymbol{B}_{[0,1)} .$$

As in Section 2, f is a random variable and m is its distribution. In particular, m is a measure. From (2.8.9) and the fact that the individual sequences have 0 measure, we have that

$$m([k2^{-n}, (k+1)2^{-n})) = P(c(b^n)) = 2^{-n} \text{ , } k = 1,2, \cdots, n-1 .$$

From the continuity of probability, this implies that for any interval $[a,b)$, $1 > b > a > 0$, $m([a,b)) = b - a$, as desired. Since this agrees with the original description of the measure desired on intervals, it must also agree on the field generated by intervals. Thus there exists a probability measure agreeing with (2.8.1) and hence with the additive extension of (2.8.1) to the field generated by the intervals. By the uniqueness of extension, the measure we have constructed must be the unique extension of (2.8.1). This measure is called the *Lebesgue measure* on the unit interval.

Thus even without a basis or an isomorphism, we can construct probability measures if we have a measurable mapping from a standard space to the given space.

REFERENCES

1. O. J. Bjornsson, "A note on the characterization of standard borel spaces," *Math. Scand.*, vol. 47, pp. 135-136, 1980.
2. N. Bourbaki, *Elements de Mathematique, Livre VI, Integration,* Hermann, Paris, 1956-1965.
3. J. P. R. Christensen, *Topology and Borel Structure,* Mathematics Studies 10, North-Holland/American Elsevier, New York, 1974.
4. D. C. Cohn, *Measure Theory,* Birkhauser, New York, 1980.
5. V. I. Levenshtein, "Binary codes capable of correcting deletions, insertaions, and reversals," *Sov. Phys. -Dokl.*, vol. 10, pp. 707-710, 1966.
6. G. Mackey, "Borel structures in groups and their duals," *Trans. Am. Math. Soc.*, vol. 85, pp. 134-165, 1957.
7. K. R. Parthasarathy, *Probability Measures on Metric Spaces,* Academic Press, New York, 1967.
8. I. Paz, *Stochastic Automata Theory,* Academic Press, New York, 1971.
9. L. Schwartz, *Radon Measures on Arbitrary Topological Spaces and Cylindrical Measures,* Oxford University Press, Oxford, 1973.
10. E. Tanaka and T. Kasai, "Synchronization and subsititution error correcting codes for the Levenshtein metric," *IEEE Transactions on Information Theory*, vol. IT-22, pp. 156-162, 1976.

3

BOREL SPACES AND POLISH ALPHABETS

We have seen that standard measurable spaces are the only measurable spaces for which all finitely additive candidate probability measures are also countably additive, and we have developed several properties and some important simple examples. In particular, sequence spaces drawn from countable alphabets and certain subspaces thereof are standard. In this chapter we develop the most important (and, in a sense, the most general) class of standard spaces–Borel spaces formed from complete separable metric spaces. We will accomplish this by showing that such spaces are isomorphic to a standard subspace of a countable alphabet sequence space and hence are themselves standard. The proof will involve a form of coding or quantization.

3.1 BOREL SPACES

In this section we further develop the structure of metric spaces and we construct σ-fields and measurable spaces from metric spaces.

Let A denote a metric space with metric d. Define for each a in A and each real $r > 0$ the *open sphere* (or open ball) with center a and radius r by $S_r(a) = \{b: b \in A, d(a,b) < r\}$. For example, an open sphere in the real line $\mathbf{R}$ with respect to the Euclidean metric is simply an open interval $(a,b) = \{x: a < x < b\}$. A subset F of A is called an *open set* if given any point $a \in F$ there is an $r > 0$ such that $S_r(a)$ is completely contained in F, that is, there are no points in F on the "edge" of F. Let $\boldsymbol{O}_A$ denote the class of all open sets in A. The following lemma collects several useful properties of open sets. The proofs are left as an exercise as they are easy and may be found in any text on topology or metric spaces; e.g., [9].

Lemma 3.1.1. All spheres are open sets. A and $\emptyset$ are open sets. A set is open if and only if it can be written as a union of spheres; in fact,

$$F = \bigcup_{a \in F} \bigcup_{r:\, r < d(a,F^c)} S_r(a), \tag{3.1.1}$$

where

$$d(a,G) = \inf_{b \in G} d(a,b).$$

If $\{F_t;\, t \in \boldsymbol{T}\}$ is any (countable or uncountable) collection of open sets, then $\bigcup_{t \in \boldsymbol{T}} F_t$ is open. If F_i, $i \in N$, is any finite collection of open sets, then $\bigcap_{i \in N} F_i$ is open.

When we have a class of open sets $\boldsymbol{V}$ such that every open set F in $\boldsymbol{O}_A$ can be written as a union of elements in $\boldsymbol{V}$, e.g., the preceding open spheres, then we say that $\boldsymbol{V}$ is a *base* for the open sets and write $\boldsymbol{O}_A = OPEN(\boldsymbol{V}) = \{\text{all unions of sets in } \boldsymbol{V}\}$.

In general, a set A together with a class of subsets $\boldsymbol{O}_A$ called open sets such that all unions and finite intersections of open sets yield open sets is called a *topology* on A. If $\boldsymbol{V}$ is a base for the open sets, then it is called a base for the topology. Observe that many different metrics may yield the same open sets and hence the same topology. For example, the metrics $d(a,b)$ and $d(a,b)/(1 + d(a,b))$ yield the same open sets and hence the same topology. Two metrics will be said to be *equivalent* on A if they yield the same topology on A.

Given a metric space A together with a class of open sets $\boldsymbol{O}_A$ of A, the corresponding *Borel field* (actually a Borel σ-field) $\boldsymbol{B}(A)$ is defined as $\sigma(\boldsymbol{O}_A)$, the σ-field generated by the open sets. The members of $\boldsymbol{B}(A)$ are called the *Borel sets* of A. A metric space A together with a Borel σ-field $\boldsymbol{B}(A)$ is called a *Borel space* $(A,\boldsymbol{B}(A))$. If this space is standard, it is naturally called a *standard Borel space.* Thus a Borel space is a measurable space where the σ-field is generated by the open sets of the alphabet with respect to some metric. Observe that equivalent metrics on a space A will yield the same Borel measurable space since they yield the same open sets.

If A is a metric space with metric d, a point a in A is said to be a *limit point* of a set F if every open sphere $S_r(a)$ contains a point $b \neq a$ such that b is in F. A set F is *closed* if it contains all of its limit points. The set F together with its limit points is called the *closure* of F and is denoted by $\overline{F}$. Given a set F, define its *diameter*

$$diam(F) = \sup_{a,b \in F} d(a,b).$$

Lemma 3.1.2. A set is closed if and only if its complement is open and hence all closed sets are in the Borel σ-field. Arbitrary intersections and finite unions of closed sets are closed. $\boldsymbol{B}(A) = \sigma($ all closed sets $)$. The closure of a set is closed. If $\overline{F}$ is the closure of F, then $diam(\overline{F}) = diam(F)$.

Proof. If F is closed and we choose x in F^c , then x is not a limit point of F and hence there is an open sphere $S_r(x)$ that is disjoint with F and hence contained in F^c. Thus F^c must be open. Conversely, suppose that F^c is open and x is a limit point of F and hence every sphere $S_r(x)$ contains a point of F. Since F^c is open this means that x cannot be in F^c and hence must be in F. Hence F is closed. This fact together with DeMorgan's laws and Lemma 3.1.1 implies that finite unions and arbitrary intersections of closed sets are closed. Since the complement of every closed set is open and vice versa, the σ-fields generated by the two classes are the same. To prove that the closure of a set is closed, first observe that if x is in $\overline{F}^c$ and hence neither a point nor a limit point of F, then there is an open sphere S centered at x which has no points in F. Since every point y in S therefore has an open sphere containing it which contains no points in F , no such y can be a limit point of F. Thus $\overline{F}^c$ contains S and hence is open. Thus its complement, the closure of F, must be closed. Any points in the closure of a set must have points in the set within an arbitrarily close distance. Hence the diameter of a set and that of its closure are the same. □

Lemma 3.1.3. A set F is closed if and only if for every sequence $\{a_n, n = 1,2,...\}$ of elements of F such that $a_n \rightarrow a$ we have that also a is in F. If d and m are two metrics on A that yield equivalent notions of convergence, that is, $d(a_n,a) \rightarrow 0$ if and only if $m(a_n,a) \rightarrow 0$ for a_n, a in A, then d and m are equivalent metrics on A, that is, they yield the same open sets and generate the same topology.

Proof. If a is a limit point of F, then one can construct a sequence of elements a_n in F such that $a = \lim_{n\to\infty} a_n$. For example, choose a_n as any point in $S_{1/n}(a)$ (there must be such an a_n since a is a limit point). Thus if a set F contains all limits of sequences of its points, it must contain its limit points and hence be closed.

Conversely, let F be closed and a_n a sequence of elements of F that converge to a point a. If for every $\varepsilon > 0$ we can find a point a_n distinct from a in $S_\varepsilon(a)$, then a is a limit point of F and hence must be in F since it is closed. If we can find no such distinct point, then all of the a_n must equal a for n greater than some value and hence a must be in F. If two metrics d and m are such that d-convergence and m-convergence are equivalent, then they must yield the same closed sets and hence by the previous lemma they must yield the same open sets. □

In addition to open and closed sets, two other kinds of sets arise often in the study of Borel spaces: Any set of the form $\bigcap_{i=1}^{\infty} F_i$ with the F_i open is called a G_δ *set* or simply a G_δ. Similarly, any countable union of closed sets is called an F_σ. Observe that G_δ and F_σ sets are not necessarily either closed or open. We have, however, the following relation.

Lemma 3.1.4. Any closed set is a G_δ, any open set is an F_σ.

Proof. If F is a closed subset of A, then

$$F = \bigcap_{n=1}^{\infty} \{x \colon d(x,F) < 1/n\} ,$$

and hence F is a G_δ. Taking complements completes the proof since any closed set is the complement of an open set and the complement of a G_δ is an F_σ. □

The added structure of Borel spaces and the previous lemma permit a different characterization of σ-fields:

Lemma 3.1.5. Given a metric space A, then $\boldsymbol{B}(A)$ is the smallest class of subsets of A that contains all the open (or closed) subsets of A, and which is closed under countable disjoint unions and countable intersections.

Proof. Let $\boldsymbol{G}$ be the given class. Clearly $\boldsymbol{G} \subset \boldsymbol{B}(A)$. Since $\boldsymbol{G}$ contains the open (or closed) sets (which generate the σ-field $\boldsymbol{B}(A)$), we need only show that it contains complements to complete the proof. Let $\boldsymbol{H}$ be the class of all sets F such that $F \in \boldsymbol{G}$ and $F^c \in \boldsymbol{G}$. Clearly $\boldsymbol{H} \subset \boldsymbol{G}$. Since $\boldsymbol{G}$ contains the open sets and countable intersections, it also contains the G_δ's and hence also the closed sets from the previous lemma. Thus both the open sets and closed sets all belong to $\boldsymbol{H}$. Suppose now that $F_i \in \boldsymbol{H}$, $i = 1,2, \cdots$ and hence both F_i and their complements belong to $\boldsymbol{G}$. Since $\boldsymbol{G}$ contains countable disjoint unions and countable intersections

$$\bigcup_{i=1}^{\infty} F_i = \bigcup_{i=1}^{\infty} (F_i \cap F_1^c \cap F_2^c \cap \cdots \cap F_{i-1}^c) \in \boldsymbol{G}$$

and

$$(\bigcup_{i=1}^{\infty} F_i)^c = \bigcap_{i=1}^{\infty} F_i^c \in \boldsymbol{G} .$$

Thus the countable unions of members of $\boldsymbol{H}$ and the complements of such sets are in $\boldsymbol{G}$. This implies that $\boldsymbol{H}$ is closed under countable unions. If $F \in \boldsymbol{H}$, then F and F^c are in $\boldsymbol{G}$, hence also $F^c \in \boldsymbol{H}$. Thus $\boldsymbol{H}$ is closed under complementation and countable unions and hence $\boldsymbol{H}$ contains both countable disjoint unions and countable intersections. Since $\boldsymbol{H}$ also contains the open sets, it must therefore contain $\boldsymbol{G}$. Since we have already argued the reverse inclusion, $\boldsymbol{H} = \boldsymbol{G}$. This implies that $\boldsymbol{G}$ is closed under complementation since $\boldsymbol{H}$ is, thus completing the proof. □

The most important mappings for Borel spaces are continuous functions. Given two Borel spaces $(A,\boldsymbol{B}(A))$ and $(B,\boldsymbol{B}(B))$ with metrics d and ρ, respectively, a function $f : A \to B$ is *continuous at a* in A if for each $\varepsilon > 0$ there exists a $\delta > 0$ such that $d(x,a) < \delta$ implies that $\rho(f(x), f(a)) < \varepsilon$. It is an easy exercise to show that if a is a limit point, then f is continuous at a if and only if $a_n \to a$ implies that $f(a_n) \to f(a)$.

A function is *continuous* if it is continuous at every point in A. The following lemma collects the main properties of continuous functions: a definition in terms of open sets and their measurability.

Lemma 3.1.6. A function f is continuous if and only if the inverse image $f^{-1}(S)$ of every open set S in B is open in A. A continuous function is Borel measurable.

Proof. Assume that f is continuous and S is an open set in B. To prove $f^{-1}(S)$ is open we must show that about each point x in $f^{-1}(S)$ there is an open sphere $S_r(x)$ contained in $f^{-1}(S)$. Since S is open we can find an s such that $\rho(f(x),y) < s$ implies that y is in S. By continuity we can find an r such that $d(x,b) < r$ implies that $d(f(x), f(b)) < s$. If $f(x)$ and $f(b)$ are in S, however, x and b are in the inverse image and hence $S_r(x)$ so obtained is also in the inverse image. Conversely, suppose that $f^{-1}(S)$ is open for every open set S in B. Given x in A and $\varepsilon > 0$ define the open sphere $S = S_\varepsilon(f(x))$ in B. By assumption $f^{-1}(S)$ is then also open, hence there is a $\delta > 0$ such that $d(x,y) < \delta$ implies that $y \in f^{-1}(S)$ and hence $f(y) \in S$ and hence $\rho(f(x), f(y)) < \varepsilon$, completing the proof. □

Since inverse images of open sets are open and hence Borel sets and since open sets generate the Borel sets, f is measurable from Lemma 1.4.1.

Exercises

1. Prove Lemma 3.1.1.
2. Is the mapping f of Section 2.8 continuous?

3.2 POLISH SPACES

Let A be a metric space. A set F is said to be *dense* in A if every point in A is a point in F or a limit point of F. A metric space A is said to be *separable* if it has a countable dense subset, that is, if there is a discrete set, say B, such that all points in A can be well approximated by points in B in the sense that all points in A are points in B or limits of points in B. For example, the rational numbers are dense in the real line $\mathbf{R}$ and are countable, and hence $\mathbf{R}$ of Example 2.5.5 is separable. Similarly, n-dimensional vectors with rational coordinates are countable and dense in $\mathbf{R}^n$, and hence Example 2.5.6 also provides a separable metric space. The discrete metric spaces of Examples 2.5.1 through 2.5.4 are trivially separable since A is countable and is dense in itself. An example of a nonseparable space is the binary sequence space $\boldsymbol{M} = \{0,1\}^{\mathbf{Z}_+}$ with the sup-norm metric

$$d(x,y) = \sup_{i \in \mathbf{Z}_+} | x_i - y_i | .$$

The principal use of separability of metric spaces is the following lemma.

Lemma 3.2.1. Let A be a metric space with Borel σ-field $\boldsymbol{B}(A)$. If A is separable, then $\boldsymbol{B}(A)$ is countably generated and it contains all of the points of the set A.

Proof. Let $\{a_i;\ i = 1,2,...\}$ be a countable dense subset of A and define the class of sets

$$\boldsymbol{V}_A = \{S_{1/n}(a_i)\ ;\ i = 1,2,...;\ n = 1,2,...\}. \tag{3.2.1}$$

Any open set F can be written as a countable union of sets in $\boldsymbol{V}_A$ analogous to (3.1.1) as

$$F = \bigcup_{i:\ a_i \in F} \ \bigcup_{n:\ n^{-1} < d(a_i, F^c)} S_{1/n}(a_i). \tag{3.2.2}$$

Thus $\boldsymbol{V}_A$ is a countable class of open sets and $\sigma(\boldsymbol{V}_A)$ contains the class of open sets $\boldsymbol{O}_A$ which generates $\boldsymbol{B}(A)$. Hence $\boldsymbol{B}(A)$ is countably generated. To see that the points of A belong to $\boldsymbol{B}(A)$, index the sets in $\boldsymbol{V}_A$ as $\{V_i;\ i = 0,1,2,...\}$ and define the sets $V_i(a)$ by V_i if a is in V_i and V_i^c if a is not in V_i. Then

$$\{a\} = \bigcap_{i=1}^{\infty} V_i(a) \in \boldsymbol{B}(A)\ . \ \Box$$

For example, the class of all open intervals $(a,b) = \{x\colon a < x < b\}$ with rational endpoints $b > a$ is a countable generating class of open sets for the Borel σ-field of the real line with respect to the usual Euclidean distance. Similar subsets of $[0,1]$ form a similar countable generating class for the Borel sets of the unit interval.

A sequence $\{a_n;\ n = 1,2,...\}$ in A is called a *Cauchy sequence* if for every $\varepsilon > 0$ there is an integer N such that $d(a_n,a_m) < \varepsilon$ if $n \geq N$ and $m \geq N$. Alternatively, a sequence is Cauchy if and only if

$$\lim_{n \to \infty} diam(\{a_n, a_{n+1}, ...\}) = 0.$$

A metric space is *complete* if every Cauchy sequence converges, that is, if $\{a_n\}$ is a Cauchy sequence, then there is an a in A for which $a = \lim_{n\to\infty} a_n$. In the words of Simmons [9], p.71, a complete metric space is one wherein "every sequence that tries to converge is successful." A standard result of elementary real analysis is that Examples 2.5.5 and 2.5.6 (R^k with the Euclidean distance) are complete (see Rudin [7], p. 46). A complete inner product space (Example 2.5.10) is called a *Hilbert space.* A complete normed space (Example 2.5.9) is called a *Banach space.*

A fundamental property of complete metric spaces is given in the following lemma, which is known as *Cantor's intersection theorem.* The corollary is a slight generalization.

Lemma 3.2.2. Given a complete metric space A, let F_n, $n = 1,2,\ldots$ be a decreasing sequence of nonempty closed subsets of A for which $diam(F_n) \to 0$. Then the intersection of all of the F_n contains exactly one point.

Proof. Since the diameter of the sets is decreasing to zero, The intersection cannot contain more than one point. If x_n is a sequence of points in F_n, then it is a Cauchy sequence and hence must converge to some point x since A is complete. If we show that x is in the intersection, the proof will be complete. Since the F_n are decreasing, for each n the sequence $\{x_k, k \geq n\}$ is in F_n and converges to x. Since F_n is closed, however, it must contain x. Thus x is in all of the F_n and hence in the intersection. □

The similarity of the preceding condition to the finite intersection property of atoms of a basis of a standard space is genuine and will be exploited soon.

Corollary 3.2.1. Given a complete metric space A, let F_n, $n = 1,2,\ldots$ be a decreasing sequence of nonempty subsets of A such that $diam(F_n) \to 0$ and such that the closure of each set is contained in the previous set, that is,

$$\overline{F}_n \subset F_{n-1}, \text{ all } n,$$

where we define $F_0 = A$. Then the intersection of all of the F_n contains exactly one point.

Proof. Since $F_n \subset \overline{F}_n \subset F_{n-1}$

$$\bigcap_{n=1}^{\infty} F_n \subset \bigcap_{n=1}^{\infty} \overline{F}_n \subset \bigcap_{n=0}^{\infty} F_n .$$

Since the F_n are decreasing, the leftmost and rightmost terms above are equal. Thus

$$\bigcap_{n=1}^{\infty} F_n = \bigcap_{n=1}^{\infty} \overline{F}_n .$$

Since the F_n are decreasing, so are the $\overline{F}_n$. Since these sets are closed, the right-hand term contains a single point from the lemma. □

A complete separable metric space is called a *Polish space* because of the pioneering work by Polish mathematicians on such spaces. Polish spaces play a central role in functional analysis and provide models for most alphabets of scalars, vectors, sequences, and functions that arise in communications applications. The best known examples are the Euclidean spaces, but we shall encounter others such as the space of square integrable functions. A Borel space $(A,\boldsymbol{B}(A))$ with A a Polish metric space will be called a *Polish Borel space.*

The remainder of this section develops some properties of Polish spaces that resemble those of standard spaces: We show that cartesian products of Polish spaces are Polish spaces, and that certain subsets of Polish spaces are also Polish.

Products of Polish Spaces

Given a Polish space A and an index set $\boldsymbol{I}$, we can define a metric $d^{\boldsymbol{I}}$ on the product space $A^{\boldsymbol{I}}$ as in Example 2.5.11. The following lemma shows that the product space is also Polish.

Lemma 3.2.3. (a) If A is a Polish space with metric d and $\boldsymbol{I}$ is a countable index set, then the product space $A^{\boldsymbol{I}}$ is a Polish space with respect to the product space metric $d^{\boldsymbol{I}}$. (b) In addition, if $\boldsymbol{B}(A)$ is the Borel σ-field of subsets of A with respect to the metric d, then the product σ-field of Section 1.4, $\boldsymbol{B}(A)^{\boldsymbol{I}} = \sigma(RECT(\boldsymbol{B}_i, i \in N))$, where the $\boldsymbol{B}_i$ are all replicas of $\boldsymbol{B}(A)$, is exactly the Borel σ-field of $A^{\boldsymbol{I}}$ with respect to $d^{\boldsymbol{I}}$. That is, $\boldsymbol{B}(A)^{\boldsymbol{I}} = \boldsymbol{B}(A^{\boldsymbol{I}})$.

Proof. (a) If A has a dense subset, say B, then a dense subset for $A^{\boldsymbol{I}}$ can be obtained as the class of all sequences taking values in B on a finite number of coordinates and taking some fixed reference value, say b, on the remaining coordinates. Hence $A^{\boldsymbol{I}}$ is separable. If a sequence $x_n^{\boldsymbol{I}} = \{x_{n,i},\ i \in N\}$ is Cauchy, then $d^{\boldsymbol{I}}(x_n^{\boldsymbol{I}}, x_m^{\boldsymbol{I}}) \to 0$ as $n, m \to \infty$. From Lemma 2.5.1 this implies that for each coordinate $d(x_{n,i}, x_{m,i}) \to 0$ as $n, m \to \infty$ and hence the sequence $x_{n,i}$ is a Cauchy sequence in A for each i. Since A is complete, the sequence has a limit, say y_i. From Lemma 2.5.1 again, this implies that $x_n^{\boldsymbol{I}} \to y^{\boldsymbol{I}}$ and hence $A^{\boldsymbol{I}}$ is complete.

(b) We next show that the product σ-field of a countably generated Borel σ-field is the Borel σ-field with respect to the product space metric; that is, both means of constructing σ-fields yield the same result. (This result need not be true for products of Borel σ-fields that are not countably generated.)

Let $\Pi_n: A^{\boldsymbol{I}} \to A$ be the coordinate or sampling function $\Pi_n(x^{\boldsymbol{I}}) = x_n$. It is easy to show that Π_n is a continuous function and hence is also measurable from Lemma 3.1.6. Thus $\Pi_n^{-1}(F) \in \boldsymbol{B}(A^{\boldsymbol{I}})$ for all $F \in \boldsymbol{B}(A)$ and $n \in N$. This implies that all finite intersections of such sets and hence all rectangles with coordinates in $\boldsymbol{B}(A)$ must be in $\boldsymbol{B}(A^{\boldsymbol{I}})$. Since these rectangles generate $\boldsymbol{B}(A)^{\boldsymbol{I}}$, we have shown that

$$\boldsymbol{B}(A)^{\boldsymbol{I}} \subset \boldsymbol{B}(A^{\boldsymbol{I}}) .$$

Conversely, let $S_r(x)$ be an open sphere in $A^{\boldsymbol{I}}$: $S_r(x) = \{y: y \in A^{\boldsymbol{I}}, d^{\boldsymbol{I}}(x,y) < r\}$. Observe that for any n

$$d^{\boldsymbol{I}}(x,y) \le \sum_{k=1}^{n} 2^{-k} \frac{d(x_k,y_k)}{1 + d(x_k,y_k)} + 2^{-n}$$

and hence we can write

$$S_r(x) = \bigcup_{n:\ 2^{-n} < r} \ \bigcup_{\{d_k\}:\ \sum_{k=1}^{n} d_k < r - 2^{-n}} \ \bigcap_{k=0}^{n} \Pi_k^{-1}(\{a: d^{\boldsymbol{I}}(x_k,a) < \frac{d_k}{1 - d_k}\}),$$

where the union over the $\{d_k\}$ is restricted to rational d_k. Every piece of the union on the right-hand side is contained in the sphere on the left, and every point in the sphere on the left is eventually included in one of the terms in the union on the right for sufficiently large n. Thus all spheres in $A^{\boldsymbol{I}}$ under $d^{\boldsymbol{I}}$ can be written as countable unions of of rectangles with coordinates in $\boldsymbol{B}(A)$, and hence these spheres are in $\boldsymbol{B}(A)^{\boldsymbol{I}}$. Since $A^{\boldsymbol{I}}$ is separable under $d^{\boldsymbol{I}}$ all open sets in A^N can be written as a countable union of spheres in A^I using (3.2.2), and hence all open sets in A^I are also in $\boldsymbol{B}(A)^I$. Since these open sets generate $\boldsymbol{B}(A^I)$, $\boldsymbol{B}(A^I) \subset \boldsymbol{B}(A)^I$, completing the proof. □

Example 3.2.1: Integer Sequence Spaces

As a simple example of some of the ideas under consideration, we consider the space of integer sequences introduced in Lemma 2.4.1. Let $B = \mathbf{Z}_+$ and $A = \boldsymbol{N} = B^B$, the space of sequences with nonnegative integer values. A discrete space such as A_0 is a Polish space with respect to the Hamming metric d_H of Example 2.5.1. This is trivial since the countable set of symbols is dense in itself and if a sequence $\{a_n\}$ is Cauchy, then choosing $\varepsilon < 1$ implies that there is an integer N such that $d(a_n, a_m) < 1$ and hence $a_n = a_m$ for all n, $m \geq N$. Note that an open sphere $S_\varepsilon(a)$ in B with $\varepsilon < 1$ is just a point and that any subset of B is a countable union of such open spheres. Thus in this case the Borel σ-field $\boldsymbol{B}(\mathbf{Z}_+)$ is just the power set, the collection of all subsets of B.

The sequence space A is a metric space with the metric of Example 2.5.11, which with a coordinate Hamming distance becomes

$$d^{\mathbf{Z}_+}(x,y) = \frac{1}{2} \sum_{i=0}^{\infty} 2^{-i} d_H(x_i, y_i) \ . \tag{3.2.3}$$

Applying Lemma 3.2.3 (b) and (3.2.3) we have

$$\boldsymbol{B}(\boldsymbol{N}) = \boldsymbol{B}(\mathbf{Z}_+^{\mathbf{Z}_+}) = \boldsymbol{B}(\mathbf{Z}_+)^{\mathbf{Z}_+} \ . \tag{3.2.4}$$

It is interesting to compare the structure used in Lemma 2.4.1 with the Borel or topological structure currently under consideration. Recall that in Lemma 2.4.1 the σ-field $\boldsymbol{B}(\mathbf{Z}_+)^{\mathbf{Z}_+}$ was obtained as the σ-field generated by the thin cylinders, the rectangles or thin cylinders having the form $\{x: x^n = a^n\}$ for some positive integer n and some n-tuple $a^n \in B^n$. For convenience abbreviate d^B to d. Suppose we know that $d(x,y) < 1/2$. Then necessarily $x_0 = y_0$ and hence both sequences must lie in a common thin cylinder, the thin cylinder $\{u: u_0 = x_0\}$. Similarly, if $d(x,y) < 1/4$, then necessarily $x^2 = y^2$ and both sequences must again lie in a common thin cylinder, $\{u: u^2 = x^2\}$. In general, if $d(x,y) < \varepsilon$ and $\varepsilon \leq 2^{-(n+1)}$, then $x^n = y^n$. Thus if $x \in A$ and $\varepsilon \leq 2^{-(n+1)}$, then

$$S_\varepsilon(x) = \{y: y^n = x^n\} \ . \tag{3.2.5}$$

This means that the thin cylinders are simply the open spheres in A! Thus the generating class used in Lemma 2.4.1 is really the same as that used to obtain a Borel σ-field for A. In other words, not only are the σ-fields of (3.2.4) all equal, the collections of sets used to generate these fields are the same.

Subspaces of Polish Spaces

The following result shows that certain subsets of Polish spaces are themselves Polish, although in some cases we need to consider an equivalent metric rather than the original metric.

Lemma 3.2.4. Let A be a complete, separable metric space with respect to a metric d. If $F = C$, C a closed subset of A, then F is itself a complete, separable metric space with respect to d. If $F = O$, where O is an open subset of A, then F is itself a complete separable metric space with respect to a metric d' such that d'-convergence and d-convergence are equivalent on F and hence d and d' are equivalent metrics on F. In fact, the following provides such a metric:

$$d'(x,y) = d(x,y) + \left| \frac{1}{d(x,O^c)} - \frac{1}{d(y,O^c)} \right|. \tag{3.2.6}$$

If F is the intersection of a closed set C with an open set O of A, then F is a complete separable metric space with respect to a metric d' such that d'-convergence and d-convergence are equivalent on F, and hence d and d' are equivalent metrics on F. Here, too, d' of (3.2.6) provides such a metric.

Proof. Suppose that F is an arbitrary subset of a separable metric space A with metric d having the countable base $\boldsymbol{V}_A = \{V_i; i = 1,2, \cdots\}$ of (3.2.2). Define the class of sets $\boldsymbol{V}_F = \{F \cap V_i \,; i = 1,2, \cdots\}$, where we eliminate all empty sets. Let y_i denote an arbitrary point in $F \cap V_i$. We claim that the set of all y_i is dense in F and hence F is separable under d. To see this let $x \in F$. Given ε there is a V_i of radius less than ε which contains x and hence $F \cap V_i$ is not empty (since it at least contains $x \in F$) and

$$d(x,y_i) \leq \sup_{y \in F \cap V_i} d(x,y) \leq \sup_{y \in V_i} d(x,y) \leq 2\varepsilon .$$

Thus for any x in F we can find a $y_i \in \mathbf{V}_F$ that is arbitrarily close to it. Thus $\mathbf{V}_F$ is dense in F and hence F is separable with respect to d. The resulting metric space is not in general complete, however, because completeness of A only ensures that Cauchy sequences converge to some point in A and hence a Cauchy sequence of points in F may converge to a point not in F. If F is closed, however, it contains all of its limit points and hence F is complete with respect to the original metric. This completes the proof of the lemma for the case of a closed subset.

If F is an open set O, then we can find a metric that makes O complete by modifying the original metric to "blow up" near the boundary and hence prevent Cauchy sequences from converging to something not in O. In other words, we modify the metric so that if a sequence $a_n \in O$ is Cauchy with respect to d but converges to a point outside O, then the sequence will not be Cauchy with respect to d'. Define the metric d' on O as in (3.2.6). Observe that by construction $d \leq d'$, and hence convergence or Cauchy convergence under d' implies the same under d. From the triangle inequality, $| d(a_n,O^c) - d(a,O^c) | \leq d(a_n,a)$. Thus provided the a_n and a are in O, d-convergence implies d'-convergence. Thus d-convergence and d'-convergence are equivalent on O and hence yield the same open sets from Lemma 3.1.3. In addition, points in O are limit points under d if and only if they are limit points under d' and hence O is separable with respect to d' since it is separable with respect to d. Since a sequence Cauchy under d' is also Cauchy under d, a d'-Cauchy sequence must have a limit since A is complete with respect to d. This point must also be in O, however, since otherwise $1/d(a_n,O^c) \to 1/d(a,O^c) = \infty$ and hence $d'(a_n,a_m)$ could not be converging to zero and hence $\{a_n\}$ could not be Cauchy. (Recall that for such a double limit to tend to zero, it must so tend regardless of the manner in which m and n go to infinity.) Thus O is complete. Since O is complete and separable, it is Polish.

If F is the intersection of a closed set C and an open set O, then the preceding metric d' yields a Polish space with the desired properties for the same reasons. In particular, d-convergence and d'-convergence are equivalent for the same reasons and a d'-Cauchy sequence must converge to a point that is in the intersection of C and O: it must be in C since C is closed and it must be in O or d' would blow up. □

It is important to point out that d-convergence and d'-convergence are only claimed to be equivalent within F; that is, a sequence a_n in F converges to a point a in F under d if and only if it converges under d'. It is possible, however, for a sequence a_n to converge to a point a not in

F under d and to not converge at all under d' (the d' distance is not even defined for points not in F). In other words, F may not be a closed subset of A with respect to d.

Example 3.2.2: Integer Sequence Spaces Revisited

We return to the integer sequence space of Example 3.2.1 to consider a simple but important special case where a subset of a Polish space is Polish. The following manipulation will resemble that of Lemma 2.3.2. Let A and d be as in that lemma and recall that A is Polish and has a Borel σ-field $\boldsymbol{B}(A) = \boldsymbol{B}(B)^B$. Suppose that we have a countable collection F_n, $n = 1,2, \cdots$ of thin cylinders with union $F = \bigcup_n F_n$. Consider the sequence space

$$\tilde{A} = A - \bigcup_{n=1}^{\infty} F_n = \bigcap_{n=1}^{\infty} F_n^c$$

formed by removing the countable collection of thin cylinders from A. Since the thin cylinders are open sets, their complements are closed. Since an arbitrary intersection of closed sets is closed, $\tilde{A}$ is a closed subset of A. Thus $\tilde{A}$ is itself Polish with respect to the same metric, d.

The Borel σ-fields of the spaces A and $\tilde{A}$ can be easily related by observing as in Example 3.2.1 that an open sphere about a sequence in either sequence space is simply a thin cylinder. Every thin cylinder in A is either contained in or disjoint from $A - F$. Thus $\boldsymbol{B}(A)$ is generated by all of the thin cylinders in A and $\boldsymbol{B}(\tilde{A})$ is generated by the thin cylinders in $\tilde{A}$ and hence by all sets of the form $\{x: x^n = a^n\} \cap (A - F)$, that is, by all the thin cylinders except those removed. Thus every open sphere in $\tilde{A}$ is also an open sphere in A and is contained in F^c; thus $\boldsymbol{B}(\tilde{A}) \subset \boldsymbol{B}(A) \cap (A - F)$. Similarly, every open sphere in A that is contained in F is also an open sphere in $\tilde{A}$, and hence the reverse inclusion holds. In summary,

$$\boldsymbol{B}(\tilde{A}) = \boldsymbol{B}(A) \cap (A - F) . \tag{3.2.7}$$

Carving

We now turn to a technique for carving up complete, separable metric spaces into pieces with a useful structure. From the previous lemma, we will then also be able to carve up certain other sets in a similar manner.

Lemma 3.2.5. (The carving lemma.) Let A be a complete, separable metric space. Given any real $r > 0$, there exists a countably infinite family of Borel sets (sets in $\boldsymbol{B}(A)$) $\{G_n, n = 1,2,...\}$ with the following properties:

(i) The $\{G_n\}$ partition A; that is, they are disjoint and their union is A.

(ii) Each $\{G_n\}$ is the intersection of the closure of an open sphere and an open set.

(iii) For all n, $diam(G_n) \le 2r$.

Proof. Let $\{a_i, 1{=}0,1,2,...\}$ be a countable dense set in A. Since A is separable, the family $\{V_i, i = 0,1,2,...\}$ of the closures of the open spheres $S_i = \{x: d(x,a_i) < r\}$ covers A; that is, its union is A. In addition, $diam(V_i) = diam(S_i) \le r$.

Construct the set $G_i(1)$ as the set V_i with all sets of lower index removed and all empty sets omitted; that is,

$$G_0(1) = V_0$$

$$G_1(1) = V_1 - V_0$$

$$\vdots$$

$$G_n(1) = V_n - \bigcup_{i<n} V_i = V_n - \bigcup_{i<n} G_i(1)$$

$$\vdots$$

where we remove all empty sets from the sequence. The sets are clearly disjoint and partition B. Furthermore, each set $G_n(1)$ has the form

$$G_n(1) = V_n \cap (\bigcap_{i<n} V_i^c)\ ;$$

that is, it is the intersection of the closure of an open sphere with an open set. Since $G_n(1)$ is a subset of V_n, it has diameter less than r. This provides the sequence with properties (i)-(iii). □

We are now ready to iterate on the previous lemmas to show that we can carve up Polish spaces into shrinking pieces that are also Polish spaces that can then be themselves carved up. This representation of a Polish space is called a *scheme*, and it is detailed in the next section.

Exercises

1. Prove the claims made in the first paragraph of this section.
2. Let $(\Omega,\boldsymbol{B},P)$ be a probability space and let $A = \boldsymbol{B}$, the collection of all events. Two events are considered to be the same if the probability of the symmetric difference is 0. Define a metric d on A as in Example 2.5.12. Is the corresponding Borel field a separable σ-field? Is A a separable metric space under d?
3. Prove that in the preceding example that all finite dimensional rectangles in A are open sets with respect to d. Since the complements of rectangles are other rectangles, this means that all rectangles (including the thin cylinders) are also closed. Thus the thin cylinders are both open and closed sets. (These unusual properties occur because the coordinate alphabets are discrete.)

3.3 POLISH SCHEMES

In this section we develop a representation of a Polish space that resembles an infinite version of the basis used to define standard spaces; the principal difference is that at each level we can now have an infinite rather that finite number of sets, but now the diameter of the sets must be decreasing as we descend through the levels. It will provide a canonical representation of Polish spaces and will provide the key element for proving that Polish spaces yield standard Borel spaces.

Recall that $\mathbf{Z}_+ = \{0,1,2,...\}$, $\mathbf{Z}_+^n$ is the corresponding finite dimensional cartesian product, and $\boldsymbol{N} = \mathbf{Z}_+^{\mathbf{Z}_+}$ is the one-sided infinite cartesian product. Given a metric space A, a *scheme* is defined as a family of sets $A(u^n)$, $n = 1,2,...$, indexed by u^n in $\mathbf{Z}_+^n$ with the following properties:

$$A(u^n) \subset A(u^{n-1}) \tag{3.3.1}$$

(the sets are decreasing),

$$A(u^n) \cap A(v^n) = \emptyset \text{ if } u^n \neq v^n \tag{3.3.2}$$

(the sets are disjoint),

$$\lim_{n\to\infty} diam(A(u^n)) = 0, \quad \text{all } u = \{u_0,u_1,...\} \in N . \tag{3.3.3}$$

If a decreasing sequence of sets $A(u^n)$, $n = 1,2,\ldots$ are all nonempty, then the limit is not empty; that is,

$$A(u^n) \neq \varnothing \text{ all } n \text{ implies } A_u \neq \varnothing, \tag{3.3.4}$$

where A_u is defined by

$$A_u = \bigcap_{n=1}^{\infty} A(u^n). \tag{3.3.5}$$

The scheme is said to be measurable if all of the sets $A(u^n)$ are Borel sets. The collection of all nonempty sets A_u, $u \in N$, of the form (3.3.5) will be called the *kernel* of the scheme. We shall let $A(u^0)$ denote the original space A.

Theorem 3.3.1. If A is a complete, separable metric space, then it is the kernel of a measurable scheme.

Proof. Construct the sequence $G_n(1)$ as in the first part of the carving lemma to satisfy properties (i)-(iii) of that lemma. Choose a radius r of, say, 1. This produces the first level sets $A(u^1)$, $u^1 \in \mathbf{Z}_+$. Each set thus produced is the intersection of the d closure $C(u^1)$ of a d-open sphere and a d–open set $O(u^1)$ in A, and hence by Lemma 3.2.4 each set $A(u^1)$ is itself a Polish space with respect to a metric d_{u^1} that within $A(u^1)$ is equivalent to d. We shall also make use of the property of the specific equivalent metric of (3.2.6) of that lemma that $d_{u^1}(a,b) \leq d(a,b)$. Henceforth let d_{u^0} denote the original metric d. The closure of $A(u^1)$ with respect to the original metric d may not be in $A(u^1)$ itself, but it is in the set $C(u^1)$.

We then repeat the procedure on these sets by applying Lemma 3.2.4 to each of the new Polish spaces with an r of 1/2 to obtain sets $A(u^2)$ that are intersections of a d_{u^1} closure $C(u^2)$ of an open sphere and a d_{u^1}-open set $O(u^2)$. Each of these sets is then itself a Polish space with respect to a metric $d_{u^2}(x,y)$ which is equivalent to the parent metric $d_{u^1}(x,y)$ within $A(u^2)$. In addition, $d_{u^2}(x,y) \geq d_{u^1}(x,y)$. Since $A(u^2) \subset A(u^1)$ and the parent metric d_{u^1} is equivalent to its parent metric d_{u^0} within $A(u^1)$, d_{u^2} is equivalent to the original metric d within $A(u^2)$.

A final crucial property is that the closure of $A(u^2)$ with respect to its parent metric d_{u^1} is contained within its parent set $A(u^1)$. This is true since $A(u^2) = C(u^2) \cap O(u^2)$ and $C(u^2)$ is a d_{u^1}-closed subset of $A(u^1)$

and therefore the d_{u^1}- closure of $A(u^2)$ lies within $C(u^2)$ and hence within $A(u^1)$.

We now continue in this way to construct succeeding levels of the scheme so that the preceding properties are retained at each level. At level n, $n = 1,2, \cdots$, we have sets

$$A(u^n) = C(u^n) \cap O(u^n), \tag{3.3.6}$$

where $C(u^n)$ is a $d_{u^{n-1}}$-closed subset of $A(u^{n-1})$ and $O(u^n)$ is a $d_{u^{n-1}}$-open subset of $A(u^{n-1})$. The set $C(u^n)$ is the $d_{u^{n-1}}$-closure of an open ball of the form

$$S(u^{n-1},i) = S_{1/n}(c_i;\ u^{n-1}) =$$

$$\{x: x \in A(u^{n-1}),\ d_{u^{n-1}}(x,c_i) < 1/n\},\ i = 0,1,2,.... \tag{3.3.7}$$

Each set $A(u^n)$ is then itself a Polish space with metric

$$d_{u^n}(x,y) =$$

$$d_{u^{n-1}}(x,y) + \left| \frac{1}{d_{u^{n-1}}(x,O(u^n)^c)} - \frac{1}{d_{u^{n-1}}(y,O(u^n)^c)} \right|. \tag{3.3.8}$$

We have by construction that for any u^n in $\mathbf{Z}_+^n$

$$d \le d_{u^1} \le d_{u^2} \le \cdots \le d_{u^n} \tag{3.3.9}$$

and hence since $A(u^n)$ is in the closure of an open sphere of the form (3.3.7) having $d_{u^{n-1}}$ radius $1/n$, we have also that the d-diameter of the sets satisfies

$$diam(A(u^n)) \le 2/n. \tag{3.3.10}$$

Furthermore, on $A(u^n)$, d_{u^n} (the complete metric) and $d_{u^{n-1}}$ yield equivalent convergence and are equivalent metrics. Since each $A(u^n)$ is contained in all of the A_{u^i} for $i = 0, \cdots, n-1$, this implies that

$$\text{given } a_k \in A(u^n),\ k = 1,2, \cdots, \text{ and } a \in A(u^n), \text{ then}$$

$$d_{u^n}(a_k,a) \to 0 \text{ iff } d_{u^j}(a_k,a) \to 0 \text{ for all } j = 0,1,2, \cdots, n-1. \tag{3.3.11}$$

This sequence of sets satisfies conditions (3.3.1)-(3.3.3). To complete the proof of the theorem we need only show that (3.3.4) is satisfied.

If we can demonstrate that the d-closure in A of the set $A(u^n)$ lies completely in its parent set $A(u^{n-1})$, then (3.3.4) will follow immediately from Corollary 3.2.1. We next provide this demonstration.

Say we have a sequence a_k; $k = 1,2, \cdots$ such that $a_k \in A(u^n)$, all k, and a point $a \in A$ such that $d(a_k,a) \to 0$ as $n \to \infty$. We will be done if we can show that necessarily $a \in A(u^{n-1})$. We accomplish this by assuming the contrary and developing a contradiction. Hence we assume that $a_k \to a$ in d, but $a \notin A(u^{n-1})$. First observe from (3.3.9) and (3.3.7) that since all of the a_k are in the closure of a sphere of $d_{u^{n-1}}$-radius $1/n$ centered at, say, c, then we must have that

$$d_{u^j}(a_k,c) \leq 1/n \ , \ j = 0,1,2,\ldots, n-1 \ ; \text{ all } k \ ; \tag{3.3.12}$$

that is, the a_k must all be close to the center $c \in A(u^{n-1})$ of the sphere used to construct $C(u^n)$ with respect to all of the previous metrics.

By assumption $a \notin A(u^{n-1})$. Suppose that l is the smallest index such that $a \notin A(u^l)$. Since the $A(u^i)$ are nested, we will then have that $a \in A(u^i)$ for $i < l$. In addition, $l \geq 1$ since $A(u^0)$ is the entire space and hence clearly $a \in A(u^0)$.

Since $a \in A(u^{l-1})$ and $a_k \in A(u^l) \subset A(u^{l-1})$, from (3.3.11) applied to $A(u^{l-1})$ d-convergence of a_k to a is equivalent to $d_{u^{l-1}}$-convergence. Since the set $C(u^l) \subset A(u^{l-1})$ used to form $A(u^l)$ via $A(u^l) = C(u^l) \cap O(u^l)$ is closed under $d_{u^{l-1}}$, we must have that $a \in C(u^l)$. Thus since

$$a \notin A(u^l) = C(u^l) \cap O(u^l)$$

and

$$a \in C(u^l),$$

then necessarily $a \in O(u^l)^c$. This, however, contradicts (3.3.12) for $j = l$ since

$$d_{u^l}(a_k,c) = d_{u^{l-1}}(a_k,c) + \left| \frac{1}{d_{u^{l-1}}(a_k,O(u^l)^c)} - \frac{1}{d_{u^{l-1}}(c,O(u^l)^c)} \right|$$

$$\geq \left| \frac{1}{d_{u^{l-1}}(a_k,O(u^l)^c)} - \frac{1}{d_{u^{l-1}}(c,O(u^l)^c)} \right| \underset{k\to\infty}{\to} \infty$$

since

$$d_{u^{l-1}}(a_k,O(u^l)^c) = \inf_{b \in O(u^l)^c} d_{u^{l-1}}(a_k,b) \leq d_{u^{l-1}}(a_k,a) \to 0 \ ,$$

where the final inequality follows since $a \in O(u^l)^c$. Thus a cannot be in $A(u^{l-1})$. But this contradicts the assumption that l is the smallest index for which $a \notin A(u_l)$ and thereby completes the proof that the given sequence of sets is indeed a scheme. □

Since at each stage every point in A must be in one of the sets $A(u^n)$, if we define $A(u^n)(x)$ as the set containing x, then

$$\{x\} = \bigcap_{n=1}^{\infty} A(u^n)(x) \qquad (3.3.13)$$

and hence every singleton set $\{x\}$ is in the kernel of the scheme. Observe that many of the sets in the scheme may be empty and not all sequences u in $\boldsymbol{N} = \mathbf{Z}_+^{\mathbf{Z}_+}$ will produce a point via (3.3.13). Since every point in A is in the kernel of a measurable scheme, (3.3.13) gives a map $f: A \to \boldsymbol{N}$ where $f(x)$ is the u for which (3.3.13) is satisfied. This scheme can be thought of as a *quantizer* mapping a continuous space A into a discrete (countable) binary representation in the following sense: Suppose that we view a point a and we encode it into a sequence of integers $u_n;\ n = 1,2,\cdots$ which is then viewed by a receiver. The receiver at time n uses the n integers to produce an estimate or reproduction of the original point, say $g_n(u^n)$. The badness or distortion of the reproduction can be measured by $d(a,g_n(u^n))$. The goal is to decrease this distortion to 0 as $n \to \infty$. The scheme provides such a code: Let the encoder at time n map x into the u^n for which $x \in A(u^n)$ and let the decoder $g_n(u^n)$ simply produce an arbitrary member of $A(u^n)$ (if it is not empty). This distortion will be bound above by $diam(A(u^n)) \leq 2/n$. This produces the sequence encoder mapping $f: A \to N$.

Unfortunately the mapping f is in general into and not onto. For example, if at level n for some n-tuple $b^n \in \mathbf{Z}_+^n$ $A(b^n)$ is empty, then the encoding will produce no sequences that begin with b^n since there can never be an x in $A(b^n)$. In other words, any sequence u with $u^n = b^n$ cannot be of the form $u = f(x)$ for any $x \in A$. Roughly speaking, not all of the integers are useful at each level since some of the cells may be empty. Alternatively, some integer sequences do not correspond to (they cannot be decoded into) any point in A. We pause to tackle the problems of empty cells.

Let N_1 denote all integers u^1 for which $A(u^1)$ is empty. Let N_2 denote the collection of all integer pairs u^2 such that $A(u^2)$ is empty but $A(u^1)$ is not. That is, once we throw out an empty cell we no longer consider its descendants. Continue in this way to form the sets N_k

consisting of all u^k for which $A(u^k)$ is empty but $A(u^{k-1})$ is not. These sets are clearly all countable. Define next the space

$$N_0 = \boldsymbol{N} - \bigcup_{n=1}^{\infty} \bigcup_{b^n \in N_n} \{u: u^n = b^n\} = \bigcap_{n=1}^{\infty} \bigcap_{b^n \in N_n} \{u: u^n = b^n\}^c \quad (3.3.14)$$

Thus N_0 is formed by removing all thin cylinders of the form $\{u: u^n = b^n\}$ for some $b^n \in N_n$, that is, all thin cylinders for which the first part of the sequence corresponds to an empty cell. This space has several convenient properties:

1. For all $x \in A$, $f(x) \in N_0$. This follows since if $x \in A$, then $x \in A(u^n)$ for some u^n for every n; that is, x can never lie in an empty cell, and hence $f(x)$ cannot be in the portion of $\boldsymbol{N}$ that was removed.

2. The set N_0 is a Borel set since it is formed by removing a countable number of thin cylinders from the entire space.

3. If we form a σ-field of N_0 in the usual way as $\boldsymbol{B}(N) \cap N_0$, then from Lemma 2.3.2 (or Lemma 2.4.2) the resulting measurable space $(N_0, \boldsymbol{B}(N) \cap N_0)$ is standard.

4. From Examples 3.2.1 and 3.2.2 and (3.2.7), the Borel field $\boldsymbol{B}(N_0)$ (using $d^{\boldsymbol{Z}_+}$) is the same as $\boldsymbol{B}(N) \cap N_0$. Thus with Property 3, $(N_0, \boldsymbol{B}(N_0))$ is standard.

5. In the reduced space N_0, all of the sequences correspond to sequences of nonempty cells in A. Because of the properties of a scheme, any descending sequence of nonempty cells must contain a point, and if the integer sequences ever differ, the points must differ. Define the mapping $g: N_0 \to A$ by

   $$g(u) = \bigcap_{n=1}^{\infty} A(u^n) .$$

   Since every point in A can be encoded uniquely into some integer sequence in N_0, the mapping g must be onto.

6. $N_0 = f(A)$; that is, N_0 is the range space of $\boldsymbol{N}$. To see this, observe first that Property 1 implies that $f(A) \subset N_0$. Property 5 implies that every point in N_0 must be of the form $f(x)$ for some $x \in A$, and hence $N_0 \subset f(A)$, thereby proving the claim. Note that with Property 2 this provides an example wherein the forward image $f(A)$ of a Borel set is also a Borel set.

By construction we have for any $x \in A$ that $g(f(x)) = x$ and that for any $u \in N_0$ $f(g(u)) = u$. Thus if we define $\hat{f}: A \to N_0$ as the mapping f considered as having a range N_0, then $\hat{f}$ is onto, the mapping g is just the inverse of the mapping $\hat{f}$, and the mappings are one-to-one. If these mappings are measurable, then the spaces $(A, \boldsymbol{B}(A))$ and $(N_0, \boldsymbol{B}(N_0))$ are isomorphic. Thus if we can show that the mappings $\hat{f}$ and g are measurable, then $(A, \boldsymbol{B}(A))$ will be proved standard by Lemma 2.3.3 and Property 4 because it is isomorphic to a standard space. We now prove this measurability and complete the proof.

First consider the mapping $g: N_0 \to A$. $\boldsymbol{N}$ and hence the subspace N_0 are metric spaces under the product space metric $d^{\mathbf{Z}_+}$ using the Hamming metric of Example 2.5.1; that is,

$$d^{\mathbf{Z}_+}(u,v) = \sum_{i=1}^{\infty} 2^{-i} d_H(u_i, v_i) ,$$

Given an $\varepsilon > 0$, find an n such that $2/n < \varepsilon$ and then choose a $\delta < 2^{-n}$. If $d^{\mathbf{Z}_+}(u,v) < \delta$, then necessarily $u^n = v^n$ which means that both $g(u)$ and $g(v)$ must be in the same $A(u^n)$ and hence $d(g(u),g(v)) < 2/n < \varepsilon$. Thus g is continuous and hence $\boldsymbol{B}(N_0)$-measurable.

Next consider $f: A \to N_0$. Let $G = c(v^n)$ denote the thin cylinder $\{u: u \in \boldsymbol{N}, u^n = v^n\}$ in N_0. Then by construction $f^{-1}(G) = g(G) = A(v^n)$, which is a measurable set. From Example 3.2.2, however, such thin cylinders generate $\boldsymbol{B}(N_0)$. Thus f is measurable on a generating class and hence, by Lemma 1.4.1, f is measurable.

We have now proved the following result.

Theorem 3.3.1. If A is a Polish space, then the Borel space $(A, \boldsymbol{B}(A))$ is standard.

Thus the real line, the unit interval, Euclidean space, and separable Hilbert spaces and Banach spaces are standard. We shall encounter several specific examples in the sequel.

The development of the theorem and some simple observations provide some subsidiary results that we now collect.

We saw that a Polish space A is the kernel of a measurable scheme and that the Borel space $(A, \boldsymbol{B}(A))$ is isomorphic to $(N_0, \boldsymbol{B}(N_0))$, a subspace of the space of integer sequences, using the metric $d^{\mathbf{Z}_+}$. In addition, the inverse mapping in the construction is continuous. The isomorphism and its properties depended only on the fact that that the set was the kernel of a measurable scheme. Thus if any set F in a metric space is the kernel of a measurable scheme, then it is isomorphic to some $(N_0,\boldsymbol{B}(N_0))$, $N_0 \subset \boldsymbol{N}$, with the metric $d^{\mathbf{Z}_+}$ on N_0, and the inverse mapping is continuous. We next show that this subspace, say N_0, is closed with respect to $d^{\mathbf{Z}_+}$. If $a_k \to a$ and $a_k \in N_0$, then the first n coordinates of a_k must match those of a for sufficiently large k and hence be contained in the thin cylinder $c(a^n)$ $= \{u: u^n = a^n\}$. But these sets are all nonempty for all n (or some a_k would not be in N_0) and hence $a \notin N_0^c$ and $a \in N_0$. Thus N_0 is closed. Since it is a closed subset of a Polish space, it is also Polish with respect to the same metric. Thus any kernel of a measurable scheme is isomorphic to a Polish space (N_0 in particular) with a continuous inverse.

Conversely, suppose that F is a subset of a Polish space and that $(F,\boldsymbol{B}(F))$ is itself isomorphic to another Polish Borel space $(B,\boldsymbol{B}(B))$ with metric m and suppose the inverse mapping is continuous. Suppose that $\{B(u^n)\}$ forms a measurable scheme for B. If $f: A \to B$ is the isomorphism and $g: B \to A$ the continuous inverse, then $A(u^n)$ defined by $f^{-1}(B(u^n)) = g(B(u^n))$ satisfy the conditions (3.3.1) and (3.3.2) because inverse images preserve set theoretic operations. Since g is continuous, $diam(B(u^n)) \to 0$ implies that also $diam(A(u^n)) \to 0$. If the $A(u^n)$ are nonempty for all n, then so are the $B(u^n)$. Since B is Polish, they must have a nonempty intersection of exactly one point, say b, but then the intersection of the $A(u^n)$ will be $g(b)$ and hence nonempty. Thus F is the kernel of a measurable scheme. We have thus proved the following lemma.

Lemma 3.3.1. Let A be a Polish space and F a Borel set in A. Then F is the kernel of a measurable scheme if and only if there is an isomorphism with continuous inverse from $(F,\boldsymbol{B}(F))$ to some Polish Borel space.

Corollary 3.3.2. Let $(A,\boldsymbol{B}(A))$ be the Borel space of a Polish space A. Let $\boldsymbol{G}$ be the class of all Borel sets F with the property that F is the kernel of a measurable scheme. Then $\boldsymbol{G}$ is closed under countable disjoint unions and countable intersections.

Proof. If F_i, $i = 1,2, \ldots$ are disjoint kernels of measurable schemes, then it is easy to say that the union $\bigcup_i F_i$ will also be the kernel of a measurable scheme. The cells of the first level are all of the first level cells for all of the F_i. They can be reindexed to have the required form. The descending levels follow exactly as in the individual F_i. Since the F_i are disjoint, so are the cells, and the remaining properties follow from those of the individual sets F_i.

Define the intersection $\underline{F} = \bigcap_{i=1}^{\infty} F_i$. From the lemma there are isomorphisms f_i: $F_i \to N_i$ with N_i being Polish subspaces of $\boldsymbol{N}$ which have continuous inverses g_i: $N_i \to F_i$. Define the space

$$N^{\infty} = \mathop{\times}_{i=1}^{\infty} N_i$$

and a subset H of N^{∞} containing all of those $(z_1, z_2, \cdots)$, $z_i \in N_i$, such that $g_1(z_1) = g_2(z_2) = \ldots$ and define a mapping g: $H \to \underline{F}$ by $g(z_1, z_2, \cdots) = g_1(z_1)$ (which equals $g_n(z_n)$ for all n). Since the g_n are continuous, H is a closed subset of N^{∞}. Since H is a closed subset of N^{∞} and since N^{∞} is Polish from Lemma 3.2.3, H is also Polish. g is continuous since the g_n are, hence it is also measurable. It is also one-to-one, its inverse being $f : \underline{F} \to H$ being defined by $f(a) = (f_1(a), f_2(a), \cdots)$. Since the mappings f_n are measurable, so is f. Thus $\underline{F}$ is isomorphic to a Polish space and has a continuous inverse and hence is the kernel of a measurable scheme. □

Theorem 3.3.2. If F is a Borel set of a Polish space A, then the Borel space $(F, \boldsymbol{B}(F))$ is standard.

Comment. It should be emphasized that the set F may not itself be Polish.

Proof. From the previous corollary the collection of all Borel sets that are kernels of measurable schemes contains countable disjoint unions and countable intersections. First, we know that each such set is isomorphic to a Polish Borel space by Lemma 3.3.1 and that each of these Polish spaces is standard by Theorem 3.3.1. Hence each set in this collection is isomorphic to a standard space and so by Lemma 2.3.3 is itself standard. To complete the proof we need to show that our collection includes all the Borel sets. Consider the closed subsets of A. Lemma 3.3.3 implies that each such subset is itself a Polish space and hence, from Theorem 3.3.1, is the kernel of a measurable scheme. Hence our collection contains all closed subsets of A. Since it also contains all countable disjoint unions and countable intersections, by Lemma 3.2.5 it contains all the Borel sets. Since our collection is itself a subset of the Borel sets, the collection must be exactly the Borel σ-field. □

This theorem gives the most general known class of standard Borel spaces. The development here departs from the traditional and elegant development of Parthasarathy [6] in a variety of ways, although the structure and use of schemes strongly resembles his use of analytic sets. The use of schemes allowed us to map a Polish space directly onto a subset of a standard space that was easily seen to be both a measurable set and standard itself. By showing that all Borel sets had such schemes, the general result was obtained. The traditional method is instead to map the original space into a subset of a known standard space that is then shown to be a Borel set and to possess a subset isomorphic to the full standard space. By showing that a set sandwiched between two isomorphic spaces is itself isomorphic, the result is proved. As a side result it is shown that all Polish spaces are actually isomorphic to $\boldsymbol{N}$ or, equivalently, to M. This development perhaps adds insight, but it requires the additional and difficult proof that the one-to-one image of a Borel set is itself a Borel set: a result that we avoided by mapping onto a subset of $\boldsymbol{N}$ formed by simply removing a countable number of thin cylinders from it. While the approach adopted here does not yield the rich harvest of related results of Parthasarathy, it is a simpler and more direct route to the desired result. Alternative but often similar developments of the properties of Polish Borel spaces may be found in Schwartz [8], Christensen [3], Bourbaki [2], Cohn [4], Mackey [5], and Bjornsson [1].

REFERENCES

1. O. J. Bjornsson, "A note on the characterization of standard borel spaces," *Math. Scand.*, vol. 47, pp. 135-136, 1980.
2. N. Bourbaki, *Elements de Mathematique, Livre VI, Integration,* Hermann, Paris, 1956-1965.
3. J. P. R. Christensen, *Topology and Borel Structure,* Mathematics Studies 10, North-Holland/American Elsevier, New York, 1974.
4. D. C. Cohn, *Measure Theory,* Birkhauser, New York, 1980.
5. G. Mackey, "Borel structures in groups and their duals," *Trans. Am. Math. Soc.*, vol. 85, pp. 134-165, 1957.

6. K. R. Parthasarathy, *Probability Measures on Metric Spaces,* Academic Press, New York, 1967.

7. W. Rudin, *Principles of Mathematical Analysis,* McGraw-Hill, New York, 1964.

8. L. Schwartz, *Radon Measures on Arbitrary Topological Spaces and Cylindrical Measures,* Oxford University Press, Oxford, 1973.

9. G. F. Simmons, *Introduction to Topology and Modern Analysis,* McGraw-Hill, New York, 1963.

4

AVERAGES

4.1 INTRODUCTION

The basic focus of classical ergodic theory was the development of conditions under which sample or time averages consisting of arithmetic means of a sequence of measurements on a random process converged to a probabilistic or ensemble average of the measurement as expressed by an integral of the measurement with respect to a probability measure. Theorems relating these two kinds of averages are called *ergodic theorems*.

In this chapter these two types of averages are defined, and several useful properties are gathered and developed. The treatment of ensemble averages or *expectations* is fairly standard, except for the emphasis on quantization and the fact that a general integral can be considered as a natural limit of simple integrals (sums, in fact) of quantized versions of the function being integrated. In this light, general (Lebesgue) integration is actually simpler to define and deal with than is the integral usually taught to most engineers, the ill-behaved and complicated Riemann integral. The only advantage of the Riemann integral is in computation, actually evaluating the integrals of specific functions. Although an important attribute for applications, it is not as important as convergence and continuity properties in developing theory.

The theory of expectation is simply a special case of the theory of integration. Excellent treatments may be found in Halmos [3], Ash [1], and Chung [2]. Many proofs in this chapter follow those in these books.

The treatment of expectation and sample averages together is uncommon, but it is useful to compare the definitions and properties of the two types of average in preparation for the ergodic theorems to come.

As we wish to form arithmetic sums and integrals of measurements, in this chapter we emphasize real-valued random variables.

4.2 DISCRETE MEASUREMENTS

We begin by focusing on measurements that take on only a finite number of possible values. This simplifies the definitions and proofs of the basic properties and provides a fundamental first step for the more general results. Let $(\Omega,\boldsymbol{B},m)$ be a probability space, e.g., the probability space $(A^{\boldsymbol{I}},B_A{}^{\boldsymbol{I}},m)$ of a directly given random process with alphabet A. A real-valued random variable $f: \Omega \to \mathbf{R}$ will also be called a *measurement* since it is often formed by taking a mapping or function of some other set of more general random variables, e.g., the outputs of some random process that might not have real-valued outputs. Measurements made on such processes, however, will always be assumed to be real.

Suppose next we have a measurement f whose range space or *alphabet* $f(\Omega) \subset \mathbf{R}$ of possible values is finite. Then f is called a *discrete random variable* or *discrete measurement* or *digital measurement* or, in the common mathematical terminology, a *simple function.* Such discrete measurements are of fundamental mathematical importance because many basic results are easy to prove for simple functions and, as we shall see, all real random variables can be expressed as limits of such functions. In addition, discrete measurements are good models for many important physical measurements: No human can distinguish all real numbers on a meter, nor is any meter perfectly accurate. Hence, in a sense, all physical measurements have for all practical purposes a finite (albeit possibly quite large) number of possible readings. Finally, many measurements of interest in communications systems are inherently discrete–binary computer data and ASCII codes being ubiquitous examples.

Given a discrete measurement f, suppose that its range space is $f(\Omega) = \{b_i, i = 1, \cdots, N\}$, where the b_i are distinct. Define the sets $F_i =$

$f^{-1}(b_i) = \{x: f(x) = b_i\}$, $i = 1, \cdots, N$. Since f is measurable, the F_i are all members of $\boldsymbol{B}$. Since the b_i are distinct, the F_i are disjoint. Since every input point in Ω must map into some b_i, the union of the F_i equals Ω. Thus the collection $\{F_i; i = 1,2, \cdots, N\}$ forms a partition of Ω. We have therefore shown that any discrete measurement f can be expressed in the form

$$f(x) = \sum_{i=1}^{M} b_i 1_{F_i}(x) , \tag{4.2.1}$$

where $b_i \in \mathbf{R}$, the $F_i \in \boldsymbol{B}$ form a partition of Ω, and 1_{F_i} is the indicator function of F_i, $i = 1, \cdots, M$. Every simple function has a unique representation in this form except possibly for sets that differ on a set of measure 0.

The *expectation* or *ensemble average* or *probabilistic average* or *mean* of a discrete measurement $f: \Omega \rightarrow \mathbf{R}$ as in (4.2.1) with respect to a probability measure m is defined by

$$E_m f = \sum_{i=0}^{M} b_i m(F_i) . \tag{4.2.2}$$

The following simple lemma shows that the expectation of a discrete measurement is well defined.

Lemma 4.2.1. Given a discrete measurement (measurable simple function) $f: \Omega \rightarrow \mathbf{R}$ defined on a probability space $(\Omega, \boldsymbol{B}, m)$, then expectation is well defined; that is, if

$$f(x) = \sum_{i=1}^{N} b_i 1_{F_i}(x) = \sum_{j=1}^{M} a_j 1_{G_j}(x)$$

then

$$E_m f = \sum_{i=1}^{N} b_i m(F_i) = \sum_{j=1}^{M} a_j m(G_j) .$$

Proof. Given the two partitions $\{F_i\}$ and $\{G_j\}$, form the new partition (called the *join* of the two partitions) $\{F_i \cap G_j\}$ and observe that

$$f(x) = \sum_{i=1}^{N} \sum_{j=1}^{M} c_{ij} 1_{F_i \cap G_j}(x)$$

where $c_{ij} = a_i = b_j$. Thus

$$\sum_{i=1}^{N} b_i m(F_i) = \sum_{i=1}^{N} b_i (\sum_{j=1}^{M} m(G_j \cap F_i)) = \sum_{i=1}^{N} \sum_{j=1}^{M} b_i m(G_j \cap F_i)$$

$$= \sum_{i=1}^{N} \sum_{j=1}^{M} c_{ij} m(G_j \cap F_i) = \sum_{j=1}^{M} \sum_{i=1}^{N} a_j m(G_j \cap F_i)$$

$$= \sum_{j=1}^{M} a_j (\sum_{i=1}^{N} m(G_j \cap F_i)) = \sum_{j=1}^{M} b_j m(G_j) . \square$$

An immediate consequence of the definition of expectation is the simple but useful fact that for any event F in the original probability space,

$$E_m 1_F = m(F) ; \tag{4.2.3}$$

that is, probabilities can be found from expectations of indicator functions.

The next lemma collects the basic properties of expectations of discrete measurements. These properties will be later seen to hold for general expectations and similar ones to hold for time averages. Thus they can be viewed as fundamental properties of averages.

Lemma 4.2.2.

(a) If $f \geq 0$ with probability 1, then $E_m f \geq 0$.

(b) $E_m 1 = 1$.

(c) Expectation is linear; that is, for any real α, β and any discrete measurements f and g,

$$E_m(\alpha f + \beta g) = \alpha E_m f + \beta E_m g .$$

(d) If f is a discrete measurement, then

$$| E_m f | \leq E_m | f | .$$

(e) Given two discrete measurements f and g for which $f \geq g$ with probability 1 ($m(x: f(x) \geq g(x)) = 1$), then

$$E_m f \geq E_m g .$$

Comments. The first three statements bear a strong resemblance to the first three axioms of probability and will be seen to follow directly from those axioms: Since probabilities are nonnegative, normalized, and additive, expectations inherit similar properties. A limiting version of additivity will be postponed. Properties (d) and (e) are useful basic integration inequalities. The following proofs are all simple, but they provide some useful practice manipulating integrals of discrete functions.

Proof.

(a) If $f \geq 0$ with probability 1, then $b_i \geq 0$ in (4.2.1) and $E_m f \geq 0$ from (4.2.2) and the nonnegativity of probability.

(b) Since $1 = 1_\Omega$, $E_m 1 = 1m(\Omega) = 1$ and the property follows from the normalization of probability.

(c) Suppose that

$$f = \sum_i b_i 1_{F_i},\ g = \sum_j a_i 1_{G_i}.$$

Again consider the join of the two partitions and write

$$f = \sum_i \sum_j b_i 1_{F_i \cap G_j}$$

$$g = \sum_i \sum_j a_j 1_{F_i \cap G_j}.$$

Then

$$\alpha f + \beta g = \sum_i \sum_j (\alpha b_i + \beta a_j) 1_{F_i \cap G_j}$$

is a discrete measurement with the partition $\{F_i \cap G_j\}$ and hence

$$E_m(\alpha f + \beta g) = \sum_i \sum_j (\alpha b_i + \beta a_j) m(F_i \cap G_j)$$

$$= \alpha \sum_i \sum_j b_i m(F_i \cap G_j) + \beta \sum_i \sum_j a_j m(F_i \cap G_j)$$

$$= \alpha \sum_i b_i m(F_i) + \beta \sum_j a_j m(G_j),$$

from the additivity of probability. This proves the claim.

(d) Let $\boldsymbol{J}$ denote those i for which b_i in (4.2.1) are nonnegative. Then

$$E_m f = \sum_i b_i m(F_i) = \sum_{i \in \boldsymbol{J}} |b_i| m(F_i) - \sum_{i \notin \boldsymbol{J}} |b_i| m(F_i)$$

$$\leq \sum_i | b_i | m(F_i) = E_m | f | .$$

(e) This follows from (a) applied to $f - g$. □

In the next section we introduce the notion of quantizing arbitrary real measurements into discrete approximations in order to extend the definition of expectation to such general measurements by limiting arguments.

4.3 QUANTIZATION

Again let $(\Omega, \boldsymbol{B}, m)$ be a probability space and $f : \Omega \to \mathbf{R}$ a measurement, that is, a real-valued random variable or measurable real-valued function. The following lemma shows that any measurement can be expressed as a limit of discrete measurements and that if the measurement is nonnegative, then the simple functions converge upward.

Lemma 4.3.1. Given a real random variable f, define the sequence of *quantizers* q_n: $\mathbf{R} \to \mathbf{R}$, $n = 1,2, \cdots$, as follows:

$$q_n(r) = \begin{cases} n & r \geq n \\ (k-1)2^{-n} & (k-1)2^{-n} \leq r < k2^{-n},\ k = 1,2, \cdots, n2^n \\ -(k-1)2^{-n} & -k2^{-n} \leq r < -(k-1)2^{-n},\ k = 1,2, \cdots, n2^n \\ -n & r < -n \ . \end{cases}$$

The sequence of quantized measurements have the following properties:

(a)

$$f(x) = \lim_{n \to \infty} q_n(f(x)) , \text{ all } x .$$

(b) If $f(x) \geq 0$, all x, then $f \geq q_{n+1}(f) \geq q_n(f)$, all n; that is, $q_n(x) \uparrow f$. More generally, $|f| \geq |q_{n+1}(f)| \geq |q_n(f)|$.

(c) If $f \geq g \geq 0$, then also $q_n(f) \geq q_n(g) \geq 0$ for all n.

(d) The magnitude quantization error $|q_n(f) - f|$ satisfies

$$|q_n(f) - f| \leq |f| \tag{4.3.1}$$

$$|q_n(f) - f| \downarrow 0 ; \tag{4.3.2}$$

that is, the magnitude error is monotonically nonincreasing and converges to 0.

(e) If f is $\boldsymbol{G}$-measurable for some sub-σ-field $\boldsymbol{G}$, than so are the f_n. Thus if f is $\boldsymbol{G}$-measurable, then there is a sequence of $\boldsymbol{G}$-measurable simple functions which converges to f.

Proof. Statements (a)-(c) are obvious from the the construction. To prove (d) observe that if x is fixed, then either f is nonnegative or it is negative. If it is nonnegative, then so is $q_n(f(x))$ and $|q_n(f) - f| = f - q_n(f)$. This is bound above by f and decreases to 0 since $q_n(f) \uparrow f$ from part (b). If $f(x)$ is negative, then so is $q_n(f)$ and $|q_n(f) - f| = q_n(f) - f$. This is bound above by $-f = |f|$ and decreases to 0 since $q_n(f) \downarrow f$ from part (b). To prove (e), suppose that f is $\boldsymbol{G}$-measurable and hence that inverse images of all Borel sets under f are in $\boldsymbol{G}$. If q_n, $n = 1,2,\cdots$ is the quantizer sequence, then the functions $q_n(f)$ are also $\boldsymbol{G}$-measurable (and hence also $\boldsymbol{B}(\mathrm{R})$ measurable) since the inverse images satisfy

$$q_n(f)^{-1}(F) = \{\omega: q_n(f)(\omega) \in F\} = \{\omega: q_n(f(\omega)) \in F\}$$
$$= \{\omega: f(\omega) \in q_n^{-1}(F)\} = f^{-1}(q_n^{-1}(F)) \in \boldsymbol{G}$$

since all inverse images under f of events F in $\boldsymbol{B}(\mathrm{R})$ are in $\boldsymbol{G}$. □

Note further that if $|f| \leq K$ for some finite K, then for $n > K$ $|f(x) - q_n(f(x))| \leq 2^{-n}$.

Measurability

As a prerequisite to the study of averages, we pause to consider some special properties of real-valued random variables. We will often be interested in the limiting properties of a sequence f_n of integrable functions. Of particular importance is the measurability of limits of measurements and conditions under which we can interchange limits and expectations. In order to state the principal convergence results, we require first a few definitions. Given a sequence a_n, $n = 1,2,\cdots$ of real numbers, define the *limit supremum* or *upper limit*

$$\limsup_{n\to\infty} a_n = \inf_n \sup_{k\geq n} a_k$$

and the *limit infimum* or *lower limit*

$$\liminf_{n\to\infty} a_n = \sup_n \inf_{k\geq n} a_k .$$

In other words, $\limsup_{n\to\infty} a_n = a$ if a is the largest number for which a_n has a subsequence, say $a_{n(k)}$, $k = 1,2,\cdots$ converging to a. Similarly, $\liminf_{n\to\infty} a_n$ is the smallest limit attainable by a subsequence. When clear from context, we shall often drop the $n\to\infty$ and simply write $\limsup a_n$ or $\liminf a_n$. The limit of a sequence a_n, say $\lim_{n\to\infty} a_n = a$, will exist if and only if $\limsup a_n = \liminf a_n = a$, that is, if all subsequences converge to the same thing–the limit of the sequence.

Lemma 4.3.2. Given a sequence of measurements f_n, $n = 1,2,\cdots$, then $\overline{f} = \limsup_{n\to\infty} f_n$ and $\underline{f} = \liminf_{n\to\infty} f_n$ are also measurements, that is, are also measurable functions. If $\lim_{n\to\infty} f_n$ exists, then it is a measurement. If f and g are measurements, then so are the functions $f+g$, $f-g$, $|f-g|$, $\min(f,g)$, and $\max(f,g)$; that is, all these functions are measurable.

Proof. To prove a real-valued function $f:\Omega\to\mathbf{R}$ measurable, it suffices from Lemma 1.4.1 to show that $f^{-1}(F)\in\boldsymbol{B}$ for all F in a generating class. From Chapter 3 $\mathbf{R}$ is generated by the countable collection of open spheres of the form $\{x: |x-r_i| < 1/n\}$; $i = 1,2,\cdots$; $n = 1,2,\cdots$, where the r_i run through the rational numbers. More simply, we can generate the Borel σ-field of $\mathbf{R}$ from the class of all sets of the form $(-\infty, r_i] = \{x: x\leq r_i\}$, with the same r_i. This is because the σ-field containing these half-open intervals also contains the open intervals since it contains the half-open intervals

$$(a,b] = (-\infty,b] - (-\infty,a]$$

with rational endpoints, the rationals themselves:

$$a = \bigcap_{n=1}^{\infty} (a - \frac{1}{n}, a] ,$$

and hence all open intervals with rational endpoints

$$(a,b) = (a,b] - b .$$

Since this includes the spheres described previously, this class generates the σ-field. Consider the limit supremum $\overline{f}$. For $\overline{f}\leq r$ we must have for any $\varepsilon > 0$ that $f_n \leq r+\varepsilon$ for all but a finite number of n. Define

$$F_n(\varepsilon) = \{x: f_n(x) \leq r + \varepsilon\} .$$

Then the collection of x for which $f_n(x) \leq r + \varepsilon$ for all but a finite number of n is

$$\underline{F}(\varepsilon) = \bigcup_{n=1}^{\infty} \bigcap_{k=n}^{\infty} F_n(\varepsilon) .$$

This set is called the *limit infimum* of the $F_n(\varepsilon)$ in analogy to the notion for sequences. As this type of set will occasionally be useful, we pause to formalize the definition: Given a sequence of sets F_n, $n = 1,2, \cdots$, define the *limit infimum* of the sequence of sets by

$$\underline{F} = \liminf_{n \to \infty} F_n = \bigcup_{n=1}^{\infty} \bigcap_{k=n}^{\infty} F_n(\varepsilon)$$

$$= \{x \; x \in F_n \text{ for all but a finite number of } n\} . \tag{4.3.3}$$

Since the f_n are measurable, $F_n(\varepsilon) \in \boldsymbol{B}$ for all n and ε and hence $\underline{F}(\varepsilon) \in \boldsymbol{B}$. Since we must be in $\underline{F}(\varepsilon)$ for all ε for $\overline{f} \leq r$,

$$F = \{x: \overline{f}(x) \leq r\} = \bigcap_{n=1}^{\infty} \underline{F}(\frac{1}{n}) .$$

Since the $\underline{F}(1/n)$ are all measurable, so is F. This proves the measurability of $\overline{f}$. That for $\underline{f}$ follows similarly. If the limit exists, then $\overline{f} = \underline{f}$ and measurability follows from the preceding argument. The remaining results are easy if the measurements are discrete. The conclusions then follow from this fact and Lemma 4.3.1. □

Exercises

1. Fill in the details of the preceding proof; that is, show that sums, differences, minima, and maxima of measurable functions are measurable.

2. Show that if f and g are measurable, then so is the product fg. Under what conditions is the division f/g measurable?

3. Prove the following corollary to the lemma:

Corollary 4.3.1. Let f_n be a sequence of measurements defined on $(\Omega, \boldsymbol{B}, m)$. Define F_f as the set of $x \in \Omega$ for which $\lim_{n \to \infty} f_n$ exists; then 1_F is a measurement and $F \in \boldsymbol{B}$.

4. Is it true that if $\underline{F}$ is the limit infimum of a sequence F_n of events, then

$$1_{\underline{F}}(x) = \liminf_{n\to\infty} 1_{F_n}(x) \ ?$$

4.4 EXPECTATION

We now define expectation for general measurements in two steps. If $f \geq 0$, then choose as in Lemma 4.3.1 the sequence of asymptotically accurate quantizers q_n and define

$$E_m f = \lim_{n\to\infty} E_m(q_n(f)) \ . \tag{4.4.1}$$

Since the q_n are discrete measurements on f, the $q_n(f)$ are discrete measurements on Ω ($q_n(f)(x) = q_n(f(x))$ is a simple function), and hence the individual expectations are well defined. Since the $q_n(f)$ are nondecreasing, so are the $E_m(q_n(f))$ from Lemma 4.2.2(e). Thus the sequence must either converge to a finite limit or grow without bound, in which case we say it converges to ∞. In both cases the expectation $E_m f$ is well defined, although it may be infinite. It is easily seen that if f is a discrete measurement, then (4.4.1) is consistent with the previous definition (4.2.1).

If f is an arbitrary real random variable, define its positive and negative parts $f^+(x) = \max(f(x),0)$ and $f^-(x) = -\min(f(x),0)$ so that $f(x) = f^+(x) - f^-(x)$ and set

$$E_m f = E_m f^+ - E_m f^- \tag{4.4.2}$$

provided this does not have the form $+\infty - \infty$, in which case the expectation does not exist. Observe that an expectation exists for all nonnegative random variables, but that it may be infinite. The following lemma provides an alternative definition of expectation as well as a limiting theorem that shows that there is nothing magic about the particular sequence of discrete measurements used in the definition.

Lemma 4.4.1.

(a) If $f \geq 0$, then

$$E_m f = \sup_{\text{discrete } g:\, g \leq f} E_m g \, .$$

(b) Suppose that $f \geq 0$ and f_n is any nondecreasing sequence of discrete measurements for which $\lim_{n\to\infty} f_n(x) = f(x)$, then

$$E_m f = \lim_{n\to\infty} E_m f_n \, .$$

Comments. Part (a) states that the expected value of a nonnegative measurement is the supremum of the expected values of discrete measurements that are no greater than f. This property is often used as a definition of expectation. Part (b) states that any nondecreasing sequence of asymptotically accurate discrete approximations to f can be used to evaluate the expectation. This is a special case of the monotone convergence theorem to be considered shortly.

Proof. Since the $q_n(f)$ are all discrete measurements less than or equal to f, clearly

$$E_m f = \lim_{n\to\infty} E_m q_n(f) \leq \sup_g E_m g \, ,$$

where the supremum is over all discrete measurements less than or equal to f. For $\varepsilon > 0$ choose a discrete measurement g with $g \leq f$ and

$$E_m g \geq \sup_g E_m g - \varepsilon \, .$$

From Lemmas 4.3.1 and 4.2.2

$$E_m f = \lim_{n\to\infty} E_m q_n(f) \geq \lim_{n\to\infty} E_m q_n(g) = E_m g \geq \sup_g E_m g - \varepsilon \, .$$

Since ε is arbitrary, this completes the proof of (a).

To prove (b), use (a) to choose a discrete function $g = \sum_i g_i 1_{G_i}$ that nearly gives $E_m f$; that is, fix $\varepsilon > 0$ and choose $g \leq f$ so that

$$E_m g = \sum_i g_i m(G_i) \geq E_m f - \frac{\varepsilon}{2} \, .$$

Define for all n the set

$$F_n = \{x: f_n(x) \geq g(x) - \frac{\varepsilon}{2}\} \, .$$

Since $g(x) \leq f(x)$ and $f_n(x) \uparrow f(x)$ by assumption, $F_n \uparrow \Omega$, and hence the continuity of probability from below implies that for any event H,

$$\lim_{n\to\infty} m(F_n \cap H) = m(H) . \tag{4.4.3}$$

Clearly $1 \geq 1_{F_n}$ and for each $x \in F_n$ $f_n \geq g - \varepsilon$. Thus

$$f_n \geq 1_{F_n} f_n \geq 1_{F_n}(g - \varepsilon) ,$$

and all three measurements are discrete. Application of Lemma 4.2.2(e) therefore gives

$$E_m f_n \geq E_m(1_{F_n} f_n) \geq E_m(1_{F_n}(g - \varepsilon)) .$$

which, from the linearity of expectation for discrete measurements (Lemma 4.2.2(c)), implies that

$$E_m f_n \geq E_m(1_{F_n} g) - \frac{\varepsilon}{2} m(F_n) \geq E_m(1_{F_n} g) - \frac{\varepsilon}{2}$$

$$= E_m(1_{F_n} \sum_i g_i 1_{G_i}) - \frac{\varepsilon}{2} = E_m(\sum_i g_i 1_{G_i \cap F_n}) - \frac{\varepsilon}{2}$$

$$= \sum_i g_i m(F_n \cap G_i) - \frac{\varepsilon}{2} .$$

Taking the limit as $n \to \infty$, using (4.4.3) for each of the finite number of probabilities in the sum,

$$\lim_{n\to\infty} E_m f_n \geq \sum_i g_i m(G_i) - \frac{\varepsilon}{2} = E_m g - \frac{\varepsilon}{2} \geq E_m f - \varepsilon ,$$

which completes the proof. □

With the preceding alternative characterizations of expectation in hand, we can now generalize the fundamental properties of Lemma 4.2.2 from discrete measurements to more general measurements.

Lemma 4.4.2.

(a) If $f \geq 0$ with probability 1, then $E_m f \geq 0$.

(b) $E_m 1 = 1$.

(c) Expectation is linear; that is, for any real α, β and any measurements f and g,

$$E_m(\alpha f + \beta g) = \alpha E_m f + \beta E_m g \ .$$

(d) $E_m f$ exists and is finite if and only if $E_m \mid f \mid$ is finite and

$$\mid E_m f \mid \leq E_m \mid f \mid .$$

(e) Given two measurements f and g for which $f \geq g$ with probability 1, then

$$E_m f \geq E_m g \ .$$

Proof. (a) follows by defining $q_n(f)$ as in Lemma 4.3.1 and applying Lemma 4.2.2(a) to conclude $E_m(q_n(f)) \geq 0$, which in the limit proves (a). (b) follows from Lemma 4.2.2(b). To prove (c), first consider nonnegative functions by using Lemma 4.2.2(c) to conclude that

$$E_m(\alpha q_n(f) + \beta q_n(g)) = \alpha E_m(q_n(f)) + \beta E_m(q_n(G))$$

The right-hand side converges upward to $\alpha E_m f + \beta E_m g$. By the linearity of limits, the sequence $\alpha q_n(f) + \beta q_n(g)$ converges upward to $f + g$, and hence, from Lemma 4.4.1, the left-hand side converges to $E_m(\alpha f + \beta g)$, proving the result for nonnegative functions. Note that it was to prove this piece that we first considered Lemma 4.4.1 and hence focused on a nontrivial limiting property before proving the fundamental properties of this lemma. The result for general measurements then follows by breaking the functions into positive and negative parts and using the nonnegative measurement result.

Part (d) follows from the fact that $f = f^+ - f^-$ and $\mid f \mid = f^+ + f^-$. The expectation $E_m \mid f \mid$ exists and is finite if and only if both positive and negative parts have finite expectations, this implying that $E_m f$ exists and is also finite. Conversely, if $E_m f$ exists and is finite, then both positive and negative parts must have finite integrals, this implying that $\mid f \mid$ must be finite. If the integrals exist, clearly

$$E_m \mid f \mid = E_m f^+ + E_m f^- \geq E_m f^+ \geq E_m f^+ - E_m f^- = E_m f \ .$$

Part (e) follows immediately by application of part (a) to $f - g$. □

Integration

The expectation is also called an *integral* and is denoted by any of the following:

$$E_m f = \int f \, dm = \int f(x) dm(x) = \int f(x) m(dx) \ .$$

The subscript m denoting the measure with respect to which the expectation is taken will occasionally be omitted if it is clear from context.

A measurement f is said to be *integrable* or *m-integrable* if $E_m f$ exists and is finite. Define $L^1(m)$ to be the space of all m-integrable functions. Given any m-integrable f and an event B, define

$$\int_B f\,dm = \int f(x)1_B(x)dm(x).$$

Two random variables f and g are said to be equal m-almost-everywhere or equal m-a.e. or equal with m-probability one if $m(f = g) = m(\{x: f(x) = g(x)\}) = 1$. The m- is dropped if it is clear from context.

The following corollary to the definition of expectation and its linearity characterizes the expectation of an arbitrary integrable random variable as the limit of the expectations of the simple measurements formed by quantizing the random variables. This is an example of an integral limit theorem, a result giving conditions under which the integral and limit can be interchanged.

Corollary 4.4.1. Let q_n be the quantizer sequence of Lemma 4.3.1 If f is m-integrable, then

$$E_m f = \lim_{n\to\infty} E_m q_n(f)$$

and

$$\lim_{n\to\infty} E_m |f - q_n(f)| = 0.$$

Proof. If f is nonnegative, the two results are equivalent and follow immediately from the definition of expectation and the linearity of Lemma 4.4.2(c). For general f $q_n(f) = q_n(f^+) - q_n(f^-)$ and $q_n(f^+) \uparrow f^+$ and $q_n(f^-) \uparrow f^-$ by construction, and hence from the nonnegative f result and the linearity of expectation that

$$E|f - q_n| = E(f^+ - q_n(f^+)) + E(f^- - q_n(f^-)) \underset{n\to\infty}{\to} 0$$

and

$$E(f - q_n) = E(f^+ - q_n(f^+)) - E(f^- - q_n(f^-)) \underset{n\to\infty}{\to} 0 . \ \square$$

The lemma can be used to prove that expectations have continuity properties with respect to the underlying probability measure:

Corollary 4.4.2. If f is m-integrable, then

$$\lim_{m(F)\to 0} \int_F f\,dm = 0\;;$$

that is, given $\varepsilon > 0$ there is a $\delta > 0$ such that if $m(F) < \delta$, then

$$\int_F f\,dm < \varepsilon\,.$$

Proof. If f is a simple function of the form (4.4.1), then

$$\int_F f\,dm = E(1_F \sum_{i=1}^{M} b_i 1_{F_i}) = \sum_{i=1}^{M} b_i E 1_{F \cap F_i}$$

$$= \sum_{i=1}^{M} b_i m(F \cap F_i) \le \sum_{i=1}^{M} b_i m(F) \underset{m(F)\to 0}{\to} 0.$$

If f is integrable, let q_n be the preceding quantizer sequence. Given $\varepsilon > 0$ use the previous corollary to choose k large enough to ensure that

$$E_m |f - q_k(f)| < \varepsilon\,.$$

From the preceding result for simple functions

$$|\int_F f\,dm| = |\int f 1_F\,dm - \int q_k(f) 1_F\,dm + \int q_k(f) 1_F\,dm|$$

$$\le |\int (f - q_k(f)) 1_F\,dm| + |\int q_k(f) 1_F\,dm|$$

$$\le \int |f - q_k(f)| 1_F\,dm + |\int_F q_k(f)\,dm|$$

$$\le \varepsilon + |\int_F q_k(f)\,dm| \underset{m(F)\to 0}{\to} \varepsilon\,.$$

Since ε is arbitrary, the result is proved. □

The preceding proof is typical of many results involving expectations: the result is easy for discrete measurements and it is extended to general functions via limiting arguments. This continuity property has two immediate consequences, the proofs of which are left as exercises.

Corollary 4.4.3. If f is nonnegative m-a.e., m-integrable, and $E_m f \neq 0$, then the set function P defined by

$$P(F) = \frac{\int_F f\, dm}{E_m f}$$

is a probability measure.

Corollary 4.4.4. If f is m-integrable, then

$$\lim_{r \to \infty} \int_{f \geq r} f\, dm = 0.$$

Corollary 4.4.5. Suppose that $f \geq 0$ m-a.e. Then $\int f\, dm = 0$ if and only if $f = 0$ m-a.e.

Corollary 4.4.5 shows that we can consider $L^1(m)$ to be a metric space with respect to the metric $d(f, g) = \int |f - g|\, dm$ if we consider measurements identical if $m(f = g) = 0$.

The following lemma collects some simple inequalities for expectations.

Lemma 4.4.3.

(a) (The Markov Inequality) Given a random variable f with $f \geq 0$ m-a.e., then

$$m(f \geq a) \leq \frac{E_m f}{a} \text{ for } a > 0.$$

Furthermore, for $p > 0$

$$m(f \geq a) \leq \frac{E_m(f^p)}{a^p}) \text{ for } a > 0 .$$

(b) (Tchebychev's Inequality) Given a random variable f with expected value $E f$, then

$$m(|f - E f| \geq a) \leq \frac{E_m(|f - E f|^2)}{a^2}, \text{ for } a > 0 .$$

(c) If $E(f^2) = 0$, them $m(f = 0) = 1$.

(d) (The Cauchy-Schwarz Inequality) Given random variables f and g, then

$$| E(fg) | \leq E(f^2)^{1/2} E(g^2)^{1/2} .$$

Proof. (a) Let $1_{\{f \geq a\}}$ denote the indicator function of the event $\{x: f(x) \geq a\}$. We have that

$$E f = E(f(1_{\{f \geq a\}} + 1_{\{f < a\}})) = E(f 1_{\{f \geq a\}}) + E(f 1_{\{f < a\}})$$
$$\geq aE1_{\{f \geq a\}} = am(f \geq a) ,$$

proving Markov's inequality. The generalization follows by substituting f^p for f and using the fact that for $p > 0$, $f^p) > a^p$ if and only if $f > a$.

(b) Tchebychev's inequality follows by replacing f by $| f - E f |^2$.

(c) From the general Markov inequality, if $E(f^2) = 0$, then for any $\varepsilon > 0$,

$$m(f^2) \geq \varepsilon^2) \leq \frac{E(f^2)}{\varepsilon^2} = 0$$

and hence $m(|f| \leq 1/n) = 1$ for all n. Since $\{\omega: |f(\omega)| \leq 1/n\} \downarrow \{\omega: f(\omega) = 0\}$, from the continuity of probability, $m(f = 0) = 1$.

(d) To prove the Cauchy-Schwarz inequality set $a = E(f^2)^{1/2}$ and $b = E(g^2)^{1/2}$. If $a = 0$, then $f = 0$ a.e. from (c) and the inequality is trivial. Hence we can assume that both a and b are nonzero. In this case

$$0 \leq E(\frac{f}{a} \pm \frac{g}{b})^2 = 2 \pm 2\frac{E(fg)}{ab} . \square$$

A sequence f_n is said to be *uniformly integrable* if the limit of Corollary 4.4.4 holds uniformly in n; that is, if

$$\lim_{r \to \infty} \int_{|f_n| \geq r} |f_n| dm = 0$$

uniformly in n, i.e., if given $\varepsilon > 0$ there exists an R such that

$$\int_{|f_n| \geq r} |f_n| dm < \varepsilon, \text{ all } r \geq R \text{ and all } n .$$

Alternatively, the sequence f_n is uniformly integrable if

$$\lim_{r \to \infty} \sup_n \int_{|f_n| \geq r} |f_n| dm = 0.$$

The following lemma collects several examples of uniformly integrable sequences.

Lemma 4.4.4. If there is an integrable function g (e.g., a constant) for which $|f_n| \le g$, all n, then the f_n are uniformly integrable. If $a > 0$ and $\sup_n E|f_n|^{1+a} < \infty$, then the f_n are uniformly integrable. If $|f_n| \le |g_n|$ for all n and the g_n are uniformly integrable, then so are the f_n. If a sequence f_n^2 is uniformly integrable, then so is the sequence f_n.

Proof. The first statement is immediate from the definition and from Corollary 4.4.3. The second follows from Markov's inequality. The third example follows directly from the definition. The fourth example follows from the definition and the Cauchy-Schwarz inequality. □

The following lemma collects the basic asymptotic integration results on *integrating to the limit.*

Lemma 4.4.5.

(a) (Monotone Convergence Theorem)

If f_n, $n = 1,2, \cdots$, is a sequence of measurements such that $f_n \uparrow f$ and $f_n \ge 0$, m-a.e., then

$$\lim_{n \to \infty} E f_n = E f . \tag{4.4.4}$$

In particular, if $\lim_{n \to \infty} E f_n < \infty$, then $E(\lim_{n \to \infty} f_n)$ exists and is finite.

(b) (Extended Monotone Convergence Theorem)

If there is a measurement h such that

$$\int h \, dm > -\infty \tag{4.4.5}$$

and if f_n, $n = 1,2, \cdots$, is a sequence of measurements such that $f_n \uparrow f$ and $f_n \ge h$, m-a.e., then (4.4.4) holds. Similarly, if there is a function h such that

$$\int h \, dm < \infty \tag{4.4.6}$$

and if f_n, $n = 1,2, \cdots$, is a sequence of measurements such that $f_n \downarrow f$ and $f_n \le h$, m-a.e., then (4.4.4) holds.

(c) (Fatou's Lemma)

If f_n is a sequence of measurements and $f_n \geq h$, where h satisfies (4.4.5), then

$$\int (\liminf_{n\to\infty} f_n)\, dm \leq \liminf_{n\to\infty} \int f_n \, dm \,. \tag{4.4.7}$$

If f_n is a sequence of measurements and $f_n \leq h$, where h satisfies (4.4.6), then

$$\limsup_{n\to\infty} \int f_n \, dm \leq \int (\limsup_{n\to\infty} f_n)\, dm \,. \tag{4.4.8}$$

(d) (Extended Fatou's Lemma) If f_n, $n = 1,2, \cdots$ is a sequence of uniformly integrable functions, then

$$E(\liminf_{n\to\infty} f_n) \leq \liminf_{n\to\infty} E f_n \leq \limsup_{n\to\infty} E f_n \leq E(\limsup_{n\to\infty} f_n) \tag{4.4.9}$$

Thus, for example, if the limit of a sequence f_n of uniformly integrable functions exists a.e., then that limit is integrable and

$$E(\lim_{n\to\infty} f_n) = \lim_{n\to\infty} E f_n \,.$$

(e) (Dominated Convergence Theorem) If f_n, $n = 1,2, \cdots$, is a sequence of measurable functions that converges almost everywhere to f and if h is an integrable function for which $|f_n| \leq h$, then f is integrable and (4.4.4) holds.

Proof.

(a) Part (a) follows in exactly the same manner as the proof of Lemma 4.4.1(b) except that the measurements f_n need not be discrete and Lemma 4.4.2 is invoked instead of Lemma 4.2.2.

(b) If $\int h\, dm$ is infinite, then so is $\int f_n\, dm$ and $\int f\, dm$. Hence assume it is finite. Then $f_n - h \geq 0$ and $f_n - h \uparrow f - h$ and the result follows from (a). The result for decreasing functions follows by a similar manipulation.

(c) Define $g_n = \inf_{k \geq n} f_k$ and $g = \liminf f_n$. Then $g_n \geq h$ for all n, h satisfies (4.4.5), and $g_n \uparrow g$. Thus from the fact that $g_n \leq f_n$ and part (b)

$$\int f_n \, dm \geq \int g_n \, dm \uparrow \int g \, dm = \int (\liminf_{n\to\infty} f_n)\, dm \,,$$

which proves (4.4.7). Eq. (4.4.8) follows in like manner.

(d) For any $r > 0$,

$$\int f_n \, dm = \int_{f_n < -r} f_n \, dm + \int_{f_n \geq -r} f_n \, dm \, .$$

Since the f_n are uniformly integrable, given $\varepsilon > 0$ r can be chosen large enough to ensure that the first integral on the right has magnitude less than ε. Thus

$$\int f_n \, dm \geq -\varepsilon + \int 1_{f_n \geq -r} f_n \, dm \, .$$

The integrand on the left is bound below by $-r$, which is integrable, and hence from part (c)

$$\liminf_{n\to\infty} \int f_n \, dm \geq -\varepsilon + \int (\liminf_{n\to\infty} 1_{f_n \geq -r} f_n) \, dm$$

$$\geq -\varepsilon + \int (\liminf_{n\to\infty} f_n) \, dm \, ,$$

where the last inequality follows since $1_{f_n \geq -r} f_n \geq f_n$. Since ε is arbitrary, this proves the limit infimum part of (4.4.9). The limit supremum part follows similarly.

(e) Follows immediately from (d) and Lemma 4.4.4. □

To complete this section we present several additional properties of expectation that will be used later. The proofs are simple and provide practice with manipulating the preceding fundamental results. The first result provides a characterization of uniformly integrable functions.

Lemma 4.4.6. A sequence of integrable functions f_n is uniformly integrable if and only if the following two conditions are satisfied: (i) The expectations $E(|f_n|)$ are bounded; that is, there is some finite K that bounds all of these expectations from above. (ii) The continuity in the sense of Corollary 4.4.4 of the integrals of f_n is uniform; that is, given $\varepsilon > 0$ there is a $\delta > 0$ such that if $m(F) < \delta$, then

$$\int_F |f_n| \, dm < \varepsilon \, , \text{ all } n.$$

Proof. First assume that (i) and (ii) hold. From Markov's inequality and (i) it follows that

$$m(|f_n| \geq r) \leq r^{-1} \int |f_n| \, dm \leq \frac{K}{r} \underset{r\to\infty}{\to} 0$$

uniformly in n. This plus uniform continuity then implies uniform integrability. Conversely, assume that the f_n are uniformly integrable. Fix $\varepsilon > 0$ and choose r so large that

$$\int_{|f_n| \geq r} |f_n| dm < \varepsilon/2 \text{ , all } n \text{ .}$$

Then

$$E|f_n| = \int_{|f_n| < r} |f_n| dm + \int_{|f_n| \geq r} |f_n| dm \leq \varepsilon/2 + r$$

uniformly in n, proving (i). Choosing F so that $m(F) < \varepsilon/(2r)$,

$$\int_F |f_n| dm = \int_{F \cap \{|f_n| \geq r\}} |f_n| dm + \int_{F \cap \{|f_n| < r\}} |f_n| dm$$

$$\leq \int_{|f_n| \geq r} |f_n| dm + rm(F) \leq \varepsilon/2 + \varepsilon/2 = \varepsilon \text{ ,}$$

proving (ii) since ε is arbitrary. □

Corollary 4.4.6. If a sequence f_n is uniformly integrable, then so is the sequence

$$n^{-1} \sum_{i=0}^{n-1} f_i \text{ .}$$

Proof. If the f_n are uniformly integrable, then there exists a finite K such that $E(|f_n|) < K$, and hence

$$E(|\frac{1}{n} \sum_{i=0}^{n-1} f_i|) \leq \frac{1}{n} \sum_{i=0}^{n-1} E|f_i| < K \text{ .}$$

The f_n are also uniformly continuous and hence given ε there is a δ such that if $m(F) \leq \delta$, then

$$\int_F f_i \, dm \leq \varepsilon \text{ , all } i \text{ ,}$$

and hence

$$\int_F |n^{-1} \sum_{i=0}^{n-1} f_i| \, dm \leq n^{-1} \sum_{i=0}^{n-1} \int_F |f_i| \, dm \leq \varepsilon \text{ , all } n \text{ .}$$

Applying the previous lemma then proves the corollary. □

The final lemma of this section collects a few final journeyman results on expectation. The proofs are left as exercises.

Lemma 4.4.7.

(a) Given two integrable functions f and g, if

$$\int_F f\,dm \leq \int_F g dm \text{ all events } F ,$$

then $f \leq g$, m-a.e. If the preceding relation holds with equality, then $f = g$ m-a.e.

(b) (Change of Variables Formula) Given a measurable mapping f from a standard probability space $(\Omega, \boldsymbol{S}, P)$ to a standard measurable space $(\Gamma, \boldsymbol{B})$ and a real-valued random variable $g: \Gamma \rightarrow \mathbf{R}$, let Pf^{-1} denote the probability measure on $(\Gamma, \boldsymbol{B})$ defined by $Pf^{-1}(F) = P(f^{-1}F)$, all $F \in \boldsymbol{B}$. Then

$$\int g(y) dPf^{-1}(y) = \int g(f(x)) dP(x) ;$$

that is, if either integral exists so does the other and they are equal.

(c) (Chain Rule) Consider a probability measure P defined as in Corollary 4.4.3 as

$$P(F) = \frac{\int_F f\,dm}{E_m f}$$

for some m-integrable f with nonzero expectation. If g is q-integrable, then

$$E_P\, g = E_m(g f) .$$

Exercises

1. Prove that (4.4.1) yields (4.2.1) as a special case if f is a discrete measurement.
2. Prove Corollary 4.4.3.
3. Prove Corollary 4.4.4.
4. Prove Corollary 4.4.5.

5. Prove Lemma 4.4.8.

   Hint for proof of (b) and (c): first consider simple functions and then limits.

6. Show that for a nonnegative random variable X

$$\sum_{n=1}^{\infty} Pr(X \geq n) \leq EX .$$

7. Verify the following relations:

$$m(F \Delta G) = E_m[(1_F - 1_G)^2]$$

$$m(F \cap G) = E_m(1_F 1_G)$$

$$m(F \cup G) = E_m[\max (1_F, 1_G)] .$$

8. Verify the following:

$$m(F \cap G) \leq \sqrt{m(F)m(G)}$$

$$m(F \Delta G) \geq | \sqrt{m(F)} - \sqrt{m(G} |^2 .$$

9. Given disjoint events F_n, $n = 1,2, \cdots$, and a measurement f, prove that integrals are countably additive in the following sense:

$$\int_{\bigcup_{n=1}^{\infty} F_n} f\, dm = \sum_{n=1}^{\infty} \int_{F_n} f\, dm$$

10. Prove that if $(\int_F)f\, dm)/m(F) \in [a,b]$ for all events F, then $f \in [a,b]$ m-a.e.

11. Show that if f, $g \in L^2(m)$, then

$$\int | f g | \, dm \leq \frac{1}{2} \int f^2 \, dm + \frac{1}{2} \int g^2 dm .$$

12. Show that if $f \in L^2(m)$, then

$$\int_{|f| > r} |f| dm \leq (\int f^2 \, dm)^{1/2} (m(\{ |f| > r\}))^{1/2} .$$

4.5 TIME AVERAGES

Given a random process $\{X_n\}$ we will often wish to apply a given measurement repeatedly at successive times, that is, to form a sequence of outputs of a given measurement. For example, if the random process has a real alphabet corresponding to a voltage reading, we might wish to record the successive outputs of the meter and hence the successive random process values X_n. Similarly, we might wish to record successive values of the energy in a unit resistor: X_n^2. We might wish to estimate the autocorrelation of lag or delay k via successive values of the form $X_n X_{n+k}$. Although all such sequences are random and hence themselves random processes, we shall see that under certain circumstances the arithmetic or Césaro means of such sequences of measurements will converge. Such averages are called *time averages* or *sample averages* and this section is devoted to their definition and to a summary of several possible forms of convergence.

A useful general model for a sequence of measurements is provided by a dynamical system $(A^{\boldsymbol{I}}, B_A{}^{\boldsymbol{I}}, m, T)$, that is, a probability space, $(A^{\boldsymbol{I}}, B_A{}^{\boldsymbol{I}}, m)$ and a transformation $T: A^{\boldsymbol{I}} \to A^{\boldsymbol{I}}$ (such as the shift), together with a real-valued measurement $f: A^{\boldsymbol{I}} \to \mathbf{R}$. Define the corresponding sequence of measurements fT^i, $i = 0,1,2, \cdots$ by $fT^i(x) = f(T^i x)$; that is, fT^i corresponds to making the measurement f at time i. For example, assume that the dynamical system corresponds to a real-alphabet random process with distribution m and that T is the shift on $A^{\mathbf{Z}}$. If we are given a sequence $x = (x_0, x_1, \cdots)$ in $A^{\mathbf{Z}}$, where $\mathbf{Z} = \{0,1,2, \cdots\}$, if $f(x) = \Pi_0(x) = x_0$, the "zero time" coordinate function, then $f(T^i x) = x_i$, the outputs of the directly given process with distribution m. More generally, f could be a filtered or coded version of an underlying process. The key idea is that a measurement f made on a sample sequence at time i corresponds to taking f on the ith shift of the sample sequence.

Given a measurement f on a dynamical system $(\Omega, \boldsymbol{B}, m, T)$, the nth order *time average* or *sample average* $<f>_n$ is defined as the arithmetic mean or Césaro mean of the measurements at times $0,1,2, \cdots, n-1$; that is,

$$<f>_n(x) = n^{-1} \sum_{i=0}^{n-1} f(T^i x) , \tag{4.5.1}$$

or, more briefly,

$$<f>_n = n^{-1} \sum_{i=0}^{n-1} fT^i .$$

For example, in the case with f being the time-zero coordinate function,

$$<f>_n(x) = n^{-1} \sum_{i=0}^{n-1} x_i ,$$

the *sample mean* or *time-average mean* of the random process. If $f(x) = x_0 x_k$, then

$$<f>_n(x) = n^{-1} \sum_{i=0}^{n-1} x_i x_{i+k} ,$$

the *sample autocorrelation of lag* k of the process. If $k = 0$, this is the time average mean of the process since if the x_k represent voltages, x_k^2 is the energy across a resistor at time k.

Finite order sample averages are themselves random variables, but the following lemma shows that they share several fundamental properties of expectations.

Lemma 4.5.1. Sample averages have the following properties for all n:

(a) If $fT^i \geq 0$ with probability 1, $i = 0,1, \cdots, n-1$ then $<f>_n \geq 0$.

(b) $< 1 >_n = 1$.

(c) For any real α, β and any measurements f and g,

$$< \alpha f + \beta g >_n = \alpha < f >_n + \beta < g >_n .$$

(d)

$$| < f >_n | \leq < | f | >_n .$$

(e) Given two measurements f and g for which $f \geq g$ with probability 1, then

$$< f >_n \geq < g >_n .$$

Proof. The properties all follow from the properties of summations. Note that for (a) we must strengthen the requirement that f is nonnegative a.e. to the requirement that all of the fT^i are nonnegative a.e. Alternatively, this is a special case of Lemma 4.4.1 in which for a particular x we define a probability measure on measurements taking values $f(T^k x)$ with probability $1/n$ for $k = 0,1, \cdots, n-1$. □

We will usually be interested not in the finite order sample averages, but in the long term or asymptotic sample averages obtained as $n \to \infty$. Since the finite order sample averages are themselves random variables, we need to develop notions of convergence of random variables (or measurable functions) in order to make such limits precise. There are, in fact, several possible approaches. In the remainder of this section we consider the notion of convergence of random variables closest to the ordinary definition of a limit of a sequence of real numbers. In the next section three other forms of convergence are considered and compared.

Given a measurement f on a dynamical system $(\Omega, \boldsymbol{B}, m, T)$, let F_f denote the set of all $x \in \Omega$ for which the limit

$$\lim_{n \to \infty} \frac{1}{n} \sum_{k=0}^{n-1} f(T^k x)$$

exists. For all such x we define the *time average* or *limiting time average* or *sample average* $<f>$ by this limit; that is,

$$<f> = \lim_{n \to \infty} <f>_n .$$

Before developing the properties of $<f>$, we show that the set F_f is indeed measurable. This follows from the measurability of limits of measurements, but it is of interest to prove it directly. Let f_n denote $<f>_n$ and define the events

$$F_n(\varepsilon) = \{x: |f_n(x) - f(x)| \leq \varepsilon\} \tag{4.5.2}$$

and define as in (4.3.3)

$$\begin{aligned} \underline{F}(\varepsilon) &= \liminf_{n \to \infty} F_n(\varepsilon) = \bigcup_{n=1}^{\infty} \bigcap_{k=n}^{\infty} F_k(\varepsilon) \\ &= \{x: x \in F_n(\varepsilon) \text{ for all but a finite number of } n\} . \end{aligned} \tag{4.5.3}$$

Then the set

$$F_f = \bigcap_{k=1}^{\infty} \underline{F}(\frac{1}{k}) \tag{4.5.4}$$

is measurable since it is formed by countable set theoretic operations on measurable sets. The sequence of sample averages $<f>_n$ will be said to converge to a limit $<f>$ *almost everywhere* or *with probability one* if $m(F_f) = 1$.

The following lemma shows that limiting time averages exhibit behavior similar to the expectations of Lemma 4.4.2.

Lemma 4.5.2. Time averages have the following properties when they are well defined; that is, when the limits exist:

(a) If $fT^i \geq 0$ with probability 1 for $i = 0,1, \cdots$, then $<f> \geq 0$.

(b) $< 1 > = 1$.

(c) For any real α, β and any measurements f and g,

$$< \alpha f + \beta g > = \alpha < f > + \beta < g > .$$

(d)

$$| < f > | \leq < | f | > .$$

(e) Given two measurements f and g for which $f \geq g$ with probability 1, then

$$< f > \geq < g > .$$

Proof. The proof follows from Lemma 4.5.1 and the properties of limits. In particular, (a) is proved as follows: If $fT^i \geq 0$ a.e. for each i, then the event $\{fT^i \geq 0, \text{ all } i\}$ has probability one and hence so does the event $\{<f>_n \geq 0 \text{ all } i\}$. (The countable intersection of a collection of sets with probability one also has probability one.) □

Note that unlike Lemma 4.5.1, Lemma 4.5.2 cannot be viewed as being a special case of Lemma 4.4.2.

4.6 CONVERGENCE OF RANDOM VARIABLES

We have introduced the notion of almost everywhere convergence of a sequence of measurements. We now further develop the properties of such convergence along with three other notions of convergence.

Suppose that $\{f_n ; n = 1,2, \cdots\}$ is a sequence of measurements (random variables, measurable functions) defined on a probability space with probability measure m.

Definitions.

1. The sequence $\{f_n\}$ is said to converge *with probability one* or *with certainty* or *m-almost surely* (m-a.s.) or m-a.e. to a random variable f if

$$m(\{x: \lim_{n\to\infty} f_n(x) = f(x)\}) = 1 ;$$

that is, if with probability one the usual limit of the sequence $f_n(x)$ exists and equals $f(x)$. In this case we write $f = \lim_{n\to\infty} f_n$, m–*a.e.*

2. The $\{f_n\}$ are said to *converge in probability* to a random variable f if for every $\varepsilon > 0$

$$\lim_{n\to\infty} m(\{x: |f_n(x) - f(x)| > \varepsilon\}) = 0 ,$$

in which case we write

$$f_n \underset{m}{\to} f$$

or

$$f = m \lim_{n\to\infty} f_n .$$

If for all positive real r

$$\lim_{n\to\infty} m(\{x: f_n(x) \le r\}) = 0 ,$$

we write

$$f_n \underset{m}{\to} \infty \text{ or } m \lim_{n\to\infty} f_n = \infty .$$

3. The space $L^1(m)$ of all m-integrable functions is a pseudo-normed space with pseudo-norm $\| f \|_1 = E\,|f|$, and hence it is a pseudo-metric space with metric

$$d(f,g) = \int |f - g| \, dm .$$

We can consider the space to be a normed metric space if we identify or consider equal functions for which $d(f,g)$ is zero. Note that $d(f,g) = 0$ if and only if $f = g$ m-a.e. and hence the metric space really has as elements equivalence classes of functions rather than functions. A sequence $\{f_n\}$ of measurements is said to converge to a measurement f in $L^1(m)$ if f is integrable and if $\lim_{n\to\infty} \| f_n - f \|_1 = 0$, in which case we write

$$f_n \underset{L^1(m)}{\longrightarrow} f .$$

Convergence in $L^1(m)$ has a simple but useful consequence. From Lemma 4.2.2(d) for any event F

$$\left| \int_F f_n \, dm - \int_F f \, dm \right| = \left| \int_F (f_n - f) \, dm \right| \leq \int_F |f_n - f| \, dm$$

$$\leq \| f_n - f \|_1 \underset{n \to \infty}{\longrightarrow} 0$$

and therefore

$$\lim_{n \to \infty} \int_F f_n \, dm = \int_F f \, dm \ \text{ if } f_n \to f \text{ in } L^1(m) . \tag{4.6.1}$$

In particular, if F is the entire space, then $\lim_{n \to \infty} E f_n = E f$.

4. Denote by $L^2(m)$ the space of all square-integrable functions, that is, the space of all f such that $E(f^2)$ is finite. This is an inner product space with inner product $(f,g) = E(fg)$ and hence also a normed space with norm $\| f \|_2 = (E(f^2))^{1/2}$ and hence also a metric space with metric $d(f,g) = (E[(f - g)^2])^{1/2}$, provided that we identify functions f and g for which $d(f,g) = 0$. A sequence of square-integrable functions $\{f_n\}$ is said to converge in $L^2(m)$ or to converge in mean square or to converge in quadratic mean to a function f if f is square-integrable and $\lim_{n \to \infty} \| f_n - f \|_2 = 0$, in which case we write

$$f_n \underset{L^2(m)}{\longrightarrow} f$$

or

$$f = \underset{n \to \infty}{\text{l.i.m.}} f_n ,$$

where *l.i.m.* stands for "limit in the mean."

Although we shall focus on L^1 and L^2, we note in passing that they are special cases of general L^p norms defined as follows: Let L^p denote the space of all measurements f for which $|f|^p$ is integrable. Then

$$\| f \|_p = \left(\int |f|^p \, dm \right)^{\frac{1}{p}}$$

is a pseudo-norm on this space. It is in fact a norm if we consider measurements to be the same if they are equal almost everywhere. We say that a sequence of measurements f_n converges to f in L^p if

$\| f_n - f \|_p \to 0$ as $n \to \infty$. As an example of such convergence, the following lemma shows that the quantizer sequence of Lemma 4.3.1 converges to the original measurement in L^p provided the f is in L^p.

Lemma 4.6.1. If $f \in L^p(m)$ and $q_n(f)$, $n = 1,2, \cdots$, is the quantizer sequence of Lemma 4.3.1, then

$$q_n(f) \underset{L^p}{\to} f.$$

Proof. From Lemma 4.3.1 $|q_n(f) - f|$ is bound above by $|f|$ and monotonically nonincreasing and goes to 0 everywhere. Hence $|q_n(f) - f|^p$ also is monotonically nonincreasing, tends to 0 everywhere, and is bound above by the integrable function $|f|^p$. The lemma then follows from the dominated convergence theorem (Lemma 4.4.5(e)). □

The following lemmas provide several properties and alternative characterizations of and some relations among the various forms of convergence.

Of all of the forms of convergence, only almost everywhere convergence relates immediately to the usual notion of convergence of sequences of real numbers. Hence almost everywhere convergence inherits several standard limit properties of elementary real analysis. The following lemma states several of these properties without proof. (Proofs may be found, e.g., in Rudin [4].)

Lemma 4.6.2. If $\lim_{n\to\infty} f_n = f$, m-a.e. and $\lim_{n\to\infty} g_n = g$, m–a.e., then $\lim_{n\to\infty}(f_n + g_n) = f + g$ and $\lim_{n\to\infty} f_n g_n = fg$, m-a.e.. If also $g \neq 0$ m–a.e. then $\lim_{n\to\infty} f_n/g_n = f/g$, m-a.e. If h: **R**→**R** is a continuous function, then $\lim_{n\to\infty} h(f_n) = h(f)$, m-a.e.

The next lemma and corollary provide useful tools for proving and interpreting almost everywhere convergence. The lemma is essentially the classic Borel-Cantelli lemma of probability theory.

Lemma 4.6.3. Given a probability space $(\Omega, \boldsymbol{B}, m)$ and a sequence of events $G_n \in \boldsymbol{B}$, $n = 1,2, \ldots$, define the set $\{G_n \text{ i.o.}\}$ as the set $\{\omega\colon \omega \in G_n$ for an infinite number of $n\}$; that is,

$$\{G_n \text{ i.o.}\} = \bigcap_{n=1}^{\infty} \bigcup_{j=n}^{\infty} G_j = \limsup_{n\to\infty} G_n .$$

Then $m(G_n \text{ i.o.}) = 0$ if and only if

$$\lim_{n\to\infty} m(\bigcup_{k=n}^{\infty} G_k) = 0.$$

A sufficient condition for $m(G_n \text{ i.o.}) = 0$ is that

$$\sum_{n=1}^{\infty} m(G_n) < \infty . \tag{4.6.2}$$

Proof. From the continuity of probability,

$$m(G_n \text{ i.o.}) = m(\bigcap_{n=1}^{\infty} \bigcup_{j=n}^{\infty} G_j) = \lim_{n\to\infty} m(\bigcup_{j=n}^{\infty} G_j) ,$$

which proves the first statement. From the union bound,

$$m(\bigcup_{j=n}^{\infty} G_n) \le \sum_{j=n}^{\infty} m(G_n) ,$$

which goes to 0 as $n\to\infty$ if (4.6.2) holds. □

Corollary 4.6.1. $\lim_{n\to\infty} f_n = f$ m-a.e. if and only if for every $\varepsilon > 0$

$$\lim_{n\to\infty} m(\bigcup_{k=n}^{\infty} \{x: | f_k(x) - f(x) | > \varepsilon\}) = 0 .$$

If for any $\varepsilon > 0$

$$\sum_{n=0}^{\infty} m(| f_n | > \varepsilon) < \infty , \tag{4.6.3}$$

then $f_n \to 0$ m-a.e.

Proof. Define $F_n(\varepsilon)$ and $\underline{F}(\varepsilon)$ as in (4.5.2)-(4.5.3). Applying the lemma to $G_n = F_n(\varepsilon)^c$ shows that the condition of the corollary is equivalent to $m(F_n(\varepsilon)^c \text{ i.o.}) = 0$ and hence $m(\underline{F}(\varepsilon)) = 1$ and therefore

$$m(\{x: \lim_{n\to\infty} f_n(x) = f(x)\}) = m(\bigcap_{k=1}^{\infty} \underline{F}(\frac{1}{k})) = 1 .$$

Conversely, if

$$m(\bigcap_k \underline{F}(\frac{1}{k})) = 1 ,$$

then since $\bigcap_{k \geq 1}\underline{F}(1/k) \subset \underline{F}(\varepsilon)$ for all $\varepsilon > 0$, $m(\underline{F}(\varepsilon)) = 1$ and therefore $\lim_{n\to\infty} m(\bigcup_{k \geq n} F_k(\varepsilon)^c) = 0$. Eq. (4.6.3) is (4.6.2) for $G_n = \{x: f_n(x) > \varepsilon\}$, and hence this condition is sufficient from the lemma. □

Note that almost everywhere convergence requires that the probability that $|f_k - f| > \varepsilon$ for *any* $k \geq n$ tend to zero as $n\to\infty$, whereas convergence in probability requires only that the probability that $|f_n - f| > \varepsilon$ tend to zero as $n\to\infty$.

Lemma 4.6.4:. Given random variables f_n, $n = 1,2, \cdots$, and f, then

$$f_n \underset{L^2(m)}{\rightarrow} f \Rightarrow f_n \underset{L^1(m)}{\rightarrow} f \Rightarrow f_n \underset{m}{\rightarrow} f ,$$

$$\lim_{n\to\infty} f_n = f \ m\text{–a.e.} \Rightarrow f_n \underset{m}{\rightarrow} f .$$

Proof. From the Cauchy-Schwarz inequality $\| f_n - f \|_1 \leq \| f_n - f \|_2$ and hence $L^2(m)$ convergence implies $L^1(m)$ convergence. From Markov's inequality

$$m(\{| f_n - f | \geq \varepsilon\}) \leq \frac{E_m(|f_n - f|)}{\varepsilon} ,$$

and hence $L^1(m)$ convergence implies convergence in m-probability. If $f_n \rightarrow f$ m-a.e., then

$$m(\{| f_n - f | > \varepsilon\}) \leq m(\bigcup_{k=n}^{\infty} \{| f_k - f | > \varepsilon\}) \underset{n\to\infty}{\rightarrow} 0$$

and hence $m\lim_{n\to\infty} f_n = f$. □

Corollary 4.6.2. If $f_n \rightarrow f$ and $f_n \rightarrow g$ in any of the given forms of convergence, then $f = g$, m-a.e.

Proof. Fix $\varepsilon > 0$ and define $F_n(\delta) = \{x: | f_n(x) - f(x) | \leq \delta\}$ and $G_n(\delta) = \{x: | f_n(x) - g(x) | \leq \delta\}$, then

$$F_n(\varepsilon/2) \cap G_n(\varepsilon/2) \subset \{x: | f(x) - g(x) | \leq \varepsilon\}$$

and hence

$$m(\{| f - g| > \varepsilon\}) \leq m(F_n(\frac{\varepsilon}{2})^c \cup G_n(\frac{\varepsilon}{2})^c)$$

$$\leq m(F_n(\frac{\varepsilon}{2})^c) + m(G_n(\frac{\varepsilon}{2})^c) \underset{n\to\infty}{\to} 0 .$$

Since this is true for all ε, $m(\,|f-g\,|>0)=0$, so the corollary is proved for convergence in probability. From the lemma, convergence in probability is implied by the other forms, and hence the proof is complete. □

The following lemma shows that the metric spaces $L^1(m)$ and $L^2(m)$ are each Polish spaces (when we consider two measurements to be the same if they are equal m-a.e.).

Lemma 4.6.5. The spaces $L^p(m)$, $k = 1,2,\cdots$ are complete, separable metric spaces, i.e., they are Polish spaces.

Proof. The spaces are separable since from Lemma 4.6.1 the collection of all simple functions with rational values is dense. Let f_n be a Cauchy sequence, that is, $\| f_m - f_n \| \to_{n,m\to\infty} 0$, where $\| \,.\, \|$ denotes the $L^p(m)$ norm. Choose an increasing sequence of integers n_l, $l = 1,2,\cdots$ so that

$$\| f_m - f_n \| \leq (\frac{1}{4})^l \text{ ; for } m,\ n \geq n_l .$$

Let $g_l = f_{n_l}$ and define the set $G_n = \{x\colon |\, g_n(x) - g_{n+1}(x)\,| \geq 2^{-n}\}$. From the Markov inequality

$$m(G_n) \leq 2^{nk} \| g_n - g_{n+1} \|^k \leq 2^{-nk}$$

and therefore from Lemma 4.6.4

$$m(\limsup_{n\to\infty} G_n) = 0 .$$

Thus with probability 1 $x \notin \limsup_n G_n$. If $x \notin \limsup_n G_n$, however, then $|\, g_n(x) - g_{n+1}(x)\,| < 2^{-n}$ for large enough n. This means that for such x $g_n(x)$ is a Cauchy sequence of real numbers, which must have a limit since **R** is complete under the Euclidean norm. Thus g_n converges on a set of probability 1 to a function g. The proof is completed by showing that $f_n \to g$ in L^p. We have that

$$\| f_n - g \|^k = \int |\, f_n - g\,|^k \, dm = \int \liminf_{l\to\infty} |\, f_n - g_l\,|^k \, dm$$

$$\leq \liminf_{l\to\infty} \int |\, f_n - f_{n_l}\,|^k \, dm = \liminf_{l\to\infty} \| f_n - f_{n_l} \|^k$$

using Fatou's lemma and the definition of g_l. The right-most term, however, can be made arbitrarily small by choosing n large, proving the required convergence. □

Lemma 4.6.6. If $f_n \to_m f$ and the f_n are uniformly integrable, then f is integrable and $f_n \to f$ in $L^1(m)$. If $f_n \to_m f$ and the f_n^2 are uniformly integrable, then $f \in L^2(m)$ and $f_n \to f$ in $L^2(m)$.

Proof. We only prove the L^1 result as the other is proved in a similar manner. Fix $\varepsilon > 0$. Since the $\{f_n\}$ are uniformly integrable we can choose from Lemma 4.4.6 a $\delta > 0$ sufficiently small to ensure that if $m(F) < \delta$, then

$$\int_F |f_n| \, dm < \frac{\varepsilon}{4}, \text{ all } n.$$

Since the f_n converge to f in probability, we can choose N so large that if $n > N$, then $m(\{|f_n - f| > \varepsilon/4\}) < \delta/2$ for $n > N$. Then for all $n,m>N$ (using the $L^1(m)$ norm)

$$\|f_n - f_m\| = \int_{\substack{|f_n - f| \le \frac{\varepsilon}{4} \\ \text{and} \\ |f_m - f| \le \frac{\varepsilon}{4}}} |f_n - f_m| \, dm + \int_{\substack{|f_n - f| > \frac{\varepsilon}{4} \\ \text{or} \\ |f_m - f| > \frac{\varepsilon}{4}}} |f_n - f_m| \, dm .$$

If f_n and f_m are each within $\varepsilon/4$ of f, then they are within $\varepsilon/2$ of each other from the triangle inequality. Also using the triangle inequality, the right-most integral is bound by

$$\int_F |f_n| \, dm + \int_F |f_m| \, dm$$

where $F = \{|f_n - f| > \varepsilon/4 \text{ or } |f_m - f| > \varepsilon/4\}$, which, from the union bound, has probability less than $\delta/2 + \delta/2 = \delta$, and hence the preceding integrals are each less than $\varepsilon/4$. Thus $\|f_n - f_m\| \le \varepsilon/2 + \varepsilon/4 + \varepsilon/4 = \varepsilon$, and hence $\{f_n\}$ is a Cauchy sequence in $L^1(m)$, and it therefore must have a limit, say g, since $L^1(m)$ is complete. From Lemma 4.6.4, however, this means that also the f_n converge to g in probability, and hence from Corollary 4.6.2 $f = g$, m-a.e., completing the proof. □

The lemma provides a simple generalization of Corollary 4.4.1 and an alternative proof of a special case of Lemma 4.6.1.

Corollary 4.6.3. If q_n is the quantizer sequence of Lemma 4.3.1 and f is square-integrable, then $q_n(f) \to f$ in L^2.

Proof. From Lemma 4.4.1, $q_n(f(x)) \to f(x)$ for all x and hence also in probability. Since for all n $|q_n(f)| \leq |f|$ by construction, $|q_n(f)|^2) \leq |f|^2$, and hence from Lemma 4.4.4 the $q_n(f)^2$ are uniformly integrable and the conclusion follows from the lemma. □

The next lemma uses uniform integrability ideas to provide a partial converse to the convergence implications of Lemma 4.6.4.

Lemma 4.6.7. If a sequence of random variables f_n converges in probability to a random variable f, then the following statements are equivalent (where $p = 1$ or 2):

(a)

$$f_n \underset{L^p}{\to} f .$$

(b)

The f_n^p are uniformly integrable.

(c)

$$\lim_{n \to \infty} E(|f_n|^p) = E(|f|^p) .$$

Proof. Lemma 4.6.6 implies that (b) => (a). If (a) is true, then the triangle inequality implies that $|\|f_n\| - \|f\|| \leq \|f_n - f\| \to 0$, which implies (c). Finally, assume that (c) holds. Consider the integral

$$\int_{|f_n|^k \geq r} |f_n|^k \, dm = \int |f_n|^k \, dm - \int_{|f_n|^k < r} f_n \, dm$$

$$= \int |f_n|^k \, dm - \int \min(|f_n|^k, r) dm + r \int_{|f_n|^k \geq r} dm . \qquad (4.6.4)$$

The first term on the right tends to $\int |f|^k \, dm$ by assumption. We consider the limits of the remaining two terms. Fix $\delta > 0$ small and observe that

$$E(|\min(|f_n|^k, r) - \min(|f|^k, r)|)$$

$$\leq \delta m(\, \|f_n\,|^k - |f|^k| \leq \delta) + rm(\, \|f_n\,|^k - |f|^k| > \delta)\,.$$

Since $f_n \to f$ in probability, it must also be true that $|f_n|^k \to |f|^k$ in probability, and hence the preceding expression is bound above by δ as $n \to \infty$. Since δ is arbitrarily small

$$\lim_{n \to \infty} |\, E(\min(\,|f_n|^k, r)) - E(\min(\,|f|^k, r))\,| \leq$$

$$\lim_{n \to \infty} E(\,|\min(\,|f_n|^k, r) - \min(\,|f|^k, r)|) = 0\,.$$

Thus the middle term on the right of (4.6.4) converges to $E(\min(\,|f|^k, r))$. The remaining term is

$$rm(\,|f_n|^k \geq r)\,.$$

For any $\varepsilon > 0$

$$m(|f_n|^k \geq r) = m(\,|f_n|^k \geq r \text{ and } \|f_n|^k - |f|^k| \leq \varepsilon)$$
$$+\, m(\,|f_n|^k \geq r \text{ and } \|f_n|^k - |f|^k| > \varepsilon)\,.$$

The right-most term tends to 0 as $n \to \infty$, and the remaining term is bound above by

$$m(\,|f|^k \geq r - \varepsilon)\,.$$

From the continuity of probability, this can be made arbitrarily close to $m(\,|f|^k \geq r)$ by choosing ε small enough, e.g., choosing $F_n = \{|f|^k \in [r - 1/n, \infty)\}$, then F_n decreases to the set $F = \{|f|^k \in [r, \infty)\}$ and $\lim_{n \to \infty} m(F_n) = m(F)$.

Combining all of the preceding facts

$$\lim_{n \to \infty} \int_{|f_n|^k \geq r} |f_n|^k \, dm = \int_{|f|^k \geq r} |f|^k \, dm$$

From Corollary 4.4.4 given $\varepsilon > 0$ we can choose R so large that for $r > R$ the right-hand side is less than $\varepsilon/2$ and then choose N so large that

$$\int_{|f_n|^k > R} |f_n|^k \, dm \leq \varepsilon\,,\ n > N\,. \tag{4.6.5}$$

Then

$$\limsup_{r \to \infty} \sup_n \int_{|f_n|^k > r} |f_n|^k \, dm \leq$$

$$\lim_{r\to\infty} \sup_{n \le N} \int_{|f_n|^k > r} |f_n|^k \, dm + \lim_{r\to\infty} \sup_{n > N} \int_{|f_n|^k > r} |f_n|^k \, dm. \quad (4.6.6)$$

The first term to the right of the inequality is zero since the limit of each of the finite terms in the sequence is zero from Corollary 4.3.4. To bound the second term, observe that for $r > R$ each of the integrals inside the supremum is bound above by a similar integral over the larger region $\{|f_n|^k > R\}$. From (4.6.5) each of these integrals is bound above by ε, and hence so is the entire right-most term. Since the left-hand side of (4.6.6) is bound above by any $\varepsilon > 0$, it must be zero, and hence the $|f_n|^k$ are uniformly integrable. □

Note as a consequence of the Lemma that if $f_n \to f$ in L^p for $p = 1,2, \cdots$ then necessarily the f_n^p are uniformly integrable.

Exercises

1. Prove Lemma 4.6.2.
2. Show that if $f_k \in L^1(m)$ for $k = 1,2, \cdots$, then there exists a subsequence n_k, $k = 1,2, \cdots$, such that with probability one

$$\lim_{k\to\infty} \frac{f_{n_k}}{n_k} = 0$$

Hint: Use the Borel-Cantelli lemma.

3. Suppose that f_n converges in probability to f. Show that given ε, $\delta > 0$ there is an N such that if

$$F_{n,m}(\varepsilon) = \{x: |f_n(x) - f_m(x) > \varepsilon\} ,$$

then $m(F_{n,m}) \le \delta$ for n, $m \ge N$; that is,

$$\lim_{n,m\to\infty} m(F_{n,m}(\varepsilon)) = 0 , \text{ all } \varepsilon > 0 .$$

In words, if f_n converges in probability, then it is a Cauchy sequence in probability.

4. Show that if $f_n \to f$ and $g_n - f_n \to 0$ in probability, then $g_n \to f$ in probability. Show that if $f_n \to f$ and $g_n \to g$ in L^1, then $f_n + g_n \to f + g$ in L^1. Show that if $f_n \to f$ and $g_n \to g$ in L^2 and $\sup_n \| f_n \|_2 < \infty$, then $f_n g_n \to f g$ in L^1.

4.7 STATIONARY AVERAGES

We have seen in the discussions of ensemble averages and time averages that uniformly integrable sequences of functions have several nice convergence properties: In particular, we can compare expectations of limits with limits of expectations and we can infer $L^1(m)$ convergence from m-a.e. convergence from such sequences. As time averages of the form $< f >_n$ are the most important sequences of random variables for ergodic theory, a natural question is whether or not such sequences are uniformly integrable if f is integrable. Of course in general time averages of integrable functions need not be uniformly integrable, but in this section we shall see that in a special, but very important, case the response is affirmative.

Let $(\Omega,\boldsymbol{B},m,T)$ be a dynamical system, where as usual we keep in mind the example where the system corresponds to a directly given random process $\{X_n,\ n \in \boldsymbol{I}\}$ with alphabet A and where T is a shift.

In general the expectation of the shift $f T^i$ of a measurement f can be found from the change of variables formula of Lemma 4.4.6(b); that is, if mT^{-i} is the measure defined by

$$mT^{-i}(F) = m(T^{-i}F) \ , \text{ all events } F \ , \tag{4.7.1}$$

then

$$E_m f T^i = \int f T^i \, dm = \int f \, dmT^{-i} = E_{mT^{-i}} f. \tag{4.7.2}$$

We shall find that the sequences of measurements $f T^n$ and sample averages $< f >_n$ will have special properties if the preceding expectations do not depend on i; that is, the expectations of measurements do not depend on when they are taken. From (4.7.2), this will be the case if the probability measures mT^{-n} defined by $mT^{-n}(F) = m(T^{-n}F)$ for all events F and all n are the same. This motivates the following definition. The dynamical system is said to be *stationary* (or T-stationary or stationary with respect to T) or *invariant* (or T-invariant or invariant with respect to T) if

$$m(T^{-1}F) = m(F) \ , \text{ all events } F. \tag{4.7.3}$$

We shall also say simply that the measure m is stationary with respect to T if (4.7.3) holds. A random process $\{X_n\}$ is said to be *stationary* (with the preceding modifiers) if its distribution m is stationary with respect to T, that is, if the corresponding dynamical system with the shift T is stationary. A stationary random process is one for which the probabilities of an event are the same regardless of when it occurs. Stationary random processes will play a central, but not exclusive, role in the ergodic theorems.

Observe that (4.7.3) implies that

$$m(T^{-n}F) = m(F), \text{ all events } F, \text{ all } n = 0,1,2,\cdots \tag{4.7.4}$$

The following lemma follows immediately from (4.7.1), (4.7.2), and (4.7.4).

Lemma 4.7.1. If m is stationary with respect to T and f is integrable with respect to m, then

$$E_m f = E_m f T^n, \; n = 0,1,2,\cdots$$

Lemma 4.7.2. If f is in $L^1(m)$ and m is stationary with respect to T, then the sequences $\{fT^n; n = 0,1,2,\cdots\}$ and $\{<f>_n; n = 0,1,2,\cdots\}$, where

$$<f>_n = n^{-1}\sum_{i=0}^{n-1} fT^i,$$

are uniformly integrable. If also $f \in L^2(m)$, then in addition the sequence $\{ffT^n; n = 0,1,2,\cdots\}$ and its arithmetic mean sequence $\{w_n; n = 0,1,\cdots\}$ defined by

$$w_n(x) = n^{-1}\sum_{i=0}^{n-1} f(x)f(T^ix) \tag{4.7.5}$$

are uniformly integrable.

Proof. First observe that if $G = \{x: |f(x)| \geq r\}$, then $T^{-i}G = \{x: f(T^ix) \geq r\}$ since $x \in T^{-i}G$ if and only if $T^ix \in G$ if and only if $f(T^ix) > r$. Thus by the change of variables formula and the assumed stationarity

$$\int_{T^{-i}G} |fT^i| \, dm = \int_G |f| \, dm \text{ , independent of } i \text{ ,}$$

and therefore

$$\int_{|fT^i|>r} |fT^i| \, dm = \int_{|f|>r} |f| \, dm \underset{r\to\infty}{\to} 0 \text{ uniformly in } i \text{ .}$$

This proves that the fT^i are uniformly integrable. Corollary 4.4.6 implies that the arithmetic mean sequence is then also uniformly integrable. □

Exercises

1. A dynamical system is said to be N-stationary for an integer N if it is stationary with respect to T^N; that is,

$$m(T^{-N}F) = m(F), \text{ all } F \in \boldsymbol{B}.$$

Show that this implies that for any event F $m(T^{-nN}F) = m(F)$ and that $m(T^{-n}F)$ is a periodic function in n with period N. Does Lemma 4.7.2 hold for N-stationary processes? Define the measure m_N by

$$m_N(F) = \frac{1}{N}\sum_{i=0}^{N-1} m(T^{-i}F)$$

for all events F, a definition that we abbreviate by

$$m_N = \frac{1}{N}\sum_{i=0}^{N-1} mT^{-i}.$$

Show that m_N is stationary and that for any measurement f

$$E_{m_N} f = \frac{1}{N}\sum_{i=1}^{N-1} E_m fT^i.$$

Hint: As usual, first consider discrete measurements and then take limits.

2. A dynamical system with measure m is said to be *asymptotically mean stationary* if there is a stationary measure $\overline{m}$ for which

$$\lim_{n\to\infty}\frac{1}{n}\sum_{i=0}^{n-1} m(T^{-i}F) = \overline{m}(F); \text{ all } F \in \boldsymbol{B}.$$

Show that both stationary and N-stationary systems are also asymptotically mean stationary. Show that if f is a bounded measurement, that is, $|f| \le K$ for some finite K with probability one, then

$$E_{\overline{m}} f = \lim_{n\to\infty}\frac{1}{n}\sum_{i=0}^{n-1} E_m(fT^i).$$

REFERENCES

1. R. B. Ash, *Real Analysis and Probability,* Academic Press, New York, 1972.
2. K. L. Chung, *A Course in Probability Theory,* Academic Press, New York, 1974.
3. P. R. Halmos, *Measure Theory,* Van Nostrand Reinhold, New York, 1950.
4. W. Rudin, *Principles of Mathematical Analysis,* McGraw-Hill, New York, 1964.

5

CONDITIONAL PROBABILITY AND EXPECTATION

5.1 INTRODUCTION

We begin the chapter by exploring some relations between measurements, that is, measurable functions, and events, that is, members of a σ-field. In particular, we explore the relation between knowledge of the value of a particular measurement or class of measurements and knowledge of an outcome of a particular event or class of events. Mathematically these are relations between classes of functions and σ-fields. Such relations will be useful in developing properties of certain special functions such as limiting sample averages arising in the study of ergodic properties of information sources. In addition, they are fundamental to the development and interpretation of conditional probability and conditional expectation, that is, probabilities and expectations when we are given partial knowledge about the outcome of an experiment.

Although conditional probability and conditional expectation are both common topics in an advanced probability course, the fact that we are living in standard spaces will result in additional properties not present in the most general situation. In particular, we will find that the conditional probabilities and expectations have more structure that enables them to be interpreted and often treated much like ordinary unconditional probabilities. In technical terms, there will always be *regular versions of conditional probability and we will be able to define conditional expectations constructively as expectations with respect to conditional probability measures.*

5.2 MEASUREMENTS AND EVENTS

Say we have a measurable spaces $(\Omega,\boldsymbol{B})$ and $(A,\boldsymbol{B}(A))$ and a function $f:\Omega\to A$. Recall from Chapter 1 that f is a random variable or measurable function or a measurement if $f^{-1}(F)\in\boldsymbol{B}$ for all $F\in\boldsymbol{B}(A)$. Clearly, being told the outcome of the measurement f provides information about certain events and likely provides no information about others. In particular, if f assumes a value in $F\in\boldsymbol{B}(A)$, then we know that $f^{-1}(F)$ occurred. We might call such an event in $\boldsymbol{B}$ an f-event since it is specified by f taking a value in $\boldsymbol{B}(A)$. Consider the class $\boldsymbol{G}=f^{-1}(\boldsymbol{B}(A))=\{$all sets of the form $f^{-1}(F),\ F\in\boldsymbol{B}(A)\}$. Since f is measurable, all sets in $\boldsymbol{G}$ are also in $\boldsymbol{B}$. Furthermore, it is easy to see that $\boldsymbol{G}$ is a σ-field of subsets of Ω. Since $\boldsymbol{G}\subset\boldsymbol{B}$ and it is a σ-field, it is referred to as a *sub-σ-field.*

Intuitively $\boldsymbol{G}$ can be thought of as the σ- field or event space consisting of those events whose outcomes are determinable by observing the output of f. Hence we define the σ-field *induced* by f as $\sigma(f)=f^{-1}(\boldsymbol{B}(A))$.

Next observe that by construction $f^{-1}(F)\in\sigma(f)$ for all $F\in\boldsymbol{B}(A)$; that is, the inverse images of all output events may be found in a sub-σ-field. That is, f is measurable with respect to the smaller σ-field $\sigma(f)$ as well as with respect to the larger σ-field $\boldsymbol{B}$. As we will often have more than one σ-field for a given space, we emphasize this notion: Say we are given measurable spaces $(\Omega,\boldsymbol{B})$ and $(A,\boldsymbol{B}(A))$ and a sub-σ-field $\boldsymbol{G}$ of $\boldsymbol{B}$ (and hence $(\Omega,\boldsymbol{G})$ is also a measurable space). Then a function $f:\Omega\to A$ is said to be *measurable with respect to* $\boldsymbol{G}$ or $\boldsymbol{G}$*-measurable* or, in more detail, $\boldsymbol{G}/\boldsymbol{B}(A)$-measurable if $f^{-1}(F)\in\boldsymbol{G}$ for all $F\in\boldsymbol{B}(A)$. Since $\boldsymbol{G}\subset\boldsymbol{B}$ a $\boldsymbol{G}$-measurable function is clearly also $\boldsymbol{B}$-measurable. Usually we shall refer to $\boldsymbol{B}$-measurable functions as simply measurable functions or as measurements.

We have seen that if $\sigma(f)$ is the σ-field induced by a measurement f and that f is measurable with respect to $\sigma(f)$. Observe that if $\boldsymbol{G}$ is any other σ-field with respect to which f is measurable, then $\boldsymbol{G}$ must contain all sets of the form $f^{-1}(F)$, $F\in\boldsymbol{B}(A)$, and hence must contain $\sigma(f)$.

Thus $\sigma(f)$ is the smallest σ-field with respect to which f is measurable.

Having begun with a measurement and found the smallest σ-field with respect to which it is measurable, we can next reverse direction and begin with a σ-field and study the class of measurements that are measurable with respect to that σ-field. The short term goal will be to characterize those functions that are measurable with respect to $\sigma(f)$. Given a σ-field $\boldsymbol{G}$ let $\boldsymbol{M}(\boldsymbol{G})$ denote the class of all measurements that are measurable with respect to $\boldsymbol{G}$. We shall refer to $\boldsymbol{M}(\boldsymbol{G})$ as the measurement class induced by $\boldsymbol{G}$. Roughly speaking, this is the class of measurements whose output events can be determined from events in $\boldsymbol{G}$. Thus, for example, $\boldsymbol{M}(\sigma(f))$ is the collection of all measurements whose output events could be determined by events in $\sigma(f)$ and hence by output events of f. We shall shortly make this intuition precise, but first a comment is in order. The basic thrust here is that if we are given information about the outcome of a measurement f, in fact we then have information about the outcomes of many other measurements, e.g., f^2, $|f|$, and other functions of f. The class of functions $\boldsymbol{M}(\sigma(f))$ will prove to be exactly that class containing functions whose outcome is determinable from knowledge of the outcome of the measurement f.

The following lemma shows that if we are given a measurement $f: \Omega \rightarrow A$, then a real-valued function g is in $\boldsymbol{M}(\sigma(f))$, that is, is measurable with respect to $\sigma(f)$, if and only if $g(\omega) = h(f(\omega))$ for some measurement $h: A \rightarrow \mathbf{R}$, that is, if g depends on the underlying sample points only through the value of f.

Lemma 5.2.1. Given a measurable space $(\Omega, \boldsymbol{B})$ and a measurement $f: \Omega \rightarrow A$, then a measurement $g: \Omega \rightarrow \mathbf{R}$ is in $\boldsymbol{M}(\sigma(f))$, that is, is measurable with respect to $\sigma(f)$, if and only if there is a $\boldsymbol{B}(A)/\boldsymbol{B}(R)$-measurable function $h: A \rightarrow \mathbf{R}$ such that $g(\omega) = h(f(\omega))$, all $\omega \in \Omega$.

Proof. If there exists such an h, then for $F \in \boldsymbol{B}(\mathbf{R})$, then

$$g^{-1}(F) = \{\omega: h(f(\omega)) \in F\} = \{\omega: f(\omega) \in h^{-1}(F)\} = f^{-1}(h^{-1}(F)) \in \boldsymbol{B}$$

since $h^{-1}(F)$ is in $\boldsymbol{B}(A)$ from the measurability of h and hence its inverse image under f is in $\sigma(f)$ by definition. Thus any g of the given form is in $\boldsymbol{M}(\sigma(f))$. Conversely, suppose that g is in $\boldsymbol{M}(\sigma(f))$. Let $q_n: A \rightarrow \boldsymbol{R}$ be the sequence of quantizers of Lemma 4.3.1. Then g_n defined by $g_n(\omega) = q_n(f(\omega))$ are $\sigma(f)$-measurable simple functions that converge to g. Thus we can write g_n as

$$g_n(\omega) = \sum_{i=1}^{M} r_i 1_{F(i)}(\omega) ,$$

where the $F(i)$ are disjoint, and hence since $g_n \in \boldsymbol{M}(\sigma(f))$ that

$$g_n^{-1}(r_i) = F(i) \in \sigma(f) .$$

Since $\sigma(f) = f^{-1}(\boldsymbol{B}(A))$, there are disjoint sets $Q(i) \in \boldsymbol{B}(A)$ such that $F(i) = f^{-1}(Q(i))$. Define the function $h_n : A \to \mathbf{R}$ by

$$h_n(a) = \sum_{i=1}^{M} r_i 1_{Q(i)}(a)$$

and

$$h_n(f(\omega)) = \sum_{i=1}^{M} r_i 1_{Q(i)}(f(\omega)) = \sum_{i=1}^{M} r_i 1_{f^{-1}(Q(i))}(\omega)$$
$$= \sum_{i=1}^{M} r_i 1_{F(i)}(\omega) = g_n(\omega) .$$

This proves the result for simple functions. By construction we have that $g(\omega) = \lim_{n\to\infty} g_n(\omega) = \lim_{n\to\infty} h_n(f(\omega))$ where, in particular, the right-most limit exists for all $\omega \in \Omega$. Define the function $h(a) = \lim_{n\to\infty} h_n(a)$ where the limit exists and 0 otherwise. Then $g(\omega) = \lim_{n\to\infty} h_n(f(\omega)) = h(f(\omega))$, completing the proof. □

Thus far we have developed the properties of σ-fields induced by random variables and of classes of functions measurable with respect to σ-fields. The idea of a σ-field induced by a single random variable is easily generalized to random vectors and sequences. We wish, however, to consider the more general case of a σ-field induced by a possibly uncountable class of measurements. Then we will have associated with each class of measurements a natural σ-field and with each σ-field a natural class of measurements. Toward this end, given a class of measurements $\boldsymbol{M}$, define $\sigma(\boldsymbol{M})$ as the smallest σ-field with respect to which all of the measurements in $\boldsymbol{M}$ are measurable. Since any σ-field satisfying this condition must contain all of the $\sigma(f)$ for $f \in \boldsymbol{M}$ and hence must contain the σ-field induced by all of these sets and since this latter collection is a σ-field,

$$\sigma(\boldsymbol{M}) = \sigma(\bigcup_{f \in \boldsymbol{M}} \sigma(f)) .$$

The following lemma collects some simple relations among σ-fields induced by measurements and classes of measurements induced by σ-fields.

Lemma 5.2.2. Given a class of measurements $\boldsymbol{M}$, then

$$\boldsymbol{M} \subset \boldsymbol{M}(\sigma(\boldsymbol{M})) .$$

Given a collection of events $\boldsymbol{G}$, then

$$\boldsymbol{G} \subset \sigma(\boldsymbol{M}(\sigma(\boldsymbol{G}))) .$$

If $\boldsymbol{G}$ is also a σ-field, then

$$\boldsymbol{G} = \sigma(\boldsymbol{M}(\boldsymbol{G})) ,$$

that is, $\boldsymbol{G}$ is the smallest σ-field with respect to which all $\boldsymbol{G}$-measurable functions are measurable. If $\boldsymbol{G}$ is a σ-field and $\boldsymbol{I}(\boldsymbol{G}) = \{\text{all } 1_G, G \in \boldsymbol{G}\}$ is the collection of all indicator functions of events in $\boldsymbol{G}$, then

$$\boldsymbol{G} = \sigma(\boldsymbol{I}(\boldsymbol{G})) ,$$

that is, the smallest σ-field induced by indicator functions of sets in $\boldsymbol{G}$ is the same as that induced by all functions measurable with respect to $\boldsymbol{G}$.

Proof. If $f \in \boldsymbol{M}$, then it is $\sigma(f)$-measurable and hence in $\boldsymbol{M}(\sigma(\boldsymbol{M}))$. If $G \in \boldsymbol{G}$, then its indicator function 1_G is $\sigma(\boldsymbol{G})$-measurable and hence $1_G \in \boldsymbol{M}(\sigma(\boldsymbol{G}))$. Since $1_G \in \boldsymbol{M}(\sigma(\boldsymbol{G}))$, it must be measurable with respect to $\sigma(\boldsymbol{M}(\sigma(\boldsymbol{G})))$ and hence $1_G^{-1}(1) = G \in \sigma(\boldsymbol{M}(\sigma(\boldsymbol{G})))$, proving the second statement. If $\boldsymbol{G}$ is a σ-field, then since since all functions in $\boldsymbol{M}(\boldsymbol{G})$ are $\boldsymbol{G}$-measurable and since $\sigma(\boldsymbol{M}(\boldsymbol{G}))$ is the smallest σ-field with respect to which all functions in $\boldsymbol{M}(\boldsymbol{G})$ are $\boldsymbol{G}$-measurable, $\boldsymbol{G}$ must contain $\sigma(\boldsymbol{M}(\boldsymbol{G}))$. Since $\boldsymbol{I}(\boldsymbol{G}) \subset \boldsymbol{M}(\boldsymbol{G})$,

$$\sigma(\boldsymbol{I}(\boldsymbol{G})) \subset \sigma(\boldsymbol{M}(\boldsymbol{G})) = \boldsymbol{G} .$$

If $G \in \boldsymbol{G}$, then 1_G is in $\boldsymbol{I}(\boldsymbol{G})$ and hence must be measurable with respect to $\sigma(\boldsymbol{I}(\boldsymbol{G}))$ and hence $1_G^{-1}(1) = G \in \sigma(\boldsymbol{I}(\boldsymbol{G}))$ and hence $\boldsymbol{G} \subset \sigma(\boldsymbol{I}(\boldsymbol{G}))$, completing the proof. □

We conclude this section with a reminder of the motivation for considering classes of events and measurements. We shall often consider such classes either because a particular class is important for a particular application or because we are simply given a particular class. In both cases we may wish to study both the given events (measurements) and the related class of measurements (events). For example, given a class of measurements $\boldsymbol{M}$, $\sigma(\boldsymbol{M})$ provides a σ-field of events whose occurrence or nonoccurrence is determinable by the output events of the functions in the class. In turn, $\boldsymbol{M}(\sigma(\boldsymbol{M}))$ is the possibly larger class of functions whose output events are determinable from the occurrence or nonoccurrence of events in $\sigma(\boldsymbol{M})$ and hence by the output events of $\boldsymbol{M}$. Thus knowing all output events of measurements in $\boldsymbol{M}$ is effectively equivalent to knowing all output events of measurements in the more structured and possibly larger class $\boldsymbol{M}(\sigma(\boldsymbol{M}))$, which is in turn equivalent to knowing the occurrence or nonoccurrence of events in the σ-field $\sigma(\boldsymbol{M})$. Hence when a class $\boldsymbol{M}$ is specified, we can instead consider the more structured classes $\sigma(\boldsymbol{M})$ or $\boldsymbol{M}(\sigma(\boldsymbol{M}))$. From the previous lemma, this is as far as we can go; that is,

$$\sigma(\boldsymbol{M}(\sigma(\boldsymbol{M}))) = \sigma(\boldsymbol{M}) . \tag{5.2.1}$$

We have seen that a function g will be in $\boldsymbol{M}(\sigma(f))$ if and only if it depends on the underlying sample points through the value of f. Since we did not restrict the structure of f, it could, for example, be a random variable, vector, or sequence, that is, the conclusion is true for countable collections of measurements as well as individual measurements. If instead we have a general class $\boldsymbol{M}$ of measurements, then it is still a useful intuition to think of $\boldsymbol{M}(\sigma(\boldsymbol{M}))$ as being the class of all functions that depend on the underlying points only through the values of the functions in $\boldsymbol{M}$.

Exercises

1. Which of the following relations are true and which are false?

$$f^2 \in \sigma(f)\ , f \in \sigma(f^2)$$

$$f + g \in \sigma(f, g)\ , f \in \sigma(f + g)$$

If $f: A \to \Omega$, $g: \Omega \to \Omega$, and $g(f): A \to \Omega$ is defined by $g(f)(x) = g(f(x))$, then $g(f) \in \sigma(f)$.

2. Given a class $\boldsymbol{M}$ of measurements, is $\bigcup_{f \in \boldsymbol{M}} \sigma(f)$ a σ-field?

3. Suppose that $(A, \boldsymbol{B})$ is a measure space and $\boldsymbol{B}$ is separable with a countable generating class $\{V_n;\ n = 1,2 \cdots\}$. Describe $\sigma(1_{V_n};\ n = 1,2, \cdots)$.

5.3 RESTRICTIONS OF MEASURES

The first application of the classes of measurements or events considered in the previous section is the notion of the restriction of a probability measure to a sub-σ-field. This occasionally provides a shortcut to evaluating expectations of functions that are measurable with respect to sub-σ-fields and in comparing such functions. Given a probability space $(\Omega,\boldsymbol{B},m)$ and a sub-σ-field $\boldsymbol{G}$ of $\boldsymbol{B}$, define the *restriction of m to* $\boldsymbol{G}$, $m_{\boldsymbol{G}}$, by

$$m_{\boldsymbol{G}}(F) = m(F) \; , F \in \boldsymbol{G}.$$

Thus $(\Omega,\boldsymbol{G},m_{\boldsymbol{G}})$ is a new probability space with a smaller event space. The following lemma shows that if f is a $\boldsymbol{G}$-measurable real-valued random variable, then its expectation can be computed with respect to either m or $m_{\boldsymbol{G}}$.

Lemma 5.3.1:. Given a $\boldsymbol{G}$-measurable real-valued measurement $f \in L^1(m)$, then also $f \in L^1(m_{\boldsymbol{G}})$ and

$$\int f \, dm = \int f \, dm_{\boldsymbol{G}} \; ,$$

where $m_{\boldsymbol{G}}$ is the restriction of m to $\boldsymbol{G}$.

If f is a simple function, the result is immediate from the definition of restriction. More generally use Lemma 4.3.1(e) to infer that $q_n(f)$ is a sequence of simple $\boldsymbol{G}$-measurable functions converging to f and combine the simple function result with Corollary 4.4.1 applied to both measures.

Corollary 5.3.1. Given $\boldsymbol{G}$-measurable functions $f,g \in L^1(m)$, if

$$\int_F f \, dm \leq \int_F g \, dm \; , \text{ all } F \in \boldsymbol{G} \; ,$$

then $f \leq g$ m-a.e. If the preceding holds with equality, then $f = g$ m-a.e.

Proof. From the previous lemma, the integral inequality holds with m replaced by $m_{\boldsymbol{G}}$, and hence from Lemma 4.4.7 the conclusion holds $m_{\boldsymbol{G}}$-a.e. Thus there is a set, say G, in $\boldsymbol{G}$ with $m_{\boldsymbol{G}}$ probability one and hence also m probability one for which the conclusion is true. □

The usefulness of the preceding corollary is that it allows us to compare $\boldsymbol{G}$-measurable functions by considering only the restricted measures and the corresponding expectations.

5.4 ELEMENTARY CONDITIONAL PROBABILITY

Say we have a probability space $(\Omega,\boldsymbol{B},m)$, how is the probability measure m altered if we are told that some event or collection of events occurred? For example, how is it influenced if we are given the outputs of a measurement or collection of measurements? The notion of conditional probability provides a response to this question. In fact there are two notions of conditional probability: elementary and nonelementary.

Elementary conditional probabilities cover the case where we are given an event, say F, having nonzero probability: $m(F) > 0$. We would like to define a conditional probability measure $m(G \mid F)$ for all events $G \in \boldsymbol{B}$. Intuitively, being told an event F occurred will put zero probability on the collection of all points outside F, but it should not effect the relative probabilities of the various events inside F. In addition, the new probability measure must be renormalized so as to assign probability one to the new "certain event" F. This suggests the definition $m(G \mid F) = km(G \cap F)$, where k is a normalization constant chosen to ensure that $m(F \mid F) = km(F \cap F) = km(F) = 1$. Thus we define for any F such that $m(F) > 0$. the conditional probability measure

$$m(G \mid F) = \frac{m(G \cap F)}{m(F)}, \text{ all } G \in \boldsymbol{B}.$$

We shall often abbreviate the elementary conditional probability measure $m(\,.\mid F)$ by m^F. Given a probability space $(\Omega,\boldsymbol{B},m)$ and an event $F \in \boldsymbol{B}$ with $m(F) > 0$, then we have a new probability space $(F,\boldsymbol{B}\cap F,m^F)$, where $\boldsymbol{B}\cap F =$ {all sets of the form $G \cap F$, $G \in \boldsymbol{B}$ }. It is easy to see that we can relate expectations with respect to the conditional and unconditional measures by

$$E_{m^F}(f) = \int f dm^F = \frac{\int_F f\,dm}{m(F)} = \frac{E_m 1_F f}{m(F)}, \tag{5.4.1}$$

where the existence of either side ensures that of the other. In particular, $f \in L^1(m^F)$ if and only if $f\,1_F \in L^1(m)$. Note further that if $G = F^c$, then

$$E_m f = m(F) E_{m^F}(f) + m(G) E_{m^G}(f) . \tag{5.4.2}$$

Instead of being told that a particular event occurred, we might be told that a random variable or measurement f is discrete and takes on a specific value, say a with nonzero probability. Then the elementary definition immediately yields

$$m(F \mid f = a) = \frac{m(G \cap \{f = a\})}{m(f = a)} .$$

If, however, the measurement is not discrete and takes on a particular value with probability zero, then the preceding elementary definition does not work. One might attempt to replace the previous definition by some limiting form, but this does not yield a useful theory in general and it is clumsy. The standard alternative approach is to replace the preceding constructive definition by a descriptive definition, that is, to define conditional probabilities by the properties that they should possess. The mathematical problem is then to prove that a function possessing the desired properties exists.

In order to motivate the descriptive definition, we make several observations on the elementary case. First note that the previous conditional probability depends on the value a assumed by f, that is, it is a function of a. It will prove more useful and more general instead to consider it a function of ω that depends on ω only through $f(\omega)$, that is, to consider conditional probability as a function $m(G \mid f)(\omega)$ = $m(G \mid \{\lambda : f(\lambda) = f(\omega)\})$, or, simply, $m(G \mid f = f(\omega))$, the probability of an event G given that f assumes a value $f(\omega)$. Thus a conditional probability is a function of the points in the underlying sample space and hence is itself a random variable or measurement. Since it depends on ω only through f, from Lemma 5.2.1 the function $m(G \mid f)$ is measurable with respect to $\sigma(f)$, the σ-field induced by the given measurement. This leads to the first property:

$$\text{For any fixed } G,\ m(G \mid f) \text{ is } \sigma(f)\text{–measurable} . \tag{5.4.3}$$

Next observe that in the elementary case we can compute the probability $m(G)$ by averaging or integrating the conditional probability $m(G \mid f)$ over all possible values of f; that is,

$$\int m(G \mid f)\, dm = \sum_a m(G \mid f = a) m(f = a)$$

$$= \sum_a m(G \cap \{f = a\}) = m(G) \ . \qquad (5.4.4)$$

In fact we can and must say more about such averaging of conditional probabilities. Suppose that F is some event in $\sigma(f)$ and hence its occurrence or nonoccurrence is determinable by observation of the value of f, that is, from Lemma 5.2.1 $1_F(\omega) = h(f(\omega))$ for some function h. Thus if we are given the value of $f(\omega)$, being told that $\omega \in F$ should not add any knowledge and hence should not alter the conditional probability of an event G given f. To try to pin down this idea, assume that $m(F) > 0$ and let m^F denote the elementary conditional probability measure given F, that is, $m^F(G) = m(G \cap F)/m(F)$. Applying (5.4.4) to the conditional measure m^F yields the formula

$$\int m^F(G \mid f)\, dm^F = m^F(G) \ ,$$

where $m^F(G \mid f)$ is the conditional probability of G given the outcome of the random variable f and given that the event F occurred. But we have argued that this should be the same as $m(G \mid f)$. Making this substitution, multiplying both sides of the equation by $m(F)$, and using (5.4.1) we derive

$$\int_F m(G \mid f)\, dm = m(G \cap F) \ , \text{ all } F \in \sigma(f) \ . \qquad (5.4.5)$$

To make this plausibility argument rigorous observe that since f is assumed discrete we can write

$$f(\omega) = \sum_{a \in A} a 1_{f^{-1}(a)}(\omega)$$

and hence

$$1_F(\omega) = h(f(\omega)) = h(\sum_{a \in A} a 1_{f^{-1}(a)}(\omega))$$

and therefore

$$F = \bigcup_{a:h(a) = 1} f^{-1}(a) \ .$$

We can then write

$$\int_F m(G \mid f)\, dm = \int 1_F\, m(G \mid f)\, dm = \int h(f) m(G \mid f)\, dm$$

$$= \sum_a h(a) \frac{m(G \cap \{f = a\})}{m(\{f = a\})} m(\{f = a\}) = \sum_a h(a) m(G \cap f^{-1}(a))$$

$$= m(G \cap \bigcup_{a:h(a) = 1} f^{-1}(a)) = m(G \cap F) .$$

Although (5.4.5) was derived for F with $m(F) > 0$, the equation is trivially satisfied if $m(F) = 0$ since both sides are then 0. Hence the equation is valid for all F in $\sigma(f)$ as stated. Eq. (5.4.5) states that not only must we be able to average $m(G \mid f)$ to get $m(G)$, a similar relation must hold when we cut down the average to any $\sigma(f)$-measurable event.

Equations (5.4.3) and (5.4.5) provide the properties needed for a rigorous descriptive definition of the conditional probability of an event G given a random variable or measurement f. In fact, this conditional probability is defined as any function $m(G \mid f)(\omega)$ satisfying these two equations. At this point, however, little effort is saved by confining interest to a single conditioning measurement, and hence we will develop the theory for more general classes of measurements. In order to do this, however, we require a basic result from measure theory–the Radon-Nikodym theorem. The next two sections develop this result.

Exercises

1. Let f and g be discrete measurements on a probability space $(\Omega, \boldsymbol{B}, m)$. Suppose in particular that

$$g(\omega) = \sum_i a_i 1_{G_i}(\omega).$$

Define a new random variable $E(g \mid f)$ by

$$E(g \mid f)(\omega) = E_{m(\,.\,\mid f(\omega))}(g) = \sum_i a_i m(G_i \mid f(\omega)) .$$

This random variable is called the *conditional expectation of g given f*. Prove that

$$E_m(g) = E_m[E(g \mid f)] .$$

5.5 PROJECTIONS

One of the proofs of the Radon-Nikodym theorem is based on the projection theorem of Hilbert space theory. As this proof is more intuitive than others (at least to the author) and as the projection theorem is of interest in its own right, this section is devoted to developing the properties of projections and the projection theorem. Good references for the theory of Hilbert spaces and projections are Halmos [4] and Luenberger [5].

Recall from Chapter 3 that a Hilbert space is a complete inner product space as defined in Example 2.5.10, which is in turn a normed linear space as Example 2.5.8. In fact, all of the results considered in this section will be applied in the next section to a very particular Hilbert space, the space $L^2(m)$ of all square integrable functions with respect to a probability measure m. Since this space is Polish, it is a Hilbert space.

Let L be a Hilbert space with inner product (f,g) and norm $\|f\| = (f,f)^{1/2}$. For the case where $L = L^2(m)$, the inner product is given by

$$(f,g) = E_m(fg) .$$

A (measurable) subset M of L will be called a subspace of L if $f, g \in M$ implies that $af+bg \in M$ for all scalar multipliers a, b. We will be primarily interested in closed subspaces of L. Recall from Lemma 3.2.4 that a closed subset of a Polish space is also Polish with the same metric, and hence M is a complete space with the same norm.

A fundamentally important property of linear spaces with inner products is the *parallelogram law*

$$\|f+g\|^2 + \|f-g\|^2 = 2\|f\|^2 + 2\|g\|^2 . \tag{5.5.1}$$

This is proved simply by expanding the left-hand side as

$$(f,f) + 2(f,g) + (g,g) + (f,f) - 2(f,g) + (g,g) .$$

A simple application of this formula and the completeness of a subspace yield the following result: given a fixed element f in a Hilbert space and given a closed subspace of the Hilbert space, we can always find a unique member of the subspace that provides the "best" approximation to f in the sense of minimizing the norm of the difference.

Lemma 5.5.1. Fix an element f in a Hilbert space L and let M be a closed subspace of L. Then there is a unique element $\hat{f} \in M$ such that

$$\| f - \hat{f} \| = \inf_{g \in M} \| f - g \| , \tag{5.5.2}$$

where f and g are considered identical if $\| f - g \| = 0$. Thus, in particular, the infimum is actually a minimum.

Proof. Let Δ denote the infimum of (5.5.2) and choose a sequence of $f_n \in M$ so that $\| f - f_n \| \to \Delta$. From the parallelogram law

$$\| f_n - f_m \|^2 = 2\| f_n - f \|^2 + 2\| f_m - f \|^2 - 4\| \frac{f_n + f_m}{2} - f \|^2 .$$

Since M is a subspace, $(f_n + f_m)/2 \in M$ and hence the latter norm squared is bound below by Δ^2. This means, however, that

$$\lim_{n \to \infty, m \to \infty} \| f_n - f_m \|^2 \le 2\Delta^2 + 2\Delta^2 - 4\Delta^2 = 0 ,$$

and hence f_n is a Cauchy sequence. Since M is complete, it must have a limit, say $\hat{f}$. From the triangle inequality

$$\| f - \hat{f} \| \le \| f - f_n \| + \| f_n - \hat{f} \| \underset{n \to \infty}{\to} \Delta ,$$

which proves that there is an $\hat{f}$ in M with the desired property. Suppose next that there are two such functions, e.g., $\hat{f}$ and g. Again invoking (5.5.1)

$$\| \hat{f} - g \|^2 \le 2\| \hat{f} - f \|^2 + 2\| g - f \|^2 - 4\| \frac{\hat{f} + g}{2} - f \|^2 .$$

Since $(\hat{f} + g)/2 \in M$, the latter norm can be no smaller than Δ, and hence

$$\| \hat{f} - g \|^2 \le 2\Delta^2 + 2\Delta^2 - 4\Delta^2 = 0 ,$$

and therefore $\| \hat{f} - g \| = 0$ and the two are identical. □

Two elements f and g of L will be called *orthogonal* and we write $f \perp g$ if $(f, g) = 0$. If $M \subset L$, we say that $f \in L$ is orthogonal to M if f is orthogonal to every $g \in M$. The collection of all $f \in L$ which are orthogonal to M is called the *orthogonal complement* of M.

Lemma 5.5.2. Given the assumptions of Lemma 5.5.1, $\hat{f}$ will yield the infimum if and only if $f - \hat{f} \perp M$; that is, $\hat{f}$ is chosen so that the error $f - \hat{f}$ is orthogonal to the subspace M.

Proof. First we prove necessity. Suppose that $f - \hat{f}$ is not orthogonal to M. We will show that $\hat{f}$ cannot then yield the infimum of Lemma 5.5.1. By assumption there is a $g \in M$ such that g is not orthogonal to $f - \hat{f}$. We can assume that $\| g \| = 1$ (otherwise just normalize g by dividing by its norm). Define $a = (f - \hat{f}, g) \neq 0$. Now set $h = \hat{f} + ag$. Clearly $h \in M$ and we now show that it is a better estimate of f than $\hat{f}$ is. To see this write

$$\| f - h \|^2 = \| f - \hat{f} - ag \|^2 = \| f - \hat{f} \|^2 + a^2 \|g\|^2 - 2a(f - \hat{f}, g)$$
$$= \| f - \hat{f} \|^2 - a^2 < \| f - \hat{f} \|^2$$

as claimed. Thus orthogonality is necessary for the minimum norm estimate. To see that it is sufficient, suppose that $\hat{f} \perp M$. Let g be any other element in M. Then

$$\| f - g \|^2 = \| f - \hat{f} + \hat{f} - g \|^2$$
$$= \| f - \hat{f} \|^2 + \| \hat{f} - g \|^2 - 2(f - \hat{f}, \hat{f} - g) .$$

Since $\hat{f} - g \in M$, the inner product is 0 by assumption, proving that $\| f - g \|$ is strictly greater than $\| f - \hat{f} \|$ if $\| \hat{f} - g \|$ is not identically 0.

Combining the two lemmas we have the following famous result.

Theorem 5.5.1. (The Projection Theorem)

Suppose that M is a closed subspace of a Hilbert space L and that $f \in L$. Then there is a unique $\hat{f} \in M$ with the properties that

$$f - \hat{f} \perp M$$

and

$$\| f - \hat{f} \| = \inf_{g \in M} \| f - g \| .$$

The resulting $\hat{f}$ is called the *projection* of f onto M and will be denoted $P_M(f)$.

Exercises

1. Show that projections satisfy the following properties:

   (a) If $f \in M$, then $P_M(f) = f$.

   (b) For any scalar a $P_M(af) = aP_M(f)$.

   (c) For any $f, g \in L$ and scalars a, b:

$$P_M(af + bg) = aP_M(f) + bP_M(g) .$$

2. Consider the special case of $L^2(m)$ of all square-integrable functions with respect to a probability measure m. Let P be the projection with respect to a closed subspace M. Prove the following:

   (a) If f is a constant, say c, then $P(f) = c$.

   (b) If $f \geq 0$ with probability one, then $P(f) \geq 0$.

3. Suppose that g is a discrete measurement and hence is in $L^2(m)$. Let M be the space of all square-integrable $\sigma(g)$-measurable functions. Assume that M is complete. Given a discrete measurement f, let $E(f \mid g)$ denote the conditional expectation of exercise 5.4.1. Show that $E(f \mid g) = P_M(f)$, the projection of f onto M.

4. Suppose that $H \subset M$ are closed subspaces of a Hilbert space L. Show that for any $f \in L$,

$$\| f - P_M(f) \| \leq \| f - P_H(f) \| .$$

5.6 THE RADON-NIKODYM THEOREM

This section is devoted to the statement and proof of the Radon-Nikodym theorem, one of the fundamental results of measure and integration theory and a key result in the proof of the existence of conditional probability measures.

A measure m is said to be *absolutely continuous* with respect to another measure P on the same measurable space and we write $m \ll P$ if $P(F) = 0$ implies $m(F) = 0$, that is, m inherits all of P's zero probability events.

Theorem 5.6.1. (The Radon-Nikodym Theorem)

Given two measures m and P on a measurable space $(\Omega,\boldsymbol{B})$ such that $m \ll P$, then there is a measurable function h: $\Omega \to \mathrm{R}$ with the properties that $h \geq 0$ P-a.e. (and hence also m-a.e.) and

$$m(F\,) = \int_F h\, dP \text{ , all } F \in \boldsymbol{B} \text{ .} \tag{5.6.1}$$

If g and h both satisfy (5.6.1), then $g = h$ P-a.e. If in addition $f : \Omega \to \mathrm{R}$ is measurable, then

$$\int_F f\, dm = \int_F f\, hdP \text{ , all } F \in \boldsymbol{B} \text{ .} \tag{5.6.2}$$

The function h is called the *Radon-Nikodym derivative* of m with respect to P and is written $h = dm/dP$.

Proof. First note that (5.6.2) follows immediately from (5.6.1) for simple functions. The general result (5.6.2) then follows, using the usual approximation techniques. Hence we need prove only (5.6.1), which will take some work.

Begin by defining a mixture measure $q = (m + P)/2$, that is, $q(\,F\,) = m(\,F\,)/2 + P(\,F\,)/2$ for all events F. For any $f \in L^2(q)$, necessarily $f \in L^2(m)$, and hence $E_m f$ is finite from the Cauchy-Schwarz inequality. As a first step in the proof we will show that there is a $g \in L^2(q)$ for which

$$E_m f = \int f\, g\, dq \text{ , all } f \in L^2(q) \text{ .} \tag{5.6.3}$$

Define the class of measurements $\boldsymbol{M} = \{f : f \in L^2(q),\ E_m f = 0\}$ and observe that the properties of expectation imply that $\boldsymbol{M}$ is a closed linear subspace of $L^2(q)$. Hence the projection theorem implies that any $f \in L^2(q)$ can be written as $f = \hat{f} + f'$, where $\hat{f} \in \boldsymbol{M}$ and $f' \perp \boldsymbol{M}$. Furthermore, this representation is unique q-a.e. If all of the functions in the orthogonal complement of $\boldsymbol{M}$ were identically 0, then $f = \hat{f} \in \boldsymbol{M}$ and (5.6.3) would be trivially true with $f' = 0$ since $E_m f = 0$ for $f \in \boldsymbol{M}$ by definition. Hence we can assume that there is some $r \perp \boldsymbol{M}$ that is not identically 0. Furthermore, $\int r\, dm \neq 0$ lest we have an $r \in \boldsymbol{M}$ such that $r \perp \boldsymbol{M}$, which can only be if r is identically 0.

Define for $x \in \Omega$

$$g(x) = \frac{r(x) E_m r}{\| r \|^2} \text{ ,}$$

where here $\| r \|$ is the $L^2(q)$ norm. Clearly $g \in L^2(q)$ and by construction $g \perp \boldsymbol{M}$ since r is. We have that

$$\int f g \, dq = \int (f - r\frac{E_m f}{E_m r}) g \, dq + \int r g \frac{E_m f}{E_m r} dq .$$

The first term on the right is 0 since $g \perp \boldsymbol{M}$ and $f - r(E_m f)/(E_m r)$ has zero expectation with respect to m and hence is in $\boldsymbol{M}$. Thus

$$\int f g \, dq = \frac{E_m f}{E_m r} \int r g \, dq = \frac{E_m f}{E_m r} \frac{E_m r \, \| r \|^2}{\| r \|^2} = E_m f$$

which proves (5.6.3). Observe that if we set $f = 1_F$ for any event F, then

$$E_m 1_F = m(F) = \int_F g \, dq ,$$

a formula which resembles the desired formula (5.6.1) except that the integral is with respect to q instead of m. From the definition of q $0 \leq m(F) \leq 2q(F)$ and hence

$$0 \leq \int_F g dq \leq \int_F 2 \, dq , \text{ all events } F .$$

From Lemma 4.4.7 this implies that $g \in [0,2]$ q-a.e.. Since $m \ll q$ and $P \ll q$, this statement also holds m-a.e. and P-a.e..

As a next step rewrite (5.6.3) in terms of P and m as

$$\int f (1 - \frac{g}{2}) \, dm = \int \frac{fg}{2} \, dP , \text{ all } f \in L^2(q) . \tag{5.6.4}$$

We next argue that (5.6.4) holds more generally for all measurable f in the sense that each integral exists if and only if the other does, in which case they are equal. To see this, first assume that $f \geq 0$ and choose the usual quantizer sequence of Lemma 4.3.1. Then (5.6.4) implies that

$$\int q_n(f) (1 - \frac{g}{2}) \, dm = \int \frac{q_n(f) g}{2} \, dP , \text{ all } f \in L^2(q)$$

for all n. Since g is between 0 and 2 with both m and P probability 1, both integrands are nonnegative a.e. and converge upward together to a finite number or diverge together. In either case, the limit is $\int f (1 - g/2) \, dm = \int f g/2 \, dP$ and (5.6.4) holds. For general f apply this argument to the pieces of the usual decomposition $f = f^+ - f^-$, and the conclusion follows, both integrals either existing or not existing together.

To complete the proof, define the set $B = \{x: g(x) = 2\}$. Application of (5.6.4) to 1_B implies that

$$P(B) = \int_B dP = \int 1_B \frac{g}{2}\, dP = \int 1_B (1 - \frac{g}{2})\, dm = 0 ,$$

and hence $P(B^c) = 1$. Combining this fact with (5.6.4) with $f = (1 - g/2)^{-1}$ implies that for any event F

$$\begin{aligned} m(F) &= m(F \cap B) + m(F \cap B^c) \\ &= m(F \cap B) + \int_{F \cap B^c} \frac{1 - \frac{g}{2}}{1 - \frac{g}{2}}\, dm \\ &= m(F \cap B) + \int_{F \cap B^c} \frac{g}{1 - \frac{g}{2}} \frac{1}{2} dP . \end{aligned} \tag{5.6.5}$$

Define the function h by

$$h(x) = \frac{g(x)}{2 - g(x)} 1_{B^c}(x) . \tag{5.6.6}$$

and (5.6.5) becomes

$$m(F) = m(F \cap B) + \int_{F \cap B^c} h\, dP , \text{ all } F \in \mathbf{B} . \tag{5.6.7}$$

Note that (5.6.7) is true in general; that is, we have not yet used the fact that $m \ll P$. Taking this fact into account, $m(B) = 0$, and hence since $P(B^c) = 1$,

$$m(F) = \int_F h\, dP \tag{5.6.8}$$

for all events F. This formula combined with (5.6.6) and the fact that $g \in [0,2]$ with P and m probability 1 completes the proof. □

In fact we have proved more than just the Radon-Nikodym theorem; we have also almost proved the Lebesgue decomposition theorem. We next complete this task.

Theorem 5.6.2. (The Lebesgue Decomposition Theorem)

Given two probability measures m and P on a measurable space $(\Omega,\boldsymbol{B})$, there are unique probability measures m_a and p and a number $\lambda \in [0,1]$ such that such that $m = \lambda m_a + (1-\lambda)p$, where $m_a \ll P$ and where p and P are *singular* in the sense that there is a set $G \in \boldsymbol{B}$ with $P(G) = 1$ and $p(G) = 0$. m_a is called the part of m *absolutely continuous* with respect to P.

Proof. If $m \ll P$, then the result is just the Radon-Nikodym theorem with $m_a = m$ and $\lambda = 1$. If m is not absolutely continuous with respect to P then several cases are possible: If $m(B)$ in (5.6.8) is 0, then the result again follows as in (5.6.8). If $m(B) = 1$, then $p = m$ and P are singular and the result holds with $\lambda = 0$. If $m(B) \in (0,1)$, then (5.6.7) yields the theorem by identifying

$$m_a(F) = \frac{\int_F h \, dP}{\int_\Omega h \, dP},$$

$$p(F) = m_B(F) = \frac{m(F \cap B)}{m(B)},$$

and

$$\lambda = \int_\Omega h \, dP = m(B^c),$$

where the last equality follows from (5.6.7) with $F = B^c$. □

The Lebesgue decomposition theorem is important because it permits any probability measure to be decomposed into an absolutely continuous part and a singular part. The absolutely continuous part can always be computed by integrating a Radon-Nikodym derivative, and hence such derivatives can be thought of as probability density functions.

Exercises

1. Show that if $m \ll P$, then $P(F) = 1$ implies that $m(F) = 1$. Two measures m and P are said to be *equivalent* if $m \ll P$ and $P \ll m$. How are the Radon-Nikodym derivatives dP/dm and dm/dP related?

5.7 CONDITIONAL PROBABILITY

Assume that we are given a probability space $(\Omega,\boldsymbol{B},m)$ and class of measurements $\boldsymbol{M}$ and we wish to define a conditional probability $m(G \mid \boldsymbol{M})$ for all events G. Intuitively, if we are told the output values of all of the measurements f in the class $\boldsymbol{M}$ (which might contain only a single measurement or an uncountable family), what then is the probability that G will occur? The first observation is that we have not yet specified what the given output values of the class are, only the class itself. We can easily nail down these output values by thinking of the conditional probability as a function of $\omega \in \Omega$ since given ω all of the $f(\omega)$, $f \in \boldsymbol{M}$, are fixed. Hence we will consider a conditional probability as a function of sample points of the form $m(G \mid \boldsymbol{M})(\omega) = m(G \mid f(\omega)$, all $f \in \boldsymbol{M})$, that is, the probability of G given that $f = f(\omega)$ for all measurements f in the given collection $\boldsymbol{M}$. Analogous to (5.4.3), since $m(G \mid \boldsymbol{M})$ is to depend only on the output values of the $f \in \boldsymbol{M}$, mathematically it should be in $\boldsymbol{M}(\sigma(\boldsymbol{M}))$, that is, it should be measurable with respect to $\sigma(\boldsymbol{M})$, the σ-field induced by the conditioning class.

Analogous to (5.4.5) the conditional probability of G given $\boldsymbol{M}$ should not be changed if we are also given that an event in $\sigma(\boldsymbol{M})$ occurred and hence averaging yields

$$m(G \cap F) = \int_F m(G \mid \boldsymbol{M})\, dm\ ,\ F \in \sigma(\boldsymbol{M})\ .$$

This leads us to the following definition:

Given a class of measurements $\boldsymbol{M}$ and an event G, the *conditional probability* $m(G \mid \boldsymbol{M})$ of G given $\boldsymbol{M}$ is defined as any function such that

$$m(G \mid \boldsymbol{M}) \text{ is } \sigma(\boldsymbol{M})\text{–measurable , and} \tag{5.7.1}$$

$$m(G \cap F) = \int_F m(G \mid \boldsymbol{M})\, dm\ ,\ \text{all } F \in \sigma(\boldsymbol{M})\ . \tag{5.7.2}$$

Clearly we have yet to show that in general such a function exists. First, however, observe that the preceding definition really depends on $\boldsymbol{M}$ only through the σ-field that it induces. Hence more generally we can define a conditional probability of an event G given a sub-σ-field $\boldsymbol{G}$ as any function $m(G \mid \boldsymbol{G})$ such that

$$m(G \mid \boldsymbol{G}) \text{ is } \boldsymbol{G}\text{–measurable, and} \tag{5.7.3}$$

$$m(G \cap F) = \int_F m(G \mid \boldsymbol{G})\, dm\ , \text{ all } F \in \boldsymbol{G}\ . \qquad (5.7.4)$$

Intuitively, the conditional probability given a σ-field is the conditional probability given the knowledge of which events in the σ-field occurred and which did not or, equivalently, given the outcomes of the indicator functions for all events in the σ-field. That is, as stated in Lemma 5.2.2, if $\boldsymbol{G}$ is a σ-field and $\boldsymbol{M} = \{\text{all } 1_F;\ F \in \boldsymbol{G}\}$, then $\boldsymbol{G} = \sigma(\boldsymbol{M})$ and hence $m(G \mid \boldsymbol{G}) = m(G \mid \boldsymbol{M})$.

By construction,

$$m(G \mid \boldsymbol{M}) = m(G \mid \sigma(\boldsymbol{M}))\ . \qquad (5.7.6)$$

Furthermore, consider $m(G \mid \boldsymbol{M}(\sigma(\boldsymbol{M})))$. This is similarly given by the conditional probability given the induced class of measurements, $m(G \mid \sigma(\boldsymbol{M}(\sigma(\boldsymbol{M}))))$. From (5.2.1), however, the conditioning σ-field is exactly $\sigma(\boldsymbol{M})$. Thus

$$m(G \mid \boldsymbol{M}) = m(G \mid \sigma(\boldsymbol{M})) = m(G \mid \boldsymbol{M}(\sigma(\boldsymbol{M})))\ , \qquad (5.7.7)$$

reinforcing the intuition discussed in Section 5.2 that knowing the outcomes of functions in $\boldsymbol{M}$ is equivalent to knowing the occurrence or nonoccurrence of events in $\sigma(\boldsymbol{M})$, which is in turn equivalent to knowing the outcomes of functions in $\boldsymbol{M}(\sigma(\boldsymbol{M}))$. It will usually be more convenient to develop general results for the more general notion of a conditional probability given a σ-field.

We are now ready to demonstrate the existence and essential uniqueness of conditional probability.

Theorem 5.7.1. Given a probability space $(\Omega, \boldsymbol{B}, m)$, an event $G \in \boldsymbol{B}$, and a sub-σ-field $\boldsymbol{G}$, then there exists a version of the conditional probability $m(G \mid \boldsymbol{G})$. Furthermore, any two such versions are equal m-a.e.

Proof. If $m(G) = 0$, then (5.7.4) becomes

$$0 = \int_F m(G \mid \boldsymbol{G})\, dm\ , \text{ all } F \in \boldsymbol{G}\ ,$$

and hence if $m(G \mid \boldsymbol{G})$ is $\boldsymbol{G}$-measurable, it is 0 m-a.e. from Corollary 5.3.1, completing the proof for this case. If $m(G) \neq 0$, then define the probability measure m^G on $(\Omega, \boldsymbol{G})$ as the elementary conditional probability

$$m^G(F) = m(F \mid G) = \frac{m(F \cap G)}{m(G)}, F \in \boldsymbol{G}.$$

The restriction $m_{\boldsymbol{G}}^G$ of $m(. \mid G)$ to $\boldsymbol{G}$ is absolutely continuous with respect to $m_{\boldsymbol{G}}$, the restriction of m to $\boldsymbol{G}$. Thus $m_{\boldsymbol{G}}^G \ll m_{\boldsymbol{G}}$ and hence from the Radon-Nikodym theorem there exists an essentially unique almost everywhere positive $\boldsymbol{G}$-measurable function h such that

$$m_{\boldsymbol{G}}^G(F) = \int_F h \, dm_{\boldsymbol{G}}$$

and hence from Lemma 5.3.1

$$m_{\boldsymbol{G}}^G(F) = \int_F h \, dm$$

and hence

$$\frac{m(F \cap G)}{m(G)} = \int_F h \, dm, F \in \boldsymbol{G}$$

and hence $m(G)h$ is the desired conditional probability measure $m(G \mid \boldsymbol{G})$. □

We close this section with a useful technical property: a chain rule for Radon-Nikodym derivatives.

Lemma 5.7.3. Suppose that three measures on a common space satisfy $m \ll \lambda \ll P$. Then P-a.e.

$$\frac{dm}{dP} = \frac{dm}{d\lambda}\frac{d\lambda}{dP}.$$

Proof. From (5.6.1) and (5.6.2)

$$m(F) = \int_F \frac{dm}{d\lambda} d\lambda = \int_F \left(\frac{dm}{d\lambda}\frac{d\lambda}{dP}\right) dP,$$

which defines $\frac{dm}{d\lambda}\frac{d\lambda}{dP}$ as a version of $\frac{dm}{dP}$. □

Exercises

1. Given a probability space $(\Omega, \boldsymbol{B}, m)$, suppose that $\boldsymbol{G}$ is a finite field with atoms G_i; $i = 1,2, \cdots, N$. Is it true that

$$m(F \mid \boldsymbol{G})(\omega) = \sum_{i=1}^{N} m(F \mid G_i) 1_{G_i}(\omega) \ ?$$

5.8 REGULAR CONDITIONAL PROBABILITY

Given an event F, the elementary conditional probability $m(G \mid F)$ of G given F is itself a probability measure $m(\,.\mid F)$ when considered as a function of the events G. A natural question is whether or not a similar conclusion holds in the more general case. In other words, we have defined conditional probabilities for a fixed event G and a sub-σ-field $\boldsymbol{G}$. What if we fix the sample point ω and consider the set function $m(\,.\,|\boldsymbol{G})(\omega)$; is it a probability measure on $\boldsymbol{B}$? In general it may not be, but we shall see two special cases in this section for which a version of the conditional probabilities exists such that with probability one $m(F \mid \boldsymbol{G})(\omega)$, $F \in \boldsymbol{B}$, is a probability measure. The first case considers the trivial case of a discrete sub-σ-field of an arbitrary σ-field. The second case considers the more important (and more involved) case of an arbitrary sub-σ-field of a standard space.

Given a probability space $(\Omega, \boldsymbol{B}, m)$ and a sub-σ- field $\boldsymbol{G}$, then a *regular conditional probability given* $\boldsymbol{G}$ is a function $f : \boldsymbol{B} \times \Omega \to [0,1]$ such that

$$f(F, \omega);\ F \in \boldsymbol{B} \tag{5.8.1}$$

is a probability measure for each $\omega \in \Omega$, and

$$f(F, \omega);\ \omega \in \Omega \tag{5.8.2}$$

is a version of the conditional probability of F given $\boldsymbol{G}$.

Lemma 5.8.1. Given a probability space $(\Omega, \boldsymbol{B}, m)$ and a discrete sub-σ-field $\boldsymbol{G}$ (that is, $\boldsymbol{G}$ has a finite or countable number of members), then there exists a regular conditional probability measure given $\boldsymbol{G}$.

Proof. Let $\{F_i\}$ denote a countable collection of atoms of the countable sub-σ-field $\boldsymbol{G}$. Then we can use elementary probability to write a version of the conditional probability for each event G:

$$m(G \mid \boldsymbol{G})(\omega) = P(G \mid F_i) = \frac{m(G \cap F_i)}{m(F_i)} \; ; \omega \in F_i$$

for those F_i with nonzero probability. If ω is in a zero probability atom, then the conditional probability can be set to $p^*(G)$ for some arbitrary probability measure p^*. If we now fix ω, it is easily seen that the given conditional probability measure is regular since for each atom F_i in the countable sub-σ-field the elementary conditional probability $P(\,.\mid F_i)$ is a probability measure. □

Theorem 5.8.1. Given a probability space $(\Omega,\boldsymbol{B},m)$ and a sub-σ-field $\boldsymbol{G}$, then if $(\Omega,\boldsymbol{B})$ is standard there exists a regular conditional probability measure given $\boldsymbol{G}$.

Proof. For each event G let $m(G \mid \boldsymbol{G})(\omega)$, $\omega \in \Omega$, be a version of the conditional probability of G given $\boldsymbol{G}$. We have that

$$\int_F m(G \mid \boldsymbol{G}) \, dm = m(G \cap F) \geq 0 \text{ , all } F \in \boldsymbol{G} \text{ ,}$$

and hence from Corollary 5.3.1 $m(G \mid \boldsymbol{G}) \geq 0$ a.e. Thus for each event G there is an event, say $H(G)$, with $m(H(G)) = 1$ such that if $\omega \in H(G)$, then $m(G \mid \boldsymbol{G})(\omega) \geq 0$. In addition, $H(G) \in \boldsymbol{G}$ since the indicator function of $H(G)$ is 1 on the set of ω on which a $\boldsymbol{G}$-measurable function is nonnegative, and hence this indicator function must also be $\boldsymbol{G}$ measurable. Since $(\Omega,\boldsymbol{B})$ is standard, it has a basis $\{\boldsymbol{F}_n\}$ which generates $\boldsymbol{B}$ as in Chapter 2. Let $\boldsymbol{F}$ be the countable union of all of the sets in the basis. Define the set

$$H_0 = \bigcap_{G \in \boldsymbol{F}} H(G)$$

and hence $H_0 \in \boldsymbol{G}$, $m(H_0) = 1$, and if $\omega \in H_0$, then $m(G \mid \boldsymbol{G})(\omega) \geq 0$ for all $G \in \boldsymbol{F}$. We also have that

$$\int_F m(\Omega|\boldsymbol{G}) \, dm = m(\Omega \cap F) = m(F) = \int_F 1 \, dm \text{ , all } F \in \boldsymbol{G} \text{ ,}$$

and hence $m(\Omega|\boldsymbol{G}) = 1$ a.e. Let H_1 be the set $\{\omega\colon m(\Omega|\boldsymbol{G})(\omega) = 1\}$ and hence $H_1 \in \boldsymbol{G}$ and $m(H_1) = 1$. Similarly, let $F_1, F_2, \ldots, F_n$ be any finite collection of disjoint sets in $\boldsymbol{F}$. Then

$$\int_F m(\bigcup_{i=1}^{n} F_i|\boldsymbol{G})\,dm = m((\bigcup_{i=1}^{n} F_i) \cap F) = m(\bigcup_{i=1}^{n} F_i \cap F)$$

$$= \sum_{i=1}^{n} m(F_i \cap F) = \sum_{i=1}^{n} \int_F m(F_i|\boldsymbol{G})\,dm = \int_F (\sum_{i=1}^{n} m(F_i|\boldsymbol{G}))\,dm\,.$$

Hence

$$m(\bigcup_{i=1}^{n} F_i \mid \boldsymbol{G}) = \sum_{i=1}^{n} m(F_i \mid \boldsymbol{G}) \text{ a.e.}$$

Thus there is a set of probability one in $\boldsymbol{G}$ on which the preceding relation holds. Since there are a countable number of choices of finite collections of disjoint unions of elements of $\boldsymbol{F}$, we can find a set, say $H_2 \in \boldsymbol{G}$, such that if $\omega \in H_2$, then $m(\,.\mid \boldsymbol{G})(\omega)$ is finitely additive on $\boldsymbol{F}$ for all $\omega \in H_2$ and $m(H_2) = 1$. Note that we could not use this argument to demonstrate countable additivity on $\boldsymbol{F}$ since there are an uncountable number of choices of countable collections of $\boldsymbol{F}$. Although this is a barrier in the general case, it does not affect the standard space case. We now have a set $H = H_0 \cap H_1 \cap H_2 \in \boldsymbol{G}$ with probability one such that for any ω in this set, $m(\,.\mid \boldsymbol{G})(\omega)$ is a normalized, nonnegative, finitely additive set function on $\boldsymbol{F}$. We now construct the desired regular conditional probability measure.

Define $f(F,\omega) = m(F \mid \boldsymbol{G})(\omega)$ for all $\omega \in H$, $F \in \boldsymbol{F}$. Fix an arbitrary probability measure P and for all $\omega \notin H$ define $f(F,\omega) = P(F)$, all $F \in \boldsymbol{B}$. From Theorem 2.6.1 $\boldsymbol{F}$ has the countable extension property, and hence for each $\omega \in H$ we have that $f(\,.\,,\omega)$ extends to a unique probability measure on $\boldsymbol{B}$, which we also denote by $f(\,.\,,\omega)$. We now have a function f such that $f(\,.\,,\omega)$ is a probability measure for all $\omega \in \Omega$, as required. We will be done if we can show that if we fix an event G, then the function we have constructed is a version of the conditional probability for G given $\boldsymbol{B}$, that is, if we can show that it satisfies (5.7.3) and (5.7.4).

First observe that for $G \in \boldsymbol{F}$, $f(G,\omega) = m(G \mid \boldsymbol{G})(\omega)$ for $\omega \in H$, where $H \in \boldsymbol{G}$, and $f(G,\omega) = P(G)$ for $\omega \in H^c$. Thus $f(G,\omega)$ is a $\boldsymbol{G}$-measurable function of ω for fixed G in $\boldsymbol{F}$. This and the extension formula (2.1.1) imply that more generally $f(G,\omega)$ is a $\boldsymbol{G}$-measurable function of ω for any $G \in \boldsymbol{B}$, proving (5.7.3).

If $G \in \boldsymbol{F}$, then by construction $m(G \mid \boldsymbol{G})(\omega) = f(G,\omega)$ a.e. and hence (5.7.4) holds for $f(G,\,.\,)$. To prove that it holds for more general G, first observe that (5.7.4) is trivially true for those $F \in \boldsymbol{G}$ with

$m(F) = 0$. If $m(F) > 0$ we can divide both sides of the formula by $m(F)$ to obtain

$$m^F(G) = \int_F f(G,\omega) \frac{dm(\omega)}{m(F)} .$$

We know the preceding equality holds for all G in a generating field $\boldsymbol{F}$. To prove that it holds for all events $G \in \boldsymbol{B}$, first observe that the left-hand side is clearly a probability measure. We will be done if we can show that the right-hand side is also a probability measure since two probability measures agreeing on a generating field must agree on all events in the generated σ-field. The right-hand side is clearly nonnegative since $f(\,.\,,\omega)$ is a probability measure for all ω. This also implies that $f(\Omega,\omega) = 1$ for all ω, and hence the right-hand side yields 1 for $G = \Omega$. If G_k, $k = 1,2, \cdots$ is a countable sequence of disjoint sets with union G, then since the $f(G_i,\omega)$ are nonnegative and the $f(\,.\,,\omega)$ are probability measures,

$$\sum_{i=1}^{n} f(G_i,\omega) \uparrow \sum_{i=1}^{\infty} f(G_i,\omega) = f(G,\omega)$$

and hence from the monotone convergence theorem that

$$\sum_{i=1}^{\infty} \int_F f(G_i,\omega) \frac{dm(\omega)}{m(F)} = \lim_{n\to\infty} \sum_{i=1}^{n} \int_F f(G_i,\omega) \frac{dm(\omega)}{m(F)}$$

$$= \lim_{n\to\infty} \int_F \left(\sum_{i=1}^{n} f(G_i,\omega)\right) \frac{dm(\omega)}{m(F)} = \int_F \left(\lim_{n\to\infty} \sum_{i=1}^{n} f(G_i,\omega)\right) \frac{dm(\omega)}{m(F)}$$

$$= \int_F f(G,\omega) \frac{dm(\omega)}{m(F)} ,$$

which proves the countable additivity and completes the proof of the theorem. □

The existence of a regular conditional probability is one of the most useful aspects of probability measures on standard spaces.

We close this section by describing a variation on the regular conditional probability results. This form will be the most common use of the theorem.

Let $X: \Omega \to A_X$ and $Y: \Omega \to A_Y$ be two random variables defined on a probability space $(\Omega,\boldsymbol{B},m)$. Let $\sigma(Y) = Y^{-1}(\boldsymbol{B}_{A_Y})$ denote the σ-field generated by the random variable Y and for each $G \in \boldsymbol{B}$ let

$m(G \mid \sigma(Y))(\omega)$ denote a version of the conditional probability of F given $\sigma(Y)$. For the moment focus on an event of the form $G = X^{-1}(F)$ for $F \in \boldsymbol{B}_{A_X}$. This function must be measurable with respect to $\sigma(Y)$, and hence from Lemma 5.2.1 there must be a function, say g: $A_Y \to [0,1]$, such that $m(X^{-1}(F) \mid \sigma(Y))(\omega) = g(Y(\omega))$. Call this function $g(y) = P_{X|Y}(F \mid y)$. By changing variables we know from the properties of conditional probability that

$$P_{XY}(F \times D) = m(X^{-1}(F) \cap Y^{-1}(D)) = \int_{Y^{-1}(D)} m(X^{-1}(F) \mid \sigma(Y))(\omega) dm(\omega)$$

$$= \int_{Y^{-1}(D)} P_{X|Y}(F \mid Y(\omega)) dm(\omega) = \int_D P_{X|Y}(F \mid y) dP_Y(y) ,$$

where $P_Y = mY^{-1}$ is the induced distribution for Y. The preceding formulas simply translate the conditional probability statements from the original space and σ-fields to the more concrete example of distributions for random variables conditioned on other random variables.

Corollary 5.8.1. Let X: $\Omega \to A_X$ and Y: $\Omega \to A_Y$ be two random variables defined on a probability space $(\Omega, \boldsymbol{B}, m)$. Then if any of the following conditions is met, there exists a regular conditional probability measure $P_X(F \mid y)$ for X given Y:

(i) The alphabet A_X is discrete.

(ii) The alphabet A_Y is discrete.

(iii) Both of the alphabets A_X and A_Y are standard.

Proof. Using the correspondence between conditional probabilities given sub-σ-fields and random variables, two cases are immediate: If the alphabet A_Y of the conditioning random variable is discrete, then the result follows from Lemma 5.8.1. If the two alphabets are standard, then so is the product space, and the result follows in the same manner from Theorem 5.8.1. If A_X is discrete, then its σ-field is standard and we can mimic the proof of Theorem 5.8.1. Let $\boldsymbol{F}$ now denote the union of all sets in a countable basis for $\boldsymbol{B}_{A_X}$ and replace $m(F \mid \boldsymbol{G})(\omega)$ by $P_{X|Y}(F \mid y)$ and dm by dP_Y. The argument then goes through virtually unchanged. □

5.9 CONDITIONAL EXPECTATION

In the previous section we showed that given a standard probability space $(\Omega,\boldsymbol{B},m)$ and a sub-σ-field $\boldsymbol{G}$, there is a regular conditional probability measure given $\boldsymbol{G}$: $m(G \mid \boldsymbol{G})(\omega)$, $G \in \boldsymbol{B}, \omega \in \Omega$; that is, for each fixed ω $m(\,.\mid \boldsymbol{G})(\omega)$ is a probability measure, and for each fixed G, $m(G \mid \boldsymbol{G})(\omega)$ is a version of the conditional probability of G given $\boldsymbol{G}$, i.e., it satisfies (5.7.3) and (5.7.4). We can use the individual probability measures to define a conditional expectation

$$E(f \mid \boldsymbol{G})(\omega) = \int f(u)dm(u \mid \boldsymbol{G})(\omega) \tag{5.9.1}$$

if the integral exists. In this section we collect a few properties of conditional expectation and relate the preceding constructive definition to the more common and more general descriptive one.

Fix an event $F \in \boldsymbol{G}$ and let f be the indicator function of an event $G \in \boldsymbol{B}$. Integrating (5.9.1) over F using (5.7.4) and the fact that $E(1_G \mid \boldsymbol{G})(\omega) = m(G \mid \boldsymbol{G})(\omega)$ yields

$$\int_F E(f \mid \boldsymbol{G})\, dm = \int_F f dm \tag{5.9.2}$$

since the right-hand side is simply $m(G \cap F)$. From the linearity of expectation (by (5.9.1) a conditional expectation is simply an ordinary expectation with respect to a measure that is a regular conditional probability measure for a particular sample point) and integration, (5.9.2) also holds for simple functions. For a nonnegative measurable f, take the usual quantizer sequence $q_n(f) \uparrow f$ of Lemma 4.3.1, and (5.9.2) then holds since the two sides are equal for each n and since both sides converge up to the appropriate integral, the right-hand side by the definition of the integral of a nonnegative function and the left-hand side by virtue of the definition and the monotone convergence theorem. For general f, use the usual decomposition $f = f^+ - f^-$ and apply the result to each piece. The preceding relation then holds in the sense that if either integral exists, then so does the other and they are equal. Note in particular that this implies that if f is in $L^1(m)$, then $E(f \mid \boldsymbol{G})(\omega)$ must be finite m-a.e.

In a similar sequence of steps, the fact that for fixed G the conditional probability of G given $\boldsymbol{G}$ is a $\boldsymbol{G}$-measurable function implies that $E(f \mid \boldsymbol{G})(\omega)$ is a $\boldsymbol{G}$-measurable function if f is an indicator function, which implies that it is also $\boldsymbol{G}$-measurable for simple functions and hence also for limits of simple functions.

Thus we have the following properties: If $f \in L^1(m)$ and $h(\omega) = E(f \mid \boldsymbol{G})(\omega)$, then

$$h(\omega) \text{ is } \boldsymbol{G}\text{–measurable, and} \tag{5.9.3}$$

$$\int_F h\, dm = \int_F f\, dm\ , \text{ all } F \in \boldsymbol{G}\ . \tag{5.9.4}$$

Eq. (5.9.4) is often referred to as *iterated expectation* or *nested expectation* since it shows that the expectation of f can be found in two steps: first find the conditional expectation given a σ-field or class of measurements (in which case $\boldsymbol{G}$ is the induced σ-field) and then integrate the conditional expectation.

These properties parallel the defining descriptive properties of conditional probability and reduce to those properties in the case of indicator functions. The following simple lemma shows that the properties provide an alternative definition of conditional expectation that is valid more generally, that is, holds even when the alphabets are not standard. Toward this end we now give the general descriptive definition of conditional expectation: Given a probability space $(\Omega, \boldsymbol{B}, m)$, a sub-σ-field $\boldsymbol{G}$, and a measurement $f \in L^1(m)$, then any function h is said to be a *version of the conditional expectation* $E(f \mid \boldsymbol{G})$. From Corollary 5.3.1, any two versions must be equal a.e. If the underlying space is standard, then the descriptive definition therefore is consistent with the constructive definition. It only remains to show that the conditional expectation exists in the general case, that is, that we can always find a version of the conditional expectation even if the underlying space is not standard. To do this observe that if f is a simple function $\sum_i a_i 1_{G_i}$, then

$$h = \sum_i a_i m(G_i \mid \boldsymbol{G})$$

is $\boldsymbol{G}$-measurable and satisfies (5.9.4) from the linearity of integration and the properties of conditional expectation:

$$\int_F (\sum_i a_i m(G_i \mid \boldsymbol{G}))\, dm = \sum_i a_i \int_F m(G_i \mid \boldsymbol{G})$$

$$= \sum_i a_i \int_F 1_{G_i}\, dm = \int_F (\sum_i a_i 1_{G_i})\, dm = \int_F f\, dm\ .$$

We then proceed in the usual way. If $f \geq 0$, let q_n be the asymptotically accurate quantizer sequence of Lemma 4.3.1. Then $q_n(f) \uparrow f$ and hence

$E(q_n(f) \mid \boldsymbol{G})$ is nondecreasing (Lemma 5.3.1), and $h = \lim_{n\to\infty} E(q_n(f) \mid \boldsymbol{G})$ satisfies (5.9.3) since limits of $\boldsymbol{G}$-measurable functions are measurable. The dominated convergence theorem implies that h also satisfies (5.9.4). The result for general integrable f follows from the decomposition $f = f^+ - f^-$. We have therefore proved the following result.

Lemma 5.9.1. Given a probability space $(\Omega,\boldsymbol{B},m)$, a sub-$\sigma$- field $\boldsymbol{G}$, and a measurement $f \in L^1(m)$, then there exists a $\boldsymbol{G}$-measurable real-valued function h satisfying the formula

$$\int_F h\, dm = \int_F f\, dm \text{ , all } F \in \boldsymbol{G} \text{ ;} \tag{5.9.5}$$

that is, there exists a version of the conditional expectation $E(f \mid \boldsymbol{G})$ of f given $\boldsymbol{G}$. If the underlying space is standard, then also (5.9.1) holds a.e.; that is, the constructive and descriptive definitions are equivalent on standard spaces.

The next result shows that the remark preceding (5.9.2) holds even if the space is not standard.

Corollary 5.9.1. Given a probability space $(\Omega,\boldsymbol{B},m)$ (not necessarily standard), a sub-σ-field $\boldsymbol{G}$, and an event $G \in \boldsymbol{G}$, then with probability one

$$m(G \mid \boldsymbol{G}) = E(1_G \mid \boldsymbol{G}) \text{ .}$$

Proof. From the descriptive definition of conditional expectation

$$\int_F E(1_G \mid \boldsymbol{G})\, dm = \int_F 1_G\, dm = m(G \cap F) \text{ , for all } F \in \boldsymbol{G} \text{ ,}$$

for any $G \in \boldsymbol{B}$. But, by (5.7.4), this is just $\int_F m(G \mid \boldsymbol{G})\, dm$ for all $F \in \boldsymbol{G}$. Since both $E(1_G \mid \boldsymbol{G})$ and $m(G \mid \boldsymbol{G})$ are $\boldsymbol{G}$-measurable, Corollary 5.3.1 completes the proof. □

Corollary 5.9.2. If $f \in L^1(m)$ is $\boldsymbol{G}$-measurable, then $E(f \mid \boldsymbol{G}) = f$ m-a.e..

Proof. The proof follows immediately from (5.9.4) and Corollary 5.3.1. □

If a conditional expectation is defined on a standard space using the constructive definitions, then it inherits the properties of an ordinary expectation; e.g., Lemma 4.4.2 can be immediately extended to conditional expectations. The following lemma shows that Lemma 4.4.2 extends to conditional expectation in the more general case of the descriptive definition.

Lemma 5.9.2. Let $(\Omega,\boldsymbol{B},m)$ be a probability space, $\boldsymbol{G}$ a sub-σ-field, and f and g integrable measurements.

(a) If $f \geq 0$ with probability 1, then $E_m(f \mid \boldsymbol{G}) \geq 0$.

(b) $E_m(1 \mid \boldsymbol{G}) = 1$.

(c) Conditional expectation is linear; that is, for any real α, β and any measurements f and g,

$$E_m(\alpha f + \beta g \mid \boldsymbol{G}) = \alpha E_m(f \mid \boldsymbol{G}) + \beta E_m(g \mid \boldsymbol{G}) .$$

(d) $E_m(f \mid \boldsymbol{G})$ exists and is finite if and only if $E_m(\,|f|\mid \boldsymbol{G})$ is finite and

$$|E_m f| \leq E_m |f| .$$

(e) Given two measurements f and g for which $f \geq g$ with probability 1, then

$$E_m(f \mid \boldsymbol{G}) \geq E_m(g \mid \boldsymbol{G}) .$$

Proof.

(a) If $f \geq 0$ and

$$\int_F E(f \mid \boldsymbol{G})dm = \int_F f\, dm \text{ , all } F \in \boldsymbol{G} \text{ ,}$$

then Lemma 4.4.2 implies that the right-hand side is nonnegative for all $F \in \boldsymbol{G}$. Since $E(f \mid \boldsymbol{G})$ is $\boldsymbol{G}$-measurable, Corollary 5.3.1 then implies that $E(f \mid \boldsymbol{G}) \geq 0$.

(b) For $f = 1$, the function $h = 1$ is both $\boldsymbol{G}$-measurable and satisfies

$$\int_F h dm = \int_F f\, dm = m(F) \text{ , all } F \in \boldsymbol{G} \text{ .}$$

Since (5.9.3)-(5.9.4) are satisfied with $h = 1$, 1 must be a version of $E(1 \mid \boldsymbol{G})$.

(c) Let $E(f \mid \boldsymbol{G})$ and $E(g \mid \boldsymbol{G})$ be versions of the conditional expectations of f and g given $\boldsymbol{G}$. Then $h = \alpha E(f \mid \boldsymbol{G}) + \beta E(g \mid \boldsymbol{G})$ is $\boldsymbol{G}$-measurable and satisfies (5.9.4) with f replaced by $\alpha f + \beta g$. Hence h is a version of $E(\alpha f + \beta g \mid \boldsymbol{G})$.

(d) Let $f = f^+ - f^-$ be the usual decomposition into positive and negative parts $f^+ \geq 0$, $f^- \geq 0$. From part (c), $E(f \mid \boldsymbol{G}) = E(f^+ \mid \boldsymbol{G}) - E(f^- \mid \boldsymbol{G})$. From part (a), $E(f^+ \mid \boldsymbol{G}) \geq 0$ and $E(f^- \mid \boldsymbol{G}) \geq 0$. Thus again using part (c)

$$E(f \mid \boldsymbol{G}) \leq E(f^+ \mid \boldsymbol{G}) + E(f^- \mid \boldsymbol{G})$$
$$= E(f^+ + f^- \mid \boldsymbol{G}) = E(\,|f|\mid \boldsymbol{G}) \,.$$

(e) The proof follows from parts (a) and (c) by replacing f by $f - g$. □

The following result shows that if a measurement f is square-integrable, then its conditional expectation $E(f \mid \boldsymbol{G})$ has a special interpretation. It is the projection of the measurement onto the space of all square-integrable $\boldsymbol{G}$-measurable functions.

Lemma 5.9.3. Given a probability space $(\Omega, \boldsymbol{B}, m)$, a measurement $f \in L^2(m)$, and a sub-σ-field $\boldsymbol{G} \subset \boldsymbol{B}$. Let M denote the subset of $L^2(m)$ consisting of all square-integrable $\boldsymbol{G}$-measurable functions. Then M is a closed subspace of $L^2(m)$ and

$$E(f \mid \boldsymbol{G}) = P_M(f) \,.$$

Proof. Since sums and limits of $\boldsymbol{G}$-measurable functions are $\boldsymbol{G}$-measurable, M is a closed subspace of $L^2(m)$. From the projection theorem (Theorem 5.5.1) there exists a function $\hat{f} = P_M(f)$ with the properties that $\hat{f} \in M$ and $f - \hat{f} \perp M$. This fact implies that

$$\int f g \, dm = \int \hat{f} g \, dm \tag{5.9.5}$$

for all $g \in M$. Choosing $g = 1_G$ for any $G \in \boldsymbol{G}$ yields

$$\int_G f \, dm = \int \hat{f} \, dm \,, \text{ all } G \in \boldsymbol{G} \,.$$

Since $\hat{f}$ is $\boldsymbol{F}$-measurable, this defines $\hat{f}$ as a version of $E(f \mid \boldsymbol{G})$ and proves the lemma. □

Eq. (5.9.5) and the lemma immediately yield a generalization of the nesting property of conditional expectation.

Corollary 5.9.3. If g is $\boldsymbol{G}$-measurable and f, $g \in L^2(m)$, then

$$E(fg) = E(g\, E(f \mid \boldsymbol{G})) .$$

The next result provides some continuity properties of conditional expectation.

Lemma 5.9.4. For $i = 1,2$, if $f \in L^i(m)$, then

$$\| E(f \mid \boldsymbol{G}) \|_i \leq \| f \|_i \;, \; i = 1,2,$$

and thus if $f \in L^i(m)$, then also $E(f \mid \boldsymbol{G}) \in L^1(m)$. If f, $g \in L^i(m)$, then

$$\| E(f \mid \boldsymbol{G}) - E(g \mid \boldsymbol{G}) \|_i \leq \| f - g \|_i \; .$$

Thus for $i = 1,2$, if $f_n \to f$ in $L^i(m)$, then also $E(f_n \mid \boldsymbol{G}) \to E(f \mid \boldsymbol{G})$.

Proof. The first inequality implies the second by replacing f with $f - g$ and using linearity. For $i = 1$ from the usual decomposition $f = f^+ - f^-$, $f^+ \geq 0$, $f^- \geq 0$, the linearity of conditional expectation, and Lemma 5.9.2(c) that

$$\| E(f \mid \boldsymbol{G}) \|_i = \int | E(f \mid \boldsymbol{G}) | \, dm = \int | E(f^+ - f^- \mid \boldsymbol{G}) | \, dm$$
$$= \int | E(f^+ \mid \boldsymbol{G}) - E(f^- \mid \boldsymbol{G}) | \, dm .$$

From Lemma 5.9.2(a), however, $E(f^+ \mid \boldsymbol{G}) \geq 0$ and $E(f^- \mid \boldsymbol{G}) \geq 0$ a.e., and hence the right-hand side is bound above by

$$\int (E(f^+ \mid \boldsymbol{G}) + E(f^- \mid \boldsymbol{G})) \, dm = \int E(f^+ + f^- \mid \boldsymbol{G}) \, dm$$
$$= \int E(\, |f| \mid \boldsymbol{G}) \, dm = E(\, |f| \,) = \| f \|_1 \; .$$

For $i = 2$ observe that

$$0 \leq E[(f - E(f \mid \boldsymbol{G}))^2]$$
$$= E(f^2) + E[\, E(f \mid \boldsymbol{G})^2 \,] - 2E[\, f E(f \mid \boldsymbol{G})] \; .$$

Apply iterated expectation and Corollary 5.9.3 to write

$$E[\, f E(f \mid \boldsymbol{G})] = E(E[\, f E(f \mid \boldsymbol{G})] \mid \boldsymbol{G})$$
$$= E(E(f \mid \boldsymbol{G}) E(f \mid \boldsymbol{G})) = E[\, E(f \mid \boldsymbol{G})^2 \,] \; .$$

Combining these two equations yields

$$E(f^2) - E[\, E(f \mid \boldsymbol{G})^2 \,] \geq 0$$

or

$$\| f \|_2^2 \geq \| E(f \mid \mathbf{G}) \|_2^2 ,$$

completing the proof. □

A special case where the conditional results simplify occurs when the measurement and the conditioning are independent in a sense we next make precise. Given a probability space $(\Omega,\mathbf{B},m)$, two events F and G are said to be *independent* if $m(F \cap G) = m(F)m(G)$. Two measurements $f: \Omega \to A_f$ and $g: \Omega \to A_g$ are said to be *independent* if the events $f^{-1}(F)$ and $g^{-1}(G)$ are independent for all events F and G; that is,

$$m(f \in F \text{ and } g \in G) = m(f^{-1}(F) \cap g^{-1}(G)) = m(f^{-1}(F))m(g^{-1}(G))$$

$$m(f \in F)m(g \in G) \text{ ; all } F \in \mathbf{B}_{A_f} \text{ ; } G \in \mathbf{B}_{A_g} .$$

Exercises

1. Prove the following corollary for indicator functions and simple functions and then prove the general result.

Corollary 5.9.4. If f and g are independent measurements, then

$$E(f \mid \sigma(g)) = E(f) ,$$

and

$$E(fg) = E(f)E(g) .$$

Thus, for example, if f and g are independent, then $m(f \in F \mid g) = m(f \in F)$ a.e.

2. Prove the following generalization of Corollary 5.9.3.

Lemma 5.9.5. Suppose that $\mathbf{G}_1 \subset \mathbf{G}_2$ are two sub-σ-fields of $\mathbf{B}$ and that $f \in L^1(m)$. Prove that

$$E(E(f \mid \mathbf{G}_2) \mid \mathbf{G}_1) = E(f \mid \mathbf{G}_1) .$$

3. Given a probability space $(\Omega,\mathbf{B},m)$ and a measurement f, for what sub-σ-field $\mathbf{G} \subset \mathbf{B}$ is it true that $E(f \mid \mathbf{G}) = E(f)$ a.e.? For what sub-σ-field $\mathbf{H} \subset \mathbf{B}$ is it true that $E(f \mid \mathbf{H}) = f$ a.e.?

4. Suppose that $\boldsymbol{G}_n$ is an increasing sequence of sub-σ-fields: $\boldsymbol{G}_n \subset \boldsymbol{G}_{n+1}$ and f is an integrable measurement. Define the random variables

$$X_n = E(f \mid \boldsymbol{G}_n) . \qquad (5.9.6)$$

Prove that X_n has the properties that it is measurable with respect to $\boldsymbol{G}_n$ and

$$E(X_n \mid X_1, X_2, \cdots, X_{n-1}) = X_{n-1} ; \qquad (5.9.7)$$

that is, the conditional expectation of the next value given all the past values is just the most recent value. A sequence of random variables with this property is called a *martingale* and there is an extremely rich theory regarding the convergence and uniform integrability properties of such sequences. (See, e.g., the classic reference of Doob [3] or the treatments in Ash [1] or Breiman [2].)

We point out in passing the general form of martingales, but we will not develop the properties in any detail. A *martingale* consists of a sequence of random variables $\{X_n\}$ defined on a common probability space $(\Omega, \boldsymbol{B}, m)$ together with an increasing sequence of sub-σ-fields $\boldsymbol{B}_n$ with the following properties: for $n = 1,2, \cdots$

$$E \mid X_n \mid < \infty , \qquad (5.9.8)$$

$$X_n \text{ is } \boldsymbol{B}_n\text{–measurable} , \qquad (5.9.9)$$

and

$$E(X_{n+1} \mid \boldsymbol{B}_n) = X_n . \qquad (5.9.10)$$

$\{X_n, \boldsymbol{B}_n\}$ is instead a *submartingale* if the third relation is replaced by $E(X_{n+1} \mid \boldsymbol{B}_n) \geq X_n$ and a *supermartingale* if $E(X_{n+1} \mid \boldsymbol{B}_n) \leq X_n$. Martingales have their roots in gambling theory, in which a martingale can be considered as a fair game–the capital expected next time is given by the current capital. A submartingale represents an unfavorable game (as in a casino), and a supermartingale a favorable game (as has been the stock market over the past decade).

5. Show that $\{X_n, \boldsymbol{B}_n\}$ is a martingale if and only if for all n

$$\int_G X_{n+1} \, dm = \int_G X_n \, dm , \text{ all } G \in \boldsymbol{B}_n .$$

5.10 INDEPENDENCE AND MARKOV CHAINS

In this section we apply some of the results and interpretations of the previous sections to obtain formulas for probabilities involving random variables that have a particular relationship called the *Markov chain* property.

Let $(\Omega,\boldsymbol{B},P)$ be a probability space and let $X: \Omega \rightarrow A_X$, $Y: \Omega \rightarrow A_Y$, and $Z: \Omega \rightarrow A_Z$ be three random variables defined on this space with σ-fields $\boldsymbol{B}_{A_X}$, $\boldsymbol{B}_{A_Y}$, and $\boldsymbol{B}_{A_Z}$, respectively. As we wish to focus on the three random variables rather than on the original space, we can consider $(\Omega,\boldsymbol{B},P)$ to be the space $(A_X \times A_Y \times A_Z, \boldsymbol{B}_{A_X} \times \boldsymbol{B}_{A_Y} \times \boldsymbol{B}_{A_Z}, P_{XYZ})$.

As a preliminary, let P_{XY} denote the joint distribution of the random variables X and Y; that is,

$$P_{XY}(F \times G) = P(X^{-1}(F) \cap Y^{-1}(G)) \ ; \ F \in \boldsymbol{B}_X \ , \ G \in \boldsymbol{B}_Y \ .$$

Translating the definition of independence of measurements of Section 5.9 into distributions the random variables X and Y are independent if and only if

$$P_{XY}(F \times G) = P_X(F) P_Y(F) \ ; \ F \in \boldsymbol{B}_X \ , \ G \in \boldsymbol{B}_Y \ ,$$

where P_X and P_Y are the marginal distributions of X and Y. We can also state this in terms of joint distributions in a convenient way. Given two distributions P_X and P_Y, define the *product distribution* $P_X \times P_Y$ or $P_{X \times Y}$ (both notations are used) as the distribution on $(A_X \times A_Y, \boldsymbol{B}_{A_X} \times \boldsymbol{B}_{A_Y})$ specified by

$$P_{X \times Y}(F \times G) = P_X(F) P_Y(G) \ ; \ F \in \boldsymbol{B}_X \ , \ G \in \boldsymbol{B}_Y \ .$$

Thus the product distribution has the same marginals as the original distribution, but it forces the random variables to be independent.

Lemma 5.10.1. Random variables X and Y are independent if and only if $P_{XY} = P_{X \times Y}$. Given measurements $f: A_X \rightarrow A_f$ and $g: A_Y \rightarrow A_g$, then if X and Y are independent, so are $f(X)$ and $g(Y)$ and

$$\int f \, g dP_{X \times Y} = (\int f \, dP_X)(\int g dP_Y) \ .$$

Proof. The first statement follows immediately from the definitions of independence and product distributions. The second statement follows from the fact that

$$P_{f(X),g(Y)}(F,G) = P_{X,Y}(f^{-1}(F),g^{-1}(G))$$
$$= P_X(f^{-1}(F))P_Y(g^{-1}(G)) = P_{f(X)}(F)P_{g(Y)}(G) .$$

The final statement then follows from Corollary 5.9.4. □

We next introduce a third random variable and a form of conditional independence among random variables. As considered in Section 5.8, the conditional probability $P(F \mid \sigma(Z))$ can be written as $g(Z(\omega))$ for some function $g(z)$ and we define this function to be $P(F \mid z)$. Similarly the conditional probability $P(F \mid \sigma(Z,Y))$ can be written as $r(Z(\omega),Y(\omega))$ for some function $r(z,y)$ and we denote this function $P(F|z,y)$. We can then define the conditional distributions $P_{X|Z}(F|z) = P(X^{-1}(F) \mid z)$, $P_{Y|Z}(G|z) = P(Y^{-1}(G) \mid z)$, $P_{XY|Z}(F \times G \mid z) = P(X^{-1}(F) \cap Y^{-1}(G) \mid z)$, and $P_{X|Z,Y}(F|z,y) = P(X^{-1}(F) \mid z,y)$. We say that $Y \to Z \to X$ is a *Markov chain* if for any event $F \in \boldsymbol{B}_{A_X}$ with probability one

$$P_{X|Z,Y}(F \mid z,y) = P_{X|Z}(F \mid z) ,$$

or in terms of the original space, with probability one

$$P(X^{-1}(F) \mid \sigma(Z,Y)) = P(X^{-1}(F) \mid \sigma(Z)) .$$

The following lemma provides an equivalent form of this property in terms of a conditional independence relation.

Lemma 5.10.2. The random variables Y,Z,X form a Markov chain $Y \to Z \to X$ if and only if for any $F \in \boldsymbol{B}_{A_X}$ and $G \in \boldsymbol{B}_{A_Y}$ we know that with probability one

$$P_{XY|Z}(F \times G \mid z) = P_{X|Z}(F \mid z)P_{Y|Z}(G \mid z) ,$$

or, equivalently, with probability one

$$P(X^{-1}(F) \cap Y^{-1}(G) \mid \sigma(Z)) = P(X^{-1}(F) \mid \sigma(Z))P(Y^{-1}(G) \mid \sigma(Z)) .$$

If the random variables have the preceding property, we also say that X and Y are conditionally independent given Z.

Proof. For any $F \in \mathbf{B}_{A_X}$, $G \in \mathbf{B}_{A_Y}$, and $D \in \mathbf{B}_{A_Z}$ we know from the properties of conditional probability that

$$P(X^{-1}(F) \cap Y^{-1}(G) \cap Z^{-1}(D)) = \int_{Y^{-1}(G) \cap Z^{-1}(D)} P(X^{-1}(F) | \sigma(Z,Y)) dP$$

$$= \int_{Y^{-1}(G) \cap Z^{-1}(D)} P(X^{-1}(F) | \sigma(Z)) dP$$

using the definition of the Markov property. Using first (5.9.4) and then Corollary 5.9.1 this integral can be written as

$$\int 1_{Z^{-1}(D)} 1_{Y^{-1}(G)} P(X^{-1}(F) \mid \sigma(Z)) dP$$

$$= \int E[1_{Z^{-1}(D)} 1_{Y^{-1}(G)} P(X^{-1}(F) \mid \sigma(Z)) \mid \sigma(Z)] dP$$

$$= \int 1_{Z^{-1}(D)} E[1_{Y^{-1}(G)} \mid \sigma(Z)] P(X^{-1}(F) \mid \sigma(Z)) dP ,$$

where we have used the fact that both $1_{Z^{-1}(D)}$ and $P(X^{-1}(F) \mid \sigma(Z))$ are measurable with respect to $\sigma(Z)$ and hence come out of the conditional expectation. From Corollary 5.9.1, however,

$$E[1_{Y^{-1}(G)} \mid \sigma(Z)] = P(Y^{-1}(G) \mid \sigma(Z)) ,$$

and hence combining the preceding relations yields

$$P(X^{-1}F \cap Y^{-1}G \cap Z^{-1}D) = \int_{Z^{-1}(D)} P(Y^{-1}(G) \mid \sigma(Z)) P(X^{-1}(F) \mid \sigma(Z)) dP .$$

We also know from the properties of conditional probability that

$$P(X^{-1}F \cap Y^{-1}G \cap Z^{-1}D) = \int_{Z^{-1}(D)} P(Y^{-1}(G) \cap X^{-1}(F) \mid \sigma(Z)) dP$$

and hence

$$P(X^{-1}F \cap Y^{-1}G \cap Z^{-1}D) = \int_{Z^{-1}(D)} P(Y^{-1}(G) \mid \sigma(Z)) P(X^{-1}(F) \mid \sigma(Z)) dP$$

$$= \int_{Z^{-1}(D)} P(Y^{-1}(G) \cap X^{-1}(F) \mid \sigma(Z)) dP . \qquad (5.10.1)$$

Since the integrands in the two right-most integrals are measurable with respect to $\sigma(Z)$, it follows from Corollary 5.3.1 that they are equal almost everywhere, proving the lemma. □

As a simple example, if the random variables are discrete with probability mass function p_{XYZ}, then Y,Z,X form a Markov chain if and only if for all x,y,z

$$p_{X|Y,Z}(x|y,z) = p_{X|Z}(x|y)$$

or, equivalently,

$$p_{X,Y|Z}(x,y|z) = p_{X|Z}(x|z)p_{Y|Z}(y|z) .$$

We close this section and chapter with an alternative statement of the previous lemma. The result is an immediate consequence of the lemma and the definitions.

Corollary 5.10.1. Given the conditions of the previous lemma, let P_{XYZ} denote the distribution of the three random variables X, Y, and Z. Let $P_{X\times Y|Z}$ denote the distribution (if it exists) for the same random variables specified by the formula

$$P_{X\times Y|Z}(F\times G\times D) = \int_{Z^{-1}(D)} P(Y^{-1}(G) \mid \sigma(Z))P(X^{-1}(F) \mid \sigma(Z))dP$$

$$= \int_D P_{X|Z}(F \mid z)P_{Y|Z}(G \mid z)dP_Z(z) \ , \quad F \in \boldsymbol{B}_{A_X} ; \; G \in \boldsymbol{B}_{A_Y} ; \; D \in \boldsymbol{B}_{A_Z} ;$$

that is, $P_{X\times Y|Z}$ is the distribution on X,Y,Z which agrees with the original distribution P_{XYZ} on the conditional distributions of X given Z and Y given Z and with the marginal distribution Z, but which is such that $Y\rightarrow Z\rightarrow X$ forms a Markov chain or, equivalently, such that X and Y are conditionally independent given Z. Then $Y\rightarrow Z\rightarrow X$ forms a Markov chain if and only if $P_{XYZ} = P_{X\times Y|Z}$.

Comment: The corollary contains the technical qualification that the given distribution exists because it need not in general, that is, there is no guarantee that the given set function is indeed countably additive and hence a probability measure. If the alphabets are assumed to be standard, however, then the conditional distributions are regular and it follows that the set function $P_{X\times Y|Z}$ is indeed a distribution.

Exercises

1. Suppose that Y_n, $n = 1,2, \cdots$ is a sequence of independent random variables with the same marginal distributions $P_{Y_n} = P_Y$. Define the sum

$$X_n = \sum_{i=1}^{n} Y_i \ .$$

Show that the sequence X_n is a martingale as defined in problem 5.9.4. This points out that martingales can be viewed as one type of memory in a random process, here formed simply by adding up memoryless random variables.

REFERENCES

1. R. B. Ash, *Real Analysis and Probability,* Academic Press, New York, 1972.
2. L. Breiman, *Probability,* Addison-Wesley, Menlo Park, Calif., 1968.
3. J. L. Doob, *Stochastic Processes,* Wiley, New York, 1953.
4. P. R. Halmos, *Introduction to Hilbert Space,* Chelsea, New York, 1957.
5. D. G. Luenberger, *Optimization by Vector Space Methods,* Wiley, New York, 1969.

6

ERGODIC PROPERTIES

6.1 ERGODIC PROPERTIES OF DYNAMICAL SYSTEMS

In this chapter we formally define ergodic properties as the existence of limiting sample averages, and we study the implications of such properties. We shall see that if sample averages converge for a sufficiently large class of measurements, e.g., the indicator functions of all events, then the random process must have a property called *asymptotic mean stationarity* and that there is a stationary measure, called the *stationary mean* of the process, that has the same sample averages. In addition, it will be seen that the limiting sample averages can be interpreted as conditional probabilities or conditional expectations and that under certain conditions convergence of sample averages implies convergence of the corresponding expectations to a single expectation with respect to the stationary mean. Finally we shall define ergodicity of a process and show that it is a necessary condition for limiting sample averages to be constants instead of random variables.

Although we have seen that there are several forms of convergence of random variables and hence we could consider several types of ergodic properties, we focus on almost everywhere convergence and later consider implications for other forms of convergence. The primary goal of this chapter is the development of necessary conditions for a random process to possess ergodic properties and an understanding of the implications of such properties. In the next chapter sufficient conditions are developed.

As in the previous chapter we focus on time and ensemble averages of measurements made on a dynamical system. Let $(\Omega,\boldsymbol{B},m,T)$ be a dynamical system. The case of principal interest will be the dynamical system corresponding to a random process, that is, a dynamical system $(A^{\boldsymbol{I}},B(A)^{\boldsymbol{I}},m,T)$, where $\boldsymbol{I}$ is an index set, usually either the nonnegative integers or the set of all integers, and T is a shift on the sequence space $A^{\boldsymbol{I}}$. The sequence of coordinate random variables $\{\Pi_n, n \in \boldsymbol{I}\}$ is the corresponding random process and m is its distribution. The dynamical system may or may not be that associated with a directly given alphabet A random process $\{X_n; n \in \boldsymbol{I}\}$ described by a distribution m and having T be the shift on the sequence space $A^{\boldsymbol{I}}$.

Dynamical systems corresponding to a directly given random process will be the most common example, but we shall be interested in others as well. For example, the space may be the same, but we may wish to consider block or variable-length shifts instead of the unit time shift.

As previously, given a measurement f (i.e., a measurable mapping of Ω into $\mathbf{R}$) define the sample averages

$$<f>_n = n^{-1} \sum_{i=0}^{n-1} f(T^i x) .$$

A dynamical system will be said to have the *ergodic property with respect to a measurement* f if the sample average $<f>_n$ converges almost everywhere as $n \to \infty$. A dynamical system will be said to have *ergodic properties* with respect to a class $\boldsymbol{M}$ of measurements if the sample averages $<f>_n$ converge almost everywhere as $n \to \infty$ for all f in the given class $\boldsymbol{M}$. When unclear from context, the measure m will be explicitly given. We shall also refer to mean ergodic properties (of order i) when the convergence is in L^i.

If $\lim_{n\to\infty} <f>_n(x)$ exists, we will usually denote it by $<f>(x)$ or $\hat{f}(x)$. For convenience we consider $<f>(x)$ to be 0 if the limit does not exist. From Corollary 4.6.2 such limits are unique only in an almost everywhere sense; that is, if $<f>_n \to <f>$ and $<f>_n \to g$ a.e., then $<f> = g$ a.e. Interesting classes of functions that we will consider include the class of all indicator functions of events, the class of all bounded measurements, and the class of all integrable measurements.

The properties of limits immediately yield properties of limiting sample averages that resemble those of probabilistic averages.

Lemma 6.1.1. Given a dynamical system, let $\boldsymbol{E}$ denote the class of measurements with respect to which the system has ergodic properties.

(a) If $f \in \boldsymbol{E}$ and $fT^i \geq 0$ a.e. for all i, then also $<f> \geq 0$ a.e. (The condition is met, for example, if $f(\omega) \geq 0$ for all $\omega \in \Omega$.)

(b) The constant functions $f(\omega) = r$, $r \in \mathbf{R}$, are in $\boldsymbol{E}$. In particular, $<1> = 1$.

(c) If $f,g \in \boldsymbol{E}$ then also that $<af+bg> = a<f> + b<g>$ and hence $af + bg \in \boldsymbol{E}$ for any real a,b. Thus $\boldsymbol{E}$ is a linear space.

(d) If $f \in \boldsymbol{E}$, then also $fT^i \in \boldsymbol{E}$, $i = 1,2,\ldots$ and $<fT^i> = <f>$; that is, if a system has ergodic properties with respect to a measurement, then it also has ergodic properties with respect to all shifts of the measurement and the limit is the same.

Proof. Parts (a)-(c) follow from Lemma 4.5.2. To prove (d) we need only consider the case $i = 1$ since that implies the result for all i. If $f \in \boldsymbol{E}$ then with probability one

$$\lim_{n\to\infty} \frac{1}{n} \sum_{i=0}^{n-1} f(T^i x) = <f>(x)$$

or, equivalently,

$$|<f>_n - <f>| \underset{n\to\infty}{\to} 0 .$$

The triangle inequality shows that $<fT>_n$ must have the same limit:

$$|<fT>_n - <f>| = |(\frac{n+1}{n})<f>_{n+1} - n^{-1}f - <f>| \leq$$

$$|(\frac{n+1}{n})<f>_{n+1} - <f>_{n+1}| + |<f>_{n+1} - <f>| + n^{-1}|f| \leq$$

$$\frac{1}{n}|<f>_{n+1}| + |<f>_{n+1} - <f>| + \frac{1}{n}|f| \underset{n\to\infty}{\to} 0$$

where the middle term goes to zero by assumption, implying that the first term must also go to zero, and the right-most term goes to zero since f cannot assume ∞ as a value. □

The preceding properties provide some interesting comparisons between the space $\boldsymbol{E}$ of all measurements on a dynamical system $(\Omega,\boldsymbol{F},m,T)$ with limiting sample averages and the space L^1 of all measurements with expectations with respect to the same measure. Properties (b) and (c) of limiting sample averages are shared by expectations. Property (a) is similar to the property of expectation that $f \geq 0$ a.e. implies that $Ef \geq 0$, but here we must require that $fT^i \geq 0$ a.e. for all nonnegative integers i. Property (d) is not, however, shared by expectations in general since integrability of f does not in general imply integrability of fT^i. There is one case, however, where the property is shared: if the measure is invariant, then from Lemma 4.7.1 f is integrable if and only if fT^i is, and, if the integrals exist, they are equal. In addition, if the measure is stationary, then

$$m(\{\omega: f(T^n\omega) \geq 0\}) = m(T^{-n}\{\omega: f(\omega) \geq 0\}) = m(f \geq 0)$$

and hence in the stationary case the condition for (a) to hold is equivalent to the condition that $f \geq 0$ a.e.. Thus for stationary measures, there is an exact parallel among the properties (a)-(d) of measurements in $\boldsymbol{E}$ and those in L^1. These similarities between time and ensemble averages will be useful for both the stationary and nonstationary systems.

The proof of the lemma yields a property of limiting sample averages that will be extremely important. We formalize this fact as a corollary after we give a needed definition: Given a dynamical system $(\Omega,\boldsymbol{B},m,T)$ a function $f: \Omega \rightarrow \Lambda$ is said to be *invariant* (with respect to T) or *stationary* (with respect to T) if $f(T\omega) = f(\omega)$, all $\omega \in \Omega$.

Corollary 6.1.1. If a dynamical system has ergodic properties with respect to a measurement f, then the limiting sample average $<f>$ is an invariant function, that is, $<f>(Tx) = <f>(x)$.

Proof. If $\lim_{n\rightarrow\infty} <f>_n(x) = <f>$, then from the lemma $\lim_{n\rightarrow\infty} <f>_n(Tx)$ also exists and is the same, but the limiting time average of $fT(x) = f(Tx)$ is just $<fT>(x)$, proving the corollary. □

The corollary simply formalizes the observation that shifting x once cannot change a limiting sample average, hence such averages must be invariant.

Ideally one would like to consider the most general possible measurements at the outset, e.g., integrable or square-integrable measurements. We often, however, initially confine attention to bounded measurements as a compromise between generality and simplicity. The class of bounded measurements is simple because the measurements and their sample averages are always integrable and uniformly integrable and the class of measurements considered does not depend on the underlying measure. All of the limiting results of probability theory immediately and easily apply. This permits us to develop properties that are true of a given class of measurements independent of the actual measure or measures under consideration. Eventually, however, we will wish to demonstrate for particular measures that sample averages converge for more general measurements.

Although a somewhat limited class, the class of bounded measurements contains some important examples. In particular, it contains the indicator functions of all events, and hence a system with ergodic properties with respect to all bounded measurements will have limiting values for all relative frequencies, that is, limits of $<1_F>_n$ as $n \to \infty$. The quantizer sequence of Lemma 4.3.1 can be used to prove a form of converse to this statement: if a system possesses ergodic properties with respect to the indicator functions of events, then it also possesses ergodic properties with respect to all bounded measurable functions.

Lemma 6.1.2. A dynamical system has ergodic properties with respect to the class of all bounded measurements if and only if it has ergodic properties with respect to the class of all indicator functions of events, i.e., all measurable indicator functions.

Proof. Since indicator functions are a subclass of all bounded functions, one implication is immediate. The converse follows by a standard approximation argument, which we present for completeness. Assume that a system has ergodic properties with respect to all indicator functions of events. Let $\boldsymbol{E}$ be the class of all measurements with respect to which the system has ergodic properties. Then $\boldsymbol{E}$ contains all indicator functions of events, and, from the previous lemma, it is linear. Hence it also contains all measurable simple functions. The general limiting sample averages can then be constructed from those of the simple functions in the same manner that we used to define ensemble averages, that is, for nonnegative functions form the increasing limit of the averages of simple functions and for general functions form the nonnegative and negative parts.

Assume that f is a bounded measurable function with, say, $|f| \leq K$. Let q_l denote the quantizer sequence of Lemma 4.3.1 and assume that $l > K$ and hence $|q_l(f(\omega)) - f(\omega)| \leq 2^{-l}$ for all l and ω by construction. This also implies that $|q_l(f(T^i\omega)) - f(T^i\omega)| \leq 2^{-l}$ for all i,l,ω. It is the fact that this bound is uniform over i (because f is bounded) that makes the following construction work. First assume that f is nonnegative. Since $q_l(f)$ is a simple function, it has a limiting sample average, say $<q_l(f)>$, by assumption. We can take these limiting sample averages to be increasing in l since $j \geq l$ implies that

$$< q_j(f) >_n - < q_l(f) >_n = < q_j(f) - q_l(f) >_n$$

$$= n^{-1} \sum_{i=0}^{n-1} (q_j(fT^i) - q_l(fT^i)) \geq 0 ,$$

since each term in the sum is nonnegative by construction of the quantizer sequence. Hence from Lemma 6.1.1 we must have that

$$< q_j(f) > - < q_l(f) > = < q_j(f) - q_l(f) > \geq 0$$

and hence $< q_l(f) >(\omega) = < q_l(f(\omega)) >$ is a nondecreasing function and hence must have a limit, say $\hat{f}$. Furthermore, since f and hence $q_l(f)$ and hence $< q_l(f) >_n$ and hence $< q_l(f) >$ are all in $[0,K]$, so must $\hat{f}$ be. We now have from the triangle inequality that for any l

$$| < f >_n - \hat{f} | \leq$$

$$| < f >_n - < q_l(f) >_n | + | < q_l(f) >_n - < q_l(f) > | + | < q_l(f) > - \hat{f} | .$$

The first term on the right-hand side is bound above by 2^{-l}. Since we have assumed that $<q_l(f)>_n$ converges with probability one to $<q_l(f)>$, then with probability one we get an ω such that $< q_l(f) >_n \to_{n \to \infty} <q_l(f)>$ for all $l = 1,2, \cdots$ (The intersection of a countable number of sets with probability one also has probability one.) Assume that there is an ω in this set. Then given ε we can choose an l so large that $| < q_l(f) > - \hat{f} | \leq \varepsilon/2$ and $2^{-l} \leq \varepsilon/2$, in which case the preceding inequality yields

$$\limsup_{n \to \infty} | < f >_n - \hat{f} | \leq \varepsilon .$$

Since this is true for arbitrary ε, for the given ω it follows that $< f >_n(\omega) \to \hat{f}(\omega)$, proving the almost everywhere result for nonnegative f. The result for general bounded f is obtained by applying the nonnegative result to f^+ and f^- in the decomposition $f = f^+ - f^-$ and using the fact that for the given quantizer sequence

$$q_l(f) = q_l(f^+) - q_l(f^-) . \ \Box$$

6.2 SOME IMPLICATIONS OF ERGODIC PROPERTIES

To begin developing the implications of ergodic properties, suppose that a system possesses ergodic properties for a class of bounded measurements. Then corresponding to each bounded measurement f in the class there will be a limit function, say $<f>$, such that with probability one $<f>_n \to <f>$ as $n \to \infty$. Convergence almost everywhere and the boundedness of f imply L^1 convergence, and hence for any event G

$$| n^{-1} \sum_{i=0}^{n-1} \int_G f T^i \, dm - \int_G <f> dm | = | \int_G (n^{-1} \sum_{i=0}^{n-1} f T^i - <f>) \, dm |$$

$$\leq \int_G | n^{-1} \sum_{i=0}^{n-1} f T^i - <f> | \, dm \leq \| <f>_n - <f> \|_1 \underset{n \to \infty}{\to} 0 .$$

Thus we have proved the following lemma:

Lemma 6.2.1. If a dynamical system with measure m has the ergodic property with respect to a bounded measurement f, then

$$\lim_{n \to \infty} n^{-1} \sum_{i=0}^{n-1} \int_G f T^i \, dm = \int_G <f> dm \text{ , all } G \in \boldsymbol{B} . \qquad (6.2.1)$$

Thus, for example, if we take G as the whole space

$$\lim_{n \to \infty} n^{-1} \sum_{i=0}^{n-1} E_m(f T^i) = E_m <f> . \qquad (6.2.2)$$

Suppose that a random process has ergodic properties with respect to the class of all bounded measurements and let f be an indicator function of an event F. Application of (6.2.1) yields

$$\lim_{n \to \infty} n^{-1} \sum_{i=0}^{n-1} m(T^{-i}F \cap G) = \lim_{n \to \infty} n^{-1} \sum_{i=0}^{n-1} \int_G 1_F T^i \, dm$$

$$= E_m(<1_F> 1_G) \text{ , all } F,G \in \boldsymbol{B} . \qquad (6.2.3)$$

For example, if G is the entire space than (6.2.3) becomes

$$\lim_{n \to \infty} n^{-1} \sum_{i=0}^{n-1} m(T^{-i}F) = E_m(<1_F>) \text{ , all events } F . \qquad (6.2.4)$$

The limit properties of the preceding lemma are trivial implications of the definition of ergodic properties, yet they will be seen to have far-reaching consequences. We note the following generalization for later use.

Corollary 6.2.1. If a dynamical system with measure m has the ergodic property with respect to a measurement f (not necessarily bounded) and if the sequence $<f>_n$ is uniformly integrable, then (6.2.1) and hence also (6.2.2) hold.

Proof. The proof follows immediately from Lemma 4.6.6 since this ensures the needed L^1 convergence. □

Recall from Corollary 4.4.6 that if the fT^n are uniformly integrable, then so are the $<f>_n$. Thus, for example, if m is stationary and f is m-integrable, then Lemma 4.7.2 and the preceding lemma imply that (6.2.1) and (6.2.2) hold, even if f is not bounded.

In general, the preceding lemmas state that if the arithmetic means of a sequence of "nice" measurements converges in any sense, then the corresponding arithmetic means of the expected values of the measurements must also converge. In particular, they imply that the arithmetic mean of the measures mT^{-i} must converge, a fact that we emphasize as a corollary:

Corollary 6.2.2. If a dynamical system with measure m has ergodic properties with respect to the class of all indicator functions of events, then

$$\lim_{n \to \infty} \frac{1}{n} \sum_{i=0}^{n-1} m(T^{-i}F) \text{ exists, all events } F. \tag{6.2.5}$$

Recall that a measure m is stationary or invariant with respect to T if all of these measures are identical. Clearly (6.2.5) holds for stationary measures. More generally, say that (6.2.5) holds for a measure m and transformation T. Define the set functions m_n, $n = 1,2,\ldots$ by

$$m_n(F) = n^{-1} \sum_{i=0}^{n-1} m(T^{-i}F).$$

The m_n are easily seen to be probability measures, and hence (6.2.5) implies that there is a sequence of measures m_n such that for every event

F the sequence $m_n(F)$ has a limit. If the limit of the arithmetic mean of the transformed measures mT^{-i} exists, we shall denote it by $\overline{m}$. The following lemma shows that if a sequence of measures m_n is such that $m_n(F)$ converges for *every* event F, then the limiting set function is itself a measure. Thus if (6.2.5) holds and we define the limit as $\overline{m}$, then $\overline{m}$ is itself a probability measure.

Lemma 6.2.2. (The Vitali-Hahn-Saks Theorem)

Given a sequence of probability measures m_n and a set function m such that

$$\lim_{n \to \infty} m_n(F) = m(F) \text{ , all events } F \text{ ,}$$

then m is also a probability measure.

Proof. It is easy to see that m is nonnegative, normalized, and finitely additive. Hence we need only show that it is countably additive. Observe that it would not help here for the space to be standard as that would only guarantee that m defined on a field would have a countably additive extension, not that the given m that is defined on all events would itself be countably additive. To prove that m is countably additive we assume the contrary and show that this leads to a contradiction.

Assume that m is not countably additive and hence from (1.2.11) there must exist a sequence of disjoint sets F_j with union F such that

$$\sum_{j=1}^{\infty} m(F_j) = b < a = m(F) \text{ .} \tag{6.2.6}$$

We shall construct a sequence of integers $M(k)$, $k = 1,2,...$, a corresponding grouping of the F_j into sets B_k as

$$B_k = \bigcup_{j=M(k-1)+1}^{M(k)} F_j, \; k = 1,2, \cdots , \tag{6.2.7}$$

a set B defined as the union of all of the B_k for odd k:

$$B = B_1 \cup B_3 \cup B_5 \cup \cdots , \tag{6.2.8}$$

and a sequence of integers $N(k)$, $k = 1,2,3, \cdots$, such that $m_{N(k)}(B)$ oscillates between values of approximately b when k is odd and approximately a when k is even. This will imply that $m_n(B)$ cannot converge and hence will yield the desired contradiction.

Fix $\varepsilon > 0$ much smaller than $a - b$. Fix $M(0) = 0$.

Step 1: Choose $M(1)$ so that

$$b - \sum_{j=1}^{M} m(F_j) < \frac{\varepsilon}{8} \quad \text{if } M \geq M(1)\,. \tag{6.2.9}$$

and then choose $N(1)$ so that

$$\left| \sum_{j=1}^{M(1)} m_n(F_j) - b \right| = | m_n(B_1) - b | < \frac{\varepsilon}{4}\,,\ n \geq N(1)\,, \tag{6.2.10}$$

and

$$| m_n(F\ ) - a| \leq \frac{\varepsilon}{4}\,,\ n \geq N(1)\,. \tag{6.2.11}$$

Thus for large n, $m_n(B_1)$ is approximately b and $m_n(F\ )$ is approximately a.

Step 2: We know from (6.2.10)-(6.2.11) that for $n \geq N(1)$

$$| m_n(F - B_1) - (a - b) | \leq | m_n(F\ ) - a | + | m_n(B_1) - b | \leq \frac{\varepsilon}{2}\,,$$

and hence we can choose $M(2)$ sufficiently large to ensure that

$$\left| m_{N(1)}\Big(\bigcup_{j=M(1)+1}^{M(2)} F_j\Big) - (a - b) \right| = | m_{N(1)}(B_2) - (a - b) |$$

$$\leq \frac{3\varepsilon}{4}\,. \tag{6.2.12}$$

From (6.2.9)

$$b - \sum_{j=1}^{M(2)} m(F_j) = b - m(B_1 \cup B_2) < \frac{\varepsilon}{8}\,,$$

and hence analogous to (6.2.10) we can choose $N(2)$ large enough to ensure that

$$| b - m_n(B_1 \cup B_2) | < \frac{\varepsilon}{4}\,,\ n \geq N(2)\,. \tag{6.2.13}$$

Observe that $m_{N(1)}$ assigns probability roughly b to B_1 and roughly $(a - b)$ to B_2, leaving only a tiny amount for the remainder of F. $m_{N(2)}$, however, assigns roughly b to B_1 and to the union $B_1 \cup B_2$ and hence assigns almost nothing to B_2. Thus the compensation term $a - b$ must be made up at higher indexed B_i's.

Step 3: From (6.2.13) of Step 2 and (6.2.11)

$$| m_n(F - \bigcup_{i=1}^{2} B_i) - (a - b) | \le | m_n(F) - a| + | m_n(\bigcup_{i=1}^{2} B_i) - b |$$

$$\le \frac{\varepsilon}{4} + \frac{\varepsilon}{4} = \frac{\varepsilon}{2}, \ n \ge N(2).$$

Hence as in (6.2.12) we can choose $M(3)$ sufficiently large to ensure that

$$| m_{N(2)}(\bigcup_{j=M(2)+1}^{M(3)} F_j) - (a - b) | = | m_{N(2)}(B_3) - (a - b) | \le \frac{3\varepsilon}{4}. \tag{6.2.14}$$

Analogous to (6.2.13) using (6.2.9) we can choose $N(3)$ large enough to ensure that

$$| b - m_n(\bigcup_{i=1}^{3} B_i) | \le \frac{\varepsilon}{4}, \ n \ge N(3). \tag{6.2.15}$$

.
.
.

Step k: From step $k - 1$ for sufficiently large $M(k)$

$$| m_{N(k-1)}(\bigcup_{j=M(k-1)+1}^{M(k)} F_j) - (a - b) |$$

$$= | m_{N(k-1)}(B_k) - (a - b) | \le \frac{3\varepsilon}{4} \tag{6.2.16}$$

and for sufficiently large $N(k)$

$$| b - m_n(\bigcup_{i=1}^{k} B_i) | \le \frac{\varepsilon}{4}, \ n \ge N(k). \tag{6.2.17}$$

Intuitively, we have constructed a sequence $m_{N(k)}$ of probability measures such that $m_{N(k)}$ always puts a probability of approximately b on the union of the first k B_i and a probability of about $a - b$ on B_{k+1}. In fact, most of the probability weight of the union of the first k B_i lies in B_1 from (6.2.9). Together this gives a probability of about a and hence yields the probability of the entire set F, thus $m_{N(k)}$ puts nearly the total probability of F on the first B_1 and on B_{k+1}. Observe then that $m_{N(k)}$ keeps pushing the required difference $a - b$ to higher indexed sets B_{k+1} as k grows. If we now define the set B of (6.2.8) as the union of all of the odd B_j, then the probability of B will include b from B_1 and it will include the $a - b$ from B_{k+1} if and only if $k + 1$ is odd, that is, if k is even. The remaining sets will contribute a negligible amount. Thus we will have that $m_{N(k)}(B)$ is about $b + (a - b) = a$ for k even and b for k odd. We complete the proof by making this idea precise. If k is even, then from (6.2.10) and (6.2.16)

$$m_{N(k)}(B) = \sum_{\text{odd } j} m_{N(k)}(B_j) \geq \tag{6.2.18}$$

$$m_{N(k)}(B_1) + m_{N(k)}(B_{k+1}) \geq b - \frac{\varepsilon}{4} + (a - b) - \frac{3\varepsilon}{4} = a - \varepsilon \text{ , } k \text{ even .}$$

For odd k the compensation term $(a - b)$ is not present and we have from (6.2.11) and (6.2.16)

$$m_{N(k)}(B) = \sum_{\text{odd } j} m_{N(k)}(B_j) \leq m_{N(k)}(F) - m_{N(k)}(B_{k+1}) \leq \tag{6.2.19}$$

$$a + \frac{\varepsilon}{4} - ((a - b) - \frac{3\varepsilon}{4}) = b + \varepsilon \text{ .}$$

Eqs. (6.2.18)-(6.2.19) demonstrate that $m_{N(k)}(B)$ indeed oscillates as claimed and hence cannot converge, yielding the desired contradiction and completing the proof of the lemma. □

Say that a probability measure m and transformation T are such that the arithmetic means of the iterates of m with respect to T converge to a set function $\overline{m}$; that is,

$$\lim_{n \to \infty} n^{-1} \sum_{i=0}^{n-1} m(T^{-i} F) = \overline{m}(F) \text{ , all events } F \text{ .} \tag{6.2.20}$$

From the previous lemma, $\overline{m}$ is a probability measure. Furthermore, it follows immediately from (6.2.20) that $\overline{m}$ is stationary with respect to T, that is,

$$\overline{m}(T^{-1} F) = \overline{m}(F) \text{ , all events } F \text{ .}$$

Since $\overline{m}$ is the limiting arithmetic mean of the iterates of a measure m and since it is stationary, a probability measure m for which (6.2.5) holds, that is, for which all of the limits of (6.2.20) exist, is said to be *asymptotically mean stationary* or *AMS*. The limiting probability measure $\overline{m}$ is called the *stationary mean of m*.

Since we have shown that a random process possessing ergodic properties with respect to the class of all indicator functions of events is AMS we have proved the following result.

Theorem 6.2.1. A necessary condition for a random process to possess ergodic properties with respect to the class of all indicator functions of events (and hence with respect to any larger class such as the class of all bounded measurements) is that it be asymptotically mean stationary.

We shall see in the next chapter that the preceding condition is also sufficient.

Exercises

1. A dynamical system $(\Omega, \boldsymbol{B}, m, T)$ is N-stationary for a positive integer N if it is stationary with respect to T^N, that is, if for all F $m(T^{-i}F)$ is periodic with period N. A system is said to be *block stationary* if it is N-stationary for some N. Show that an N-stationary process is AMS with stationary mean

$$\overline{m}(F) = \frac{1}{N}\sum_{i=0}^{N-1} m(T^{-i}F) .$$

Show in this case that $\overline{m} \gg m$.

6.3 ASYMPTOTICALLY MEAN STATIONARY PROCESSES

In this section we continue to develop the implications of ergodic properties. The emphasis here is on a class of measurements and events that are intimately connected to ergodic properties and will play a basic role in further necessary conditions for such properties as well as in the demonstration of sufficient conditions. These measurements help develop the relation between an AMS measure m and its stationary mean $\overline{m}$. The theory of AMS processes follows the work of Dowker [1, 2], Rechard [8], and Gray and Kieffer [3].

As before, we have a dynamical system $(\Omega, \boldsymbol{B}, m, T)$ with ergodic properties with respect to a measurement, f and hence with probability one the sample average

$$\lim_{n \to \infty} <f>_n = \lim_{n \to \infty} \frac{1}{n} \sum_{i=0}^{n-1} f T^i \tag{6.3.1}$$

exists. From Corollary 6.1.1, the limit, say $<f>$, must be an invariant function. Define the event

$$F = \{x: \lim_{n \to \infty} n^{-1} \sum_{i=0}^{n-1} f(T^i x) \text{ exists}\}, \tag{6.3.2}$$

and observe that by assumption $m(F) = 1$. Consider next the set $T^{-1}F$. Then $x \in T^{-1}F$ or $Tx \in F$ if and only if the limit

$$\lim_{n \to \infty} n^{-1} \sum_{i=0}^{n-1} f(T^i(Tx)) = <f>(Tx)$$

exists. Provided that f is a real-valued measurement and hence cannot take on infinity as a value (that is, it is not an extended real-valued measurement), then as in Lemma 6.1.1 the left-hand limit is exactly the same as the limit

$$\lim_{n \to \infty} n^{-1} \sum_{i=0}^{n-1} f(T^i x).$$

In particular, either limit exists if and only if the other one does and hence $x \in T^{-1}F$ if and only if $x \in F$. We formalize this property in a definition: If an event F is such that $T^{-1}F = F$, then the event is said to be *invariant* or *T-invariant* or *invariant with respect to T*.

Thus we have seen that the event that limiting sample averages converge is an invariant event and that the limiting sample averages themselves are invariant functions. Observe that we can write for all x that

$$<f>(x) = 1_F(x) \lim_{n \to \infty} \frac{1}{n} \sum_{n=0}^{n-1} f(T^i x).$$

We shall see that invariant events and invariant measurements are intimately connected. When T is the shift, invariant measurements can be interpreted as measurements whose output does not depend on when the measurement is taken and invariant events are events that are unaffected by shifting.

Limiting sample averages are not the only invariant measurements; other examples are

$$\limsup_{n \to \infty} < f >_n(x) ,$$

and

$$\liminf_{n \to \infty} < f >_n(x) .$$

Observe that an event is invariant if and only if its indicator function is an invariant measurement. Observe also that for any event F, the event

$$F_{\text{i.o.}} = \limsup_{n \to \infty} T^{-n}F = \bigcap_{n=0}^{\infty} \bigcup_{k=n}^{\infty} T^{-k}F$$

$$= \{x: x \in T^{-n}F \text{ i.o.}\} = \{x: T^n x \in F \text{ i.o.}\} \tag{6.3.3}$$

is invariant, where *i.o.* means "infinitely often." To see this, observe that

$$T^{-1}F_{\text{i.o.}} = \bigcap_{n=1}^{\infty} \bigcup_{k=n}^{\infty} T^{-k}F$$

and

$$\bigcup_{k=n}^{\infty} T^{-k}F \downarrow F_{\text{i.o.}} .$$

Invariant events are "closed" to transforming, e.g., shifting: If a point begins inside (outside) an invariant event, it must remain inside (outside) that event for all future transformations or shifts. Invariant measurements are measurements that yield the same value on $T^n x$ for any n. Invariant measurements and events yield special relationships between an AMS random process and its asymptotic mean, as described in the following lemma.

Lemma 6.3.1. Given an AMS system with measure m with stationary mean $\overline{m}$, then

$$m(F) = \overline{m}(F) \text{ , all invariant } F \text{ ,}$$

and

$$E_m f = E_{\overline{m}} f \text{ , all invariant } f \text{ ,}$$

where the preceding equation means that if either integral exists, then so does the other and they are equal.

Proof. The first equality follows by direct substitution into the formula defining an AMS measure. The second formula then follows from standard approximation arguments, i.e., since it holds for indicator functions, it holds for simple functions by linearity. For positive functions take an increasing quantized sequence and apply the simple function result. The two limits must converge or diverge together. For general f, decompose it into positive and negative parts and apply the positive function result. □

In some applications we deal with measurements or events that are "almost invariant" in the sense that they equal their shift with probability one instead of being invariant in the strict sense. As there are two measures to consider here, an AMS measure and its stationary mean, a little more care is required to conclude results like Lemma 6.3.1. For example, suppose that an event F is invariant $\overline{m}$-a.e. in the sense that $\overline{m}(F \Delta T^{-1} F) = 0$. It does not then follow that F is also invariant m-a.e. or that $m(F) = \overline{m}(F)$. The following lemma shows that if one strengthens the definitions of almost invariant to include all shifts, then m and $\overline{m}$ will behave similarly for invariant measurements and events.

Lemma 6.3.2. An event F is said to be *totally invariant* m-a.e. if

$$m(F \Delta T^{-k} F) = 0 \, , \; k = 1,2, \cdots \, .$$

Similarly, a measurement f is said to be *totally invariant* m-a.e. if

$$m(f = f T^k ; \, k=1,2, \cdots) = 1 \, .$$

Suppose that m is AMS with stationary mean $\overline{m}$. If F is totally invariant m-a.e. then

$$\overline{m}(F) = m(F) \, ,$$

and if f is totally invariant m-a.e.

$$E_m f = E_{\overline{m}} f \, .$$

Furthermore, if f is totally invariant m-a.e., then the event $f^{-1}(G) = \{\omega : f(\omega) \in G\}$ is also totally invariant and hence

$$m(f^{-1}(G)) = \overline{m}(f^{-1}(G)) \, .$$

Proof. It follows from the elementary inequality

$$| m(F) - m(G) | \leq m(F \Delta G)$$

that if F is totally invariant m-a.e., then $m(F) = m(T^{-k} F)$ for all positive integers k. This and the definition of $\overline{m}$ imply the first relation. The second relation then follows from the first as in Lemma 6.3.1. If m puts probability 1 on the set D of ω for which the $f(T^k\omega)$ are all equal, then for any k

$$m(f^{-1}(G) \Delta T^{-k} f^{-1}(G)) = m(\omega: \{f(\omega) \in G\} \Delta \{f(T^k\omega) \in G\})$$

$$= m(\omega: \{\omega \in D\} \cap (\{f(\omega) \in G\} \Delta \{f(T^k\omega) \in G\})) = 0$$

since if ω is in D, $f(\omega) = f(T^k\omega)$ and the given difference set is empty. □

Lemma 6.3.1 has the following important implication:

Corollary 6.3.1. Given an AMS system $(\Omega, \boldsymbol{B}, m, T)$ with stationary mean $\overline{m}$, then the system has ergodic properties with respect to a measurement f if and only if the stationary system $(\Omega, \boldsymbol{B}, \overline{m}, T)$ has ergodic properties with respect to f. Furthermore, the limiting sample average can be chosen to be the same for both systems.

Proof. The set $F = \{x: \lim_{n \to \infty} < f >_n \text{ exists}\}$ is invariant and hence $m(F) = \overline{m}(F)$ from the lemma. Thus either both measures assign probability one to this set and hence both systems have ergodic properties with respect to f or neither does. Choosing $< f >(x)$ as the given limit for $x \in F$ and 0 otherwise completes the proof. □

Another implication of the lemma is the following relation between an AMS measure and its stationary mean. We first require the definition of an asymptotic form of domination or absolute continuity. Recall that $m \ll \eta$ means that $\eta(F) = 0$ implies $m(F) = 0$. We say that a measure η *asymptotically dominates* m (with respect to T) if $\eta(F) = 0$ implies that

$$\lim_{n \to \infty} m(T^{-n} F) = 0 .$$

Roughly speaking, if η asymptotically dominates m, then zero probability events under η will have vanishing probability under m if they are shifted far into the future. This can be considered as a form of steady state behavior in that such events may have nonzero probability as transients, but eventually their probability must tend to 0 as things settle down.

Corollary 6.3.2. If m is asymptotically mean stationary with stationary mean $\overline{m}$, then $\overline{m}$ asymptotically dominates m. If T is invertible (e.g., the two-sided shift), then asymptotic domination implies ordinary domination and hence $m \ll \overline{m}$.

Proof. From the lemma m and $\overline{m}$ place the same probability on invariant sets. Suppose that F is an event such that $\overline{m}(F) = 0$. Then the set $G = \limsup_{n \to \infty} T^{-n}F$ of (6.3.3) has $\overline{m}$ measure 0 since from the continuity of probability and the union bound

$$\overline{m}(G) = \lim_{n \to \infty} \overline{m}\Big(\bigcup_{k=n}^{\infty} T^{-n}F \Big) \le \overline{m}\Big(\bigcup_{k=1}^{\infty} T^{-n}F \Big)$$

$$\le \sum_{j=1}^{\infty} \overline{m}(T^{-n}F) = 0 .$$

In addition G is invariant, and hence from the lemma $m(G) = 0$. Applying Fatou's lemma (Lemma 4.4.5(c)) to indicator functions yields

$$\limsup_{n \to \infty} m(T^{-n}F) \le m(\limsup_{n \to \infty} T^{-n}F) = 0 ,$$

proving the first statement. Suppose that η asymptotically dominates m and that T is invertible. Say $\eta(F) = 0$. Since T is invertible the set $\bigcup_{n=-\infty}^{\infty} T^{n}F$ is measurable and as in the preceding argument has $\overline{m}$ measure 0. Since the set is invariant it also has m measure 0. Since the set contains the set F, we must have $m(F) = 0$, completing the proof. □

The next lemma pins down the connection between invariant events and invariant measurements. Let **I** denote the class of all invariant events and, as in the previous chapter, let $\boldsymbol{M}$(**I**) denote the class of all measurements ($\boldsymbol{B}$-measurable functions) that are measurable with respect to **I**. The following lemma shows that $\boldsymbol{M}$(**I**) is exactly the class of all invariant measurements.

Lemma 6.3.3. Let **I** denote the class of all invariant events. Then **I** is a σ-field and $\boldsymbol{M}$(**I**) is the class of all invariant measurements. In addition, **I** $= \sigma(\boldsymbol{M}(\mathbf{I}))$ and hence **I** is the σ-field generated by the class of all invariant measurements.

Proof. That **I** is a σ-field follows immediately since inverse images (under T) preserve set theoretic operations. For example, if F and G are invariant, then $T^{-1}(F \cup G) = T^{-1}F \cup T^{-1}G = F \cup G$ and hence the union is also invariant. The class $\boldsymbol{M}$(**I**) contains all indicator functions of invariant events, hence all simple invariant functions. For example, if $f(\omega) =$

$\sum b_i 1_{F_i}(\omega)$ is a simple function in this class, then its inverse images must be invariant, and hence $f^{-1}(b_i) = F_i \in \mathbf{I}$ is an invariant event. Thus f must be an invariant simple function. Taking limits in the usual way then implies that every function in $\boldsymbol{M}(\mathbf{I})$ is invariant. Conversely, suppose that f is a measurable invariant function. Then for any Borel set G, $T^{-1} f^{-1}(G) = (fT)^{-1}(G) = f^{-1}(G)$, and hence the inverse images of all Borel sets under f are invariant and hence in $\mathbf{I}$. Thus measurable invariant function must be in $\boldsymbol{M}(\mathbf{I})$. The remainder of the lemma follows from Lemma 5.2.2. □

The sub-σ-field of invariant events plays an important role when studying limiting averages. There is another sub-σ-field that plays an important role in studying asymptotic behavior, especially when dealing with one-sided processes or noninvertible transformations. Given a dynamical system $(\Omega, \boldsymbol{B}, m, T)$ and a random process $\{X_n\}$ with $X_n(\omega) = X_0(T^n\omega)$, define the sub-σ-fields

$$\boldsymbol{F}_n = \sigma(X_n, X_{n+1}, X_{n+2}, \cdots) \tag{6.3.4}$$

and

$$\boldsymbol{F}_\infty = \bigcap_{n=1}^{\infty} \boldsymbol{F}_n . \tag{6.3.5}$$

$\boldsymbol{F}_\infty$ is called the *tail σ-field* or the *remote σ-field*; intuitively, it contains all of those events that can be determined by looking at future outcomes arbitrarily far into the future; that is, membership in a tail event cannot be determined by any of the individual X_n, only by their aggregate behavior in the distant future. If the system is a directly given one-sided random process and T is the (noninvertible) shift, then we can write

$$\boldsymbol{F}_n = T^{-n}\boldsymbol{B}$$

$$\boldsymbol{F}_\infty = \bigcap_{n=0}^{\infty} T^{-n}\boldsymbol{B} .$$

In the two-sided process case this is not true since if T is invertible $T^{-n}\boldsymbol{B} = \boldsymbol{B}$. In the one-sided case invariant events are clearly tail events since they are in all of the $T^{-n}\boldsymbol{B}$. Not all tail events are invariant, however. A measurement f is called a *tail function* if it is measurable with respect to the tail σ-field. Intuitively, a measurement f is a tail function only if we can determine its value by looking at the outputs $X_n, X_{n+1}, \cdots$ for arbitrarily large values of n.

We have seen that if T is invertible as in a two-sided directly given random process and if m is AMS with stationary mean $\overline{m}$, then $m \ll \overline{m}$, and hence any event having 0 probability under $\overline{m}$ also has 0 probability under m. In the one-sided case this is not in general true, although we have seen that it is true for invariant events. We shall shortly see that it is also true for tail events. First, however, we need a basic property of asymptotic dominance. If $m \ll \overline{m}$, the Radon-Nikodym theorem implies that $m(F) = \int_F (dm/d\overline{m})d\overline{m}$. The following lemma provides an asymptotic version of this relation for asymptotically dominated measures.

Lemma 6.3.4. Given measures m and η, let $(mT^{-n})_a$ denote the absolutely continuous part of mT^{-n} with respect to η as in the Lebesgue decomposition theorem (Theorem 5.6.2) and define the Radon-Nikodym derivative

$$f_n = \frac{d(mT^{-n})_a}{d\eta}\ ;\ n = 1,2, \cdots .$$

If η asymptotically dominates m then

$$\lim_{n\to\infty} [\sup_{F\in \boldsymbol{B}} | m(T^{-n}F) - \int_F f_n\, d\eta\, |] = 0.$$

Proof. From the Lebesgue decomposition theorem and the Radon-Nikodym theorem (Theorems 5.6.2 and 5.6.1) for each $n = 1,2, \cdots$ there exists a $B_n \in \boldsymbol{B}$ such that

$$mT^{-n}(F) = mT^{-n}(F \cap B_n) + \int_F f_n\, d\eta\ ;\ F \in \boldsymbol{B}, \qquad (6.3.6)$$

with

$$\eta(B_n) = 0 .$$

Define

$$B = \bigcup_{n=0}^{\infty} B_n$$

Then $\eta(B) = 0$, and hence by assumption

$$0 \le mT^{-n}(F) - \int_F f_n\, d\eta = mT^{-n}(F \cap B_n)$$

$$\le mT^{-n}(F \cap B) \le mT^{-n}(B) \underset{n\to\infty}{\to} 0 .$$

Since the bound is uniform over F, the lemma is proved.

Corollary 6.3.3. Suppose that η is stationary and asymptotically dominates m (e.g., m is AMS and η is the stationary mean), then if F is a tail event ($F \in \boldsymbol{F}_\infty$) and $\eta(F) = 0$, then also $m(F) = 0$.

Proof. First observe that if T is invertible, the result follows from the fact that asymptotic dominance implies dominance. Hence assume that T is not invertible and $\boldsymbol{F}_\infty = \bigcap T^{-n}\boldsymbol{B}$. Let $F \in \boldsymbol{F}_\infty$ have η probability 0. Find $\{F_n\}$ so that $T^{-n}F_n = F$, $n = 1,2,\cdots$. (This is possible since $F \in \boldsymbol{F}_\infty$ means that also $F \in T^{-n}\boldsymbol{B}$ for all n and hence there must be an $F_n \in \boldsymbol{B}$ such that $F = T^{-n}F_n$.) From the lemma, $m(F) - \int_{F_n} f_n \, d\eta \to 0$. Since η is stationary, $\int_{F_n} f_n \, d\eta = \int_F f_n T^n \, d\eta = 0$, hence $m(F) = 0$.

Corollary 6.3.4. Let μ and η be probability measures on $(\Omega,\boldsymbol{B})$ with η stationary. Then the following are equivalent:

(a) η asymptotically dominates μ.

(b) If $F \in \boldsymbol{B}$ is invariant and $\eta(F) = 0$, then also $\mu(F) = 0$.

(c) If $F \in \boldsymbol{F}_\infty$ and $\eta(F) = 0$, then also $\mu(F) = 0$.

Proof. The previous corollary shows that (a) => (c). (c) => (b) immediately from the fact that invariant events are also tail events. To prove that (b) => (a), assume that $\eta(F) = 0$. The set $G = \limsup_{n \to \infty} T^{-n}F$ of (6.3.3) is invariant and has η measure 0. Thus as in the proof of Corollary 6.3.2

$$\limsup_{n \to \infty} m(T^{-n}F) \le m(\limsup_{n \to \infty} T^{-n}F) = 0 .$$

Corollary 6.3.5. Let μ and η be probability measures on $(\Omega,\boldsymbol{B})$. If η is stationary and has ergodic properties with respect to a measurement f and if η asymptotically dominates μ, then also μ has ergodic properties with respect to f.

Proof. The set on which $\{x: \lim_{n \to \infty} < f >_n \text{ exists}\}$ is invariant and hence if its complement (which is also invariant) has measure 0 under η, it also has measure zero under m from the previous corollary.

Exercises

1. Is the event

$$\liminf_{n\to\infty} T^{-n}F = \bigcup_{n=0}^{\infty} \bigcap_{k=n}^{\infty} T^{-k}$$

$$= \{x: x \in T^{-n}F \text{ for all but a finite number of } n\}$$

invariant? Is it a tail event?

2. Show that given an event F and $\varepsilon > 0$, the event

$$\{x: |\frac{1}{n}\sum_{i=0}^{n-1} f(T^i x)| \leq \varepsilon \text{ i.o.}\}$$

is invariant.

3. Suppose that m is stationary and let G be an event for which $T^{-1}G \subset G$. Show that $m(G - T^{-1}G) = 0$ and that there is an invariant event $\underline{G}$ with $m(G\Delta\underline{G}) = 0$.
4. Is an invariant event a tail event? Is a tail event an invariant event? Is $F_{\text{i.o.}}$ of (6.3.3) a tail event?
5. Suppose that m is the distribution of a stationary process and that G is an event with nonzero probability. Show that if G is invariant, then the conditional distribution m^G defined by $m^G(F) = m(F \mid G)$ is also stationary. Show that even if G is not invariant, m^G is AMS.
6. Suppose that T is an invertible transformation (e.g., the shift in a two-sided process) and that m is AMS with respect to T. Show that $m(F) > 0$ implies that $m(T^nF) > 0$ i.o. (infinitely often).

6.4 RECURRENCE

We have seen that AMS measures are closely related to stationary measures. In particular, an AMS measure is either dominated by a stationary measure or asymptotically dominated by a stationary measure. Since the simpler and stronger nonasymptotic form of domination is much easier to work with, the question arises as to conditions under which an AMS system will be dominated by a stationary system. We have already seen that this is the case if the transformation T is invertible, e.g., the system is a two-sided random process with shift T. In this section we provide a more general set of conditions that relates to what is sometimes called the most basic ergodic property of a dynamical system, that of *recurrence*.

The original recurrence theorem goes back to Poincaré (1899) [7]. We briefly describe it as it motivates some related definitions and results.

Given a dynamical system $(\Omega,\boldsymbol{B},m,T)$, let $G \in \boldsymbol{B}$. A point $x \in G$ is said to be *recurrent with respect to* G if there is a finite $n = n_G(x)$ such that

$$T^{n_G(x)} \in G ,$$

that is, if the point in G returns to G after some finite number of shifts. An event G having positive probability is said to be *recurrent* if almost every point in G is recurrent with respect to G. In order to have a useful definition for all events (with or without positive probability), we state it in a negative fashion: Define

$$G^* = \bigcup_{n=1}^{\infty} T^{-n} G , \tag{6.4.1}$$

the set of all sequences that land in G after one or more shifts. Then

$$B = G - G^* \tag{6.4.2}$$

is the collection of all points in G that never return to G, the bad or nonrecurrent points. We now define an event G to be *recurrent* if $m(G - G^*) = 0$. The dynamical system will be called *recurrent* if every event is a recurrent event. Observe that if the system is recurrent, then for all events G

$$\begin{aligned} m(G \cap G^*) &= m(G) - m(G \cap G^{*c}) \\ &= m(G) - m(G - G^*) = m(G) . \end{aligned} \tag{6.4.3}$$

Thus if $m(G) > 0$, then the conditional probability that the point will eventually return to G is $m(G^* \mid G) = m(G \cap G^*)/m(G) = 1$.

Before stating and proving Poincaré's recurrence theorem, we make two interesting observations about the preceding sets. First observe that if $x \in B$, then we cannot have $T^k x \in G$ for any $k \geq 1$. Since B is a subset of G, this means that we cannot have $T^k x \in B$ for any $k \geq 1$, and hence the events B and $T^{-k}B$ must be pairwise disjoint for $k = 1,2, \cdots$; that is,

$$B \cap T^{-k}B = \varnothing \ ; \ k = 1,2, \cdots$$

If these sets are empty, then so must be their inverse images with respect to T^n for any positive integer n, that is,

$$\varnothing = T^{-n}(B \cap T^{-k}B) = (T^{-n}B) \cap (T^{-(n+k)}B)$$

for all integers n and k. This in turn implies that

$$T^{-k}B \cap T^{-j}B = \varnothing \ ; \ k, j = 1,2, \cdots , k \neq j \ ; \tag{6.4.4}$$

that is, the sets $T^{-n}B$ are pairwise disjoint. We shall call any set W with the property that $T^{-n}W$ are pairwise disjoint a *wandering set.* Thus (6.4.4) states that the collection $G - G^*$ of nonrecurrent sequences is a wandering set.

As a final property, observe that G^* as defined in (6.4.1) has a simple property when it is shifted:

$$T^{-1}G^* = \bigcup_{n=1}^{\infty} T^{-n-1}G = \bigcup_{n=2}^{\infty} T^{-n}G \subset G^* ,$$

and hence

$$T^{-1}G^* \subset G^* . \tag{6.4.5}$$

Thus G^* has the property that it is *shrunk* or *compressed* by the inverse transformation. Repeating this shows that

$$T^{-k}G^* = \bigcup_{n=k+1}^{\infty} T^{-n}G \subset T^{-(k+1)}G^* . \tag{6.4.6}$$

We are now prepared to state and prove what is often considered the first ergodic theorem.

Theorem 6.4.1. (The Poincaré Recurrence Theorem) If the dynamical system is stationary, then it is recurrent.

Proof. We need to show that for any event G, the event $B = G - G^*$ has 0 measure. We have seen, however, that B is a wandering set and hence its iterates are disjoint. This means from countable additivity of probability that

$$m\left(\bigcup_{n=0}^{\infty} T^{-n}B\right) = \sum_{n=0}^{\infty} m(T^{-n}B) .$$

Since the measure is stationary, all of the terms of the sum are the same. They then must all be 0 or the sum would blow up, violating the fact that the left-hand side is less than 1. □

Although such a recurrence theorem can be viewed as a primitive ergodic property, unlike the usual ergodic properties, there is no indication of the relative frequency with which the point will return to the target set. We shall shortly see that recurrence at least guarantees infinite recurrence in the sense that the point will return infinitely often, but again there is no notion of the fraction of time spent in the target set.

We wish next to explore the recurrence property for AMS systems. We shall make use of several of the ideas developed above. We begin with several definitions regarding the given dynamical system. Recall for comparison that the system is stationary if $m(T^{-1}G) = m(G)$ for all events G. The system is said to be *incompressible* if every event G for which $T^{-1}G \subset G$ has the property that $m(G - T^{-1}G) = 0$. Note the resemblance to stationarity: For all such G $m(G) = m(T^{-1}G)$. The system is said to be *conservative* if all wandering sets have 0 measure. As a final definition, we modify the definition of recurrent. Instead of the requirement that a recurrent system have the property that if an event has nonzero probability, then points in that event must return to that event with probability 1, we can strengthen the requirement to require that the points return to the event infinitely often with probability 1. Define as in (6.3.3) the set

$$G_{\text{i.o.}} = \bigcap_{n=0}^{\infty} \bigcup_{k=n}^{\infty} T^{-k}G$$

of all points that are in G infinitely often. The dynamical system will be said to be *infinitely recurrent* if for all events G

$$m(G - G_{\text{i.o.}}) = 0 \ .$$

Note from (6.4.6) that

$$\bigcup_{k=n}^{\infty} T^{-k}G = T^{-(n+1)}G^* \ ,$$

and hence since the $T^{-n}G^*$ are decreasing

$$G_{\text{i.o.}} = \bigcap_{n=0}^{\infty} T^{-(n+1)}G^* = \bigcap_{n=0}^{\infty} T^{-n}G^* \ . \qquad (6.4.7).$$

The following theorem is due to Halmos [4], and the proof largely follows Wright [9], Petersen [6], or Krengel [5]. We do not, however, assume as they do that the system is *nonsingular* in the sense that $mT^{-1} \ll m$. When this assumption is made, T is said to be *null preserving* because shifts of the measure inherit the zero probability sets of the original measure.

Theorem 6.4.2. Suppose that $(\Omega,\boldsymbol{B},m,T)$ is a dynamical system. Then the following statements are equivalent:

(a) T is infinitely recurrent.

(b) T is recurrent.

(c) T is conservative.

In addition, any of the above equivalent conditions implies

(d) T is incompressible.

Proof. (a) => (b): In words: if we must return an infinite number of times, then we must obviously return at least once. Alternatively, application of (6.4.7) implies that

$$G - G_{\text{i.o.}} = G - \bigcap_{n=0}^{\infty} T^{-n}G^* \tag{6.4.8}$$

so that $G - G^* \subset G - G_{\text{i.o.}}$ and hence the system is recurrent if it is infinitely recurrent.

(b) => (c): Suppose that W is a wandering set. From the recurrence property $W - W^*$ must have 0 measure. Since W is wandering, $W \cap W^* = \emptyset$ and hence $m(W) = m(W - W^*) = 0$ and the system is conservative.

(c) => (d): Consider an event G with $T^{-1}G \subset G$. Then $G^* = T^{-1}G$ since all the remaining terms in the union defining G^* are further subsets. Thus $G - T^{-1}G = G - G^*$. We have seen, however, that $G - G^*$ is a wandering set and hence has 0 measure since the system is conservative.

(c) => (b): The set $G - G^*$ is wandering. Since the system is conservative, the set must have 0 measure, which implies that the system is recurrent.

(b) and (d) => a: From (6.4.8) we need to prove that

$$m(G - G_{\text{i.o.}}) = m(G - \bigcap_{n=0}^{\infty} T^{-n}G^*) = 0 .$$

Toward this end we use the fact that the $T^{-n}G^*$ are decreasing to decompose G^* as

$$G^* = (\bigcap_{n=1}^{\infty} T^{-n}G^*) \bigcup_{n=0}^{\infty} (T^{-n}G^* - T^{-(n+1)}G^*) .$$

This breakup can be thought of as a "core" $\bigcap_{n=1}^{\infty} T^{-n}G^*$ containing the sequences that shift into G^* for all shifts combined with a union of an infinity of "doughnuts," where the nth doughnut contains those sequences that when shifted n times land in G^*, but never arrive in G^* if shifted more than n times. The core and all of the doughnuts are disjoint and together yield all the sequences in G^*. Because the core and doughnuts are disjoint,

$$\bigcap_{n=1}^{\infty} T^{-n}G^* = G^* - \bigcup_{n=0}^{\infty} (T^{-n}G^* - T^{-(n+1)}G^*)$$

so that

$$G - \bigcap_{n=1}^{\infty} T^{-n}G^* = (G - G^*) \bigcup (G \bigcap (\bigcup_{n=0}^{\infty} (T^{-n}G^* - T^{-(n+1)}G^*)) \quad (6.4.9)$$

Thus since the system is recurrent we know by (6.4.9) and the union bound that

$$m(G - \bigcap_{n=1}^{\infty} T^{-n}G^*) \leq m(G - G^*) + m(G \bigcap (\bigcup_{n=0}^{\infty} (T^{-n}G^* - T^{-(n+1)}G^*))$$

$$\leq m(G - G^*) + m(\bigcup_{n=0}^{\infty} (T^{-n}G^* - T^{-(n+1)}G^*))$$

$$= m(G - G^*) + \sum_{n=0}^{\infty} m(T^{-n}G^* - T^{-(n+1)}G^*)) ,$$

using the disjointness of the doughnuts. (Actually, the union bound and an inequality would suffice here.) Since $T^{-1}G^* \subset G^*$, it is also true that $T^{-1}T^{-n}G^* = T^{-n}T^{-1}G^* \subset T^{-n}G^*$ and hence since the system is incompressible, all the terms in the previous sum are zero.

We have now shown that (a) => (b) => (c) => (d), that (c) => (b), and that (b) and (d) together imply (a). This implies that (a), (b), and (c) are equivalent and that (d) is implied by any of them. (d) alone does not imply any of the remaining three. This completes the proof. □

If one assumes (as is often done in books on ergodic theory) that the measure m is nonsingular with respect to T, then all four of the conditions become equivalent. The proof is left as an exercise.

Thus, for example, Theorem 2.4.1 implies that a stationary system has all the preceding properties. Our goal, however, is to characterize the properties for an AMS system. The following theorem characterizes recurrent AMS processes: An AMS process is recurrent if and only if it is dominated by its stationary mean. The result is a slight extension of a result for nonsingular measures of Wu Chou in unpublished work.

Theorem 6.4.3. Given an AMS dynamical system $(\Omega, \boldsymbol{B}, m, T)$ with stationary mean $\overline{m}$, then $m \ll \overline{m}$ if and only the system is recurrent.

Proof. If $m \ll \overline{m}$, then the system is obviously recurrent since $\overline{m}$ is from the previous two theorems and m inherits all of the zero probability sets from $\overline{m}$. To prove the converse implication, suppose the contrary: Say that there is a set G for which $m(G) > 0$ and $\overline{m}(G) = 0$ and that the system is recurrent. From Theorem 6.4.2 this implies that the system is also infinitely recurrent, and hence

$$m(G - G_{\text{i.o.}}) = 0$$

and therefore

$$m(G) = m(G_{\text{i.o.}}) > 0 \ .$$

We have seen, however, that $G_{i.o}$ is an invariant set in Section 6.3 and hence from Lemma 6.3.1, $\overline{m}(G_{\text{i.o.}}) > 0$. Since $\overline{m}$ is stationary and gives probability 0 to G, $\overline{m}(T^{-k}G) = 0$ for all k, and hence

$$\overline{m}(G_{\text{i.o.}}) = \overline{m}(\bigcap_{n=0}^{\infty} \bigcup_{k=n}^{\infty} T^{-n}G) \leq \overline{m}(\bigcup_{k=0}^{\infty} T^{-n}G)$$

$$\leq \sum_{k=0}^{\infty} \overline{m}(G) = 0 \ ,$$

yielding a contradiction and proving the theorem. □

If the transformation T is invertible; e.g., the two-sided shift, then we have seen that $m \ll \overline{m}$ and hence the system is recurrent. When the shift is noninvertible, the theorem proves that domination by a stationary measure is equivalent to recurrence. Thus, in particular, not all AMS processes need be recurrent.

Exercises

1. Show that if an AMS dynamical system is nonsingular ($mT^{-1} \ll m$, then m and the stationary mean $\overline{m}$ will be equivalent (have the same probability 0 events) if and only if the system is recurrent.

2. Show that if m is nonsingular (null preserving) with respect to T, then incompressibility of T implies that it is conservative and hence that all four conditions of Theorem 6.4.2 are equivalent.

6.5 ASYMPTOTIC MEAN EXPECTATIONS

We have seen that the existence of ergodic properties implies that limits of sample means of expectations exist; e.g., if $<f>_n$ converges, then so does $n^{-1}\sum_{i=0}^{n-1} E_m f T^i$. In this section we relate these limits to expectations with respect to the stationary mean $\overline{m}$.

Lemma 6.2.1 showed that the arithmetic mean of the expected values of transformed measurements converged. The following lemma shows that the limit is given by the expected value of the original measurement under the stationary mean.

Lemma 6.5.1. Let m be an AMS random process with stationary mean $\overline{m}$. If a measurement f is bounded, then

$$\lim_{n \to \infty} E_m(<f>_n) = \lim_{n \to \infty} n^{-1} \sum_{i=0}^{n-1} E_m(f T^i)$$

$$= \lim_{n \to \infty} E_{m_n} f = E_{\overline{m}}(f) . \qquad (6.5.1)$$

Proof. The result is immediately true for indicator functions from the definition of AMS. It then follows for simple functions from the linearity of limits. In general let q_k denote the quantizer sequence of Lemma 4.3.1 and observe that for each k we have

$$| E_m(<f>_n) - E_{\overline{m}}(f) | \le | E_m(<f>_n) - E_m(<q_k(f)>_n) | +$$

$$| E_m(<q_k>_n) - E_{\overline{m}}(q_k(f)) | + | E_{\overline{m}}(q_k(f)) - E_{\overline{m}}(f) | .$$

Choose k larger that the maximum of the bounded function f. The middle term goes to zero from the simple function result. The rightmost term is bound above by 2^{-k} since we are integrating a function that is everywhere less than 2^{-k}. Similarly the first term on the left is bound above by

$$n^{-1} \sum_{i=0}^{n-1} E_m | f T^i - q_k(f T^i) | \leq 2^{-k} ,$$

where we have used the fact that $q_m(f)(T^i x) = q_m(f(T^i x))$. Since k can be made arbitrarily large and hence 2×2^{-k} arbitrarily small, the lemma is proved. □

Thus the stationary mean provides the limiting average expectation of the transformed measurements. Combining the previous lemma with Lemma 6.2.1 immediately yields the following corollary.

Corollary 6.5.1. Let m be an AMS measure with stationary mean $\overline{m}$. If m has ergodic properties with respect to a bounded measurement f and if the limiting sample average is $< f >$, then

$$E_m < f > = E_{\overline{m}} f . \tag{6.5.2}$$

Thus the expectation of the measurement under the stationary mean gives the expectation of the limiting sample average under the original measure.

If we also have uniform integrability, we can similarly extend Corollary 6.2.1.

Lemma 6.5.2. Let m be an AMS random process with stationary mean $\overline{m}$ that has the ergodic property with respect to a measurement f. If the $< f >_n$ are uniformly m-integrable, then f is also $\overline{m}$ integrable and (6.5.1) and (6.5.2) hold.

Proof. From Corollary 6.2.1

$$\lim_{n \to \infty} E_m(< f >_n) = E_m < f > .$$

In particular, the right-hand integral exists and is finite. From Corollary 6.1.1 the limit $< f >$ is invariant. If the system is AMS, the invariance of $< f >$ and Lemma 6.3.1 imply that

$$E_m < f > = E_{\overline{m}} < f > .$$

Since $\overline{m}$ is stationary Lemma 4.7.2 implies that the sequence $< f >_n$ is uniformly integrable with respect to $\overline{m}$. Thus from Lemma 4.6.6 and the convergence of $< f >_n$ on a set of $\overline{m}$ probability one (Corollary 6.3.1), the right-hand integral is equal to

$$\lim_{n \to \infty} E_{\overline{m}}(n^{-1} \sum_{i=0}^{n-1} f T^i) .$$

From the linearity of expectation, the stationarity of $\overline{m}$, and Lemma 4.7.1, for each value of n the preceding expectation is simply $E_{\overline{m}} f$. □

Exercises

1. Show that if m is AMS, then for any positive integer N and all $i = 0,1, \cdots, N-1$, then the following limits exist:

$$\lim_{n \to \infty} \frac{1}{n} \sum_{j=0}^{n-1} m(T^{-i-jN} F)$$

Define the preceding limit to be $\overline{m}_i$. Show that the $\overline{m}_i$ are N-stationary and that if $\overline{m}$ is the stationary mean of m, then

$$\overline{m}(F) = \frac{1}{N} \sum_{i=0}^{N-1} \overline{m}_i(F) .$$

6.6 LIMITING SAMPLE AVERAGES

In this section limiting sample averages are related to conditional expectations. The basis for this development is the result of Section 6.1 that limiting sample averages are invariant functions and hence are measurable with respect to I, the σ-field of all invariant events.

We begin with limiting relative frequencies, that is, the limiting sample averages of indicator functions. Say we have a dynamical system $(\Omega, \boldsymbol{B}, m, T)$ that has ergodic properties with respect to the class of all indicator functions and which is therefore AMS Fix an event G and let $< 1_G >$ denote the corresponding limiting relative frequency, that is $< 1_G >_n \to < 1_G >$ as $n \to \infty$ a.e. This limit is invariant and therefore in $\boldsymbol{M}$(I) and is measurable with respect to I. Fix an event $F \in$ I and observe that from Lemma 6.3.1

$$\int_F < 1_G > dm = \int_F < 1_G > d\overline{m} . \qquad (6.6.1)$$

For invariant F, 1_F is an invariant function and hence

$$< 1_F f >_n = \frac{1}{n} \sum_{i=0}^{n-1} (1_F f)T^i = 1_F \frac{1}{n} \sum_{i=0}^{n-1} f T^i = 1_F < f >_n$$

and hence if $< f >_n \rightarrow < f >$ then

$$< 1_F \, f > = 1_F < f > . \tag{6.6.2}$$

Thus from Corollary 6.5.1 we have for invariant F that

$$\int_F < 1_G > dm = E_m < 1_F \, 1_G > = E_{\overline{m}} 1_F \, 1_G$$

$$= \int_F 1_G d\overline{m} = \overline{m}(F \cap G) ,$$

which with (6.6.1) implies that

$$\int_F < 1_G > \, d\overline{m} = \overline{m}(F \cap G) \text{ , all } F \in \mathbf{I} . \tag{6.6.3}$$

Eq. (6.6.3) together with the I-measurability of the limiting sample average yield (6.6.3) and (6.6.4) and thereby show that the limiting sample average is a version of the conditional probability $\overline{m}(G \mid \mathbf{I})$ of G given the σ-field $\mathbf{I}$ or, equivalently, given the class of all invariant measurements! We formalize this in the following theorem.

Theorem 6.6.1. Given a dynamical system $(\Omega, \boldsymbol{B}, m, T)$ let $\mathbf{I}$ denote the class of all T-invariant events. If the system is AMS and if it has ergodic properties with respect to the indicator function 1_G of an event G (e.g., if the system has ergodic properties with respect to all indicator functions of events), then with probability one under m and $\overline{m}$

$$< 1_G >_n = n^{-1} \sum_{i=0}^{n-1} 1_G T^i \underset{n \rightarrow \infty}{\rightarrow} \overline{m}(G \mid \mathbf{I}) ,$$

In other words,

$$< 1_G > = \overline{m}(G \mid \mathbf{I}) . \tag{6.6.4}$$

It is important to note that the conditional probability is defined by using the stationary mean measure $\overline{m}$ and not the original probability measure m. Intuitively, this is because it is the stationary mean that describes asymptotic events.

Eq. (6.6.4) shows that the limiting sample average of the indicator function of an event, that is, the limiting relative frequency of the event, satisfies the descriptive definition of the conditional probability of the event given the σ-field of invariant events. An interpretation of this result is that if we wish to guess the probability of an event G based on the knowledge of the occurrence or nonoccurrence of all invariant events or, equivalently, on the outcomes of all invariant measurements, then our guess is given by one particular invariant measurement: the relative frequency of the event in question.

A natural question is whether Theorem 6.6.1 holds for more general functions than indicator functions; that is, if a general sample average $<f>_n$ converges, must the limit be the conditional expectation? The following lemma shows that this is true in the case of bounded measurements.

Lemma 6.6.1. Given the assumptions of Theorem 6.6.1, if the system has ergodic properties with respect to a bounded measurement f, then with probability one under m and $\overline{m}$

$$<f>_n = n^{-1} \sum_{i=0}^{n-1} f T^i \underset{n \to \infty}{\to} E_{\overline{m}}(f \mid \mathbf{I}) .$$

In other words,

$$<f> = E_{\overline{m}}(f \mid \mathbf{I}).$$

Proof. Say that the limit of the left-hand side is an invariant function $<f>$. Apply Lemma 6.5.1 to the bounded measurement $g = f 1_F$ with an invariant event F and

$$\lim_{n \to \infty} n^{-1} \sum_{i=0}^{n-1} \int_F f T^i \, dm = \int_F f \, d\overline{m} .$$

From Lemma 6.2.1 the left-hand side is also given by

$$\int_F <f> dm .$$

The equality of the two integrals for all invariant F and the invariance of the limiting sample average imply from Lemma 6.7.1 that $<f>$ is a version of the conditional expectation of f given $\mathbf{I}$.

Replacing Lemma 6.5.1 by Lemma 6.5.2 in the preceding proof gives part of the following generalization.

Corollary 6.6.1. Given a dynamical system $(\Omega, \boldsymbol{B}, m, T)$ let I denote the class of all T-invariant events. If the system is AMS with stationary mean $\overline{m}$ and if it has ergodic properties with respect to a measurement f for which either (i) the $< f >_n$ are uniformly integrable with respect to m or (ii) f is $\overline{m}$-integrable, then with probability one under m and $\overline{m}$

$$< f >_n = n^{-1} \sum_{i=0}^{n-1} f T^i \underset{n \to \infty}{\to} E_{\overline{m}}(f \mid \mathrm{I}) .$$

In other words,

$$< f > = E_{\overline{m}}(f \mid \mathrm{I}).$$

Proof. Part (i) follows from the comment before the corollary. From Corollary 7.3.1, since m has ergodic properties with respect to f, then so does $\overline{m}$ and the limiting sample average $< f >$ is the same (up to an almost everywhere equivalence under both measures). Since f is by assumption $\overline{m}$-integrable and since $\overline{m}$ is stationary, then $< f >_n$ is uniformly $\overline{m}$-integrable and hence application of (i) to $\overline{m}$ then proves that the given equality holds $\overline{m}$-a.e.. The set on which the the equality holds is invariant, however, since both $< f >$ and $E_{\overline{m}}(f \mid \mathrm{I})$ are invariant. Hence the given equality also holds m-a.e. from Lemma 6.3.1.

Exercises

1. Let m be AMS with stationary mean $\overline{m}$. Let $\boldsymbol{M}$ denote the space of all square-integrable invariant measurements. Show that $\boldsymbol{M}$ is a closed linear subspace of $L^2(\overline{m})$. If m has ergodic properties with respect to a bounded measurement f, show that

$$< f > = P_{\boldsymbol{M}}(f) ;$$

that is, the limiting sample average is the projection of the original measurement onto the subspace of invariant measurements. From the projection theorem, this means that $< f >$ is the minimum mean-squared estimate of f given the outcome of all invariant measurements.

2. Suppose that $f \in L^2(m)$ is a measurement and $f_n \in L^2(m)$ is a sequence of measurements for which $\| f_n - f \| \to_{n \to \infty} 0$. Suppose

also that m is stationary. Prove that

$$\frac{1}{n}\sum_{i=0}^{n-1} f_i T^i \underset{L^2(m)}{\rightarrow} <f>,$$

where $<f>$ is the limiting sample average of f.

6.7 ERGODICITY

A special case of ergodic properties of particular interest occurs when the limiting sample averages are constants instead of random variables, that is, when a sample average converges with probability one to a number known a priori rather than to a random variable. This will be the case, for example, if a system is such that all invariant measurements equal constants with probability one. If this is true for all measurements, then it is true for indicator functions of invariant events. Since such functions can assume only values of 0 or 1, for such a system

$$m(F) = 0 \text{ or } 1 \text{ for all } F \text{ such that } T^{-1}F = F . \tag{6.7.1}$$

Conversely, if (6.7.1) holds every indicator function of an invariant event is either 1 or 0 with probability 1, and hence every simple invariant function is equal to a constant with probability one. Since every invariant measurement is the pointwise limit of invariant simple measurements (combine Lemma 4.3.1 and Lemma 6.3.3), every invariant measurement is also a constant with probability one (the ordinary limit of the constants equaling the invariant simple functions). Thus every invariant measurement will be constant with probability one if and only if (6.7.1) holds.

Suppose next that a system possesses ergodic properties with respect to all indicator functions or, equivalently, with respect to all bounded measurements. Since the limiting sample averages of all indicator functions of invariant events (in particular) are constants with probability one, the indicator functions themselves are constant with probability one. Then, as previously, (6.7.1) follows.

Thus we have shown that if a dynamical system possesses ergodic properties with respect to a class of functions containing the indicator functions, then the limiting sample averages are constant a.e. if and only if (6.7.1) holds. This motivates the following definition: A dynamical system (or the associated random process) is said to be *ergodic with respect to a transformation* T or *T-ergodic* or, if T is understood, simply *ergodic*, if every invariant event has probability 1 or 0. Another name for ergodic that is occasionally found in the mathematical literature is *metrically transitive*.

Given the definition, we have proved the following result.

Lemma 6.7.1. A dynamical system is ergodic if and only if all invariant measurements are constants with probability one. A necessary condition for a system to have ergodic properties with limiting sample averages being a.e. constant is that the system be ergodic.

Lemma 6.7.2. If a dynamical system $(\Omega,\boldsymbol{B},m,T)$ is AMS with stationary mean $\overline{m}$, then $(\Omega,\boldsymbol{B},m,T)$ is ergodic if and only if $(\Omega,\boldsymbol{B},\overline{m},T)$ is.

Proof. The proof follows from the definition and Lemma 6.3.1. □

Coupling the definition and properties of ergodic systems with the previously developed ergodic properties of dynamical systems yields the following results.

Lemma 6.7.3. If a dynamical system has ergodic properties with respect to a bounded measurement f and if the system is ergodic, then with probability one

$$<f>_n \underset{n\to\infty}{\to} E_m<f> = \lim_{n\to\infty} n^{-1}\sum_{i=0}^{n-1} E_m f T^i .$$

If more generally the $<f>_n$ are uniformly integrable (and hence the convergence is also in $L^1(m)$), then the preceding still holds. If the system is AMS and either the $<f>_n$ are uniformly integrable with respect to m or f is $\overline{m}$-integrable, then also with probability one

$$<f>_n \underset{n\to\infty}{\to} E_{\overline{m}} f .$$

In particular, the limiting relative frequencies are given by

$$n^{-1}\sum_{i=0}^{n-1} 1_F T^i \underset{n\to\infty}{\to} \overline{m}(F) \text{ , all } F \in \boldsymbol{B} .$$

In addition,

$$\lim_{n \to \infty} n^{-1} \sum_{i=0}^{n-1} \int_G f T^i \, dm = (E_{\overline{m}} f) m(G) \text{ , all } G \in \boldsymbol{B} \text{ ,}$$

where again the result holds for measurements for which the $<f>_n$ are uniformly integrable or f is $\overline{m}$ integrable. Letting f be the indicator function of an event F, then

$$\lim_{n \to \infty} n^{-1} \sum_{i=0}^{n-1} m(T^{-i} F \cap G) = \overline{m}(F) m(G) \text{ .}$$

Proof. If the limiting sample average is constant, then it must equal its own expectation. The remaining results then follow from the results of this chapter. □

The final result of the previous lemma yields a useful test for determining whether an AMS system is ergodic.

Lemma 6.7.4. Suppose that a dynamical system $(\Omega, \boldsymbol{B}, m, T)$ has ergodic properties with respect to the indicator functions and hence is AMS with stationary mean $\overline{m}$ and suppose that $\boldsymbol{B}$ is generated by a field $\boldsymbol{F}$ (e.g., the rectangles), then m is ergodic if and only if

$$\lim_{n \to \infty} n^{-1} \sum_{i=0}^{n-1} m(T^{-i} F \cap F) = \overline{m}(F) m(F) \text{ , all } F \in \boldsymbol{F} \text{ .}$$

Proof. Necessity follows from the previous result. To show that the preceding formula is sufficient for ergodicity, first assume that the formula holds for all events F and let F be invariant. Then the left-hand side is simply $m(F) = \overline{m}(F)$ and the right-hand side is $m(F)\overline{m}(F) = \overline{m}(F)^2$. But these two can be equal only if $\overline{m}(F)$ is 0 or 1, i.e., if it (and hence also m) is ergodic. We will be done if we can show that the preceding relation holds for all events given that it holds on a generating field. Toward this end fix $\varepsilon > 0$. From Corollary 1.5.3 we can choose a field event F_0 such that $m(F \Delta F_0) \leq \varepsilon$ and $\overline{m}(F \Delta F_0) \leq \varepsilon$. To see that we can choose a field event F_0 that provides a good approximation to F simultaneously for both measures m and $\overline{m}$, apply Corollary 1.5.3 to the mixture measure $p = (m + \overline{m})/2$ to obtain an F_0 for which $p(F \Delta F_0) \leq \varepsilon/2$. This implies that both $m(F \Delta F_0)$ and $\overline{m}(F \Delta F_0))$ must be less than ε.

From the triangle inequality

$$\left| \frac{1}{n} \sum_{i=0}^{n-1} m(T^{-i} F \cap F) - \overline{m}(F) m(F) \right| \le$$

$$\left| \frac{1}{n} \sum_{i=0}^{n-1} m(T^{-i} F \cap F) - \frac{1}{n} \sum_{i=0}^{n-1} m(T^{-i} F_0 \cap F_0) \right| +$$

$$\left| \frac{1}{n} \sum_{i=0}^{n-1} m(T^{-i} F_0 \cap F_0) - \overline{m}(F_0) m(F_0) \right| + \left| \overline{m}(F_0) m(F_0) - \overline{m}(F) m(F) \right| .$$

The middle term on the right goes to 0 as $n \to \infty$ by assumption. The rightmost term is bound above by 2ϵ. The leftmost term is bound above by

$$\frac{1}{n} \sum_{i=0}^{n-1} \left| m(T^{-i} F \cap F) - m(T^{-i} F_0 \cap F_0) \right| .$$

Since for any events D, C

$$|m(D) - m(C)| \le m(D \Delta C) \tag{6.7.2}$$

each term in the preceding sum is bound above by $m((T^{-i} F \cap F) \Delta (T^{-i} F_0 \cap F_0))$. Since for any events D,C,H $m(D \Delta C) \le m(D \Delta H) + m(H \Delta C)$, each of the terms in the sum is bound above by

$$m((T^{-i} F \cap F) \Delta (T^{-i} F_0 \cap F)) + m((T^{-i} F_0 \cap F) \Delta (T^{-i} F_0 \cap F_0))$$
$$\le m(T^{-i} (F \Delta F_0)) + m(F \Delta F_0) .$$

Thus the remaining sum term is bound above by

$$\frac{1}{n} \sum_{i=0}^{n-1} m(T^{-i} (F \Delta F_0)) + m(F \Delta F_0) \underset{n \to \infty}{\to} \overline{m}(F \Delta F_0) + m(F \Delta F_0) \le 2\epsilon,$$

which completes the proof. □

Examples

We next consider a few examples of ergodic processes and systems. Suppose that we have a random process $\{X_n\}$ with distribution m and alphabet A. The process is said to be *memoryless* or *independent and identically distributed* (*IID*) if there is a measure q on $(A, \boldsymbol{B}_A)$ such that for any rectangle $G = \times_{i \in \boldsymbol{J}} G_i$, $G_i \in \boldsymbol{B}_A$, $j \in \boldsymbol{J}$, $\boldsymbol{J} \subset \boldsymbol{I}$ a finite collection of distinct indices,

$$m(\times_{j \in \boldsymbol{J}} G_j) = \prod_{j \in \boldsymbol{J}} q(G_j) \; ;$$

that is, the random variables X_n are mutually independent. Clearly an IID process has the property that if G and F are two finite-dimensional rectangles, then there is an integer M such that for $N \geq M$

$$m(G \cap T^{-n}F) = m(G)m(F) \; ;$$

that is, shifting rectangles sufficiently far apart ensures their independence. A generalization of this idea is obtained by having a process satisfy this behavior asymptotically as $n \to \infty$. A measure m is said to be *mixing* or *strongly mixing* with respect to a transformation T if for all $F,G \in \boldsymbol{B}$

$$\lim_{n \to \infty} | m(T^{-n}F \cap G) - m(T^{-n}F)m(G) | = 0 \tag{6.7.3}$$

and *weakly mixing* if for all $F,G \in \boldsymbol{B}$

$$\lim_{n \to \infty} \frac{1}{n} \sum_{i=0}^{n-1} | m(T^{-i}F \cap G) - m(T^{-i}F)m(G) | = 0 \; . \tag{6.7.4}$$

If a measure is mixing, then it is also weakly mixing. We have not proved, however, that an IID process is mixing, since the mixing conditions must be proved for all events, not just the rectangles considered in the discussion of the IID process. The following lemma shows that proving that (6.7.3) or (6.7.4) holds on a generating field is sufficient in certain cases.

Lemma 6.7.5. If a measure m is AMS and if (6.7.4) holds for all sets in a generating field, then m is weakly mixing. If a measure m is asymptotically stationary in the sense that $\lim_{n \to \infty} m(T^{-n}F)$ exists for all events F and if (6.7.3) holds for all sets in a generating field, then m is strongly mixing.

Proof. The first result is proved by using the same approximation techniques of the proof of Lemma 6.7.4, that is, approximate arbitrary events F and G, by events in the generating field for both m and its stationary mean $\bar{m}$. In the second case, as in the theory of asymptotically mean stationary processes, the Vitali-Hahn-Saks theorem implies that there must exist a measure $\overline{m}$ such that

$$\lim_{n\to\infty} m(T^{-n}F) = \overline{m}(F),$$

and hence arbitrary events can again be approximated by field events under both measures and the result follows. □

An immediate corollary of the lemma is the fact that IID processes are strongly mixing since (6.7.3) is satisfied for all rectangles and rectangles generate the entire event space.

If the measure m is AMS, then either of the preceding mixing properties implies that the condition of Lemma 6.7.4 is satisfied since

$$\begin{aligned} &| n^{-1}\sum_{i=0}^{n-1} m(T^{-i}F \cap F) - \overline{m}(F)m(F)| \\ &\le | n^{-1}\sum_{i=0}^{n-1} (m(T^{-i}F \cap F) - m(T^{-i}F)m(F)| \\ &+ | n^{-1}\sum_{i=0}^{n-1} m(T^{-i}F)m(F)) - \overline{m}(F)m(F)| \\ &\le n^{-1}\sum_{i=0}^{n-1} | m(T^{-i}F \cap F) - m(T^{-i}F)m(F)| \\ &+ m(F)\, | n^{-1}\sum_{i=0}^{n-1} m(T^{-i}F) - \overline{m}(F)| \underset{n\to\infty}{\to} 0 . \end{aligned}$$

Thus AMS weakly mixing systems are ergodic, and hence AMS strongly mixing systems and IID processes are also ergodic. A simpler proof for the case of mixing measures follows from the observation that if F is invariant, then mixing implies that $m(F \cap F) = m(F)^2$, which in turn implies that $m(F)$ must be 0 or 1 and hence that m is ergodic.

Exercises

1. Show that if m and p are two stationary and ergodic processes with ergodic properties with respect to all indicator functions, then either they are identical or they are singular in the sense that there is an event G such that $p(G) = 1$ and $m(G) = 0$.

2. Suppose that m_i are distinct stationary and ergodic sources with ergodic properties with respect to all indicator functions. Show that the mixture

$$m(F) = \sum_i \lambda_i m_i(F),$$

where

$$\sum_i \lambda_i = 1$$

is stationary but not ergodic. Show that more generally if the m_i are AMS, then so is m.

3. A random process is N-ergodic if it is ergodic with respect to T^N, that is, if $T^{-N}F = F$ for all F implies that $m(F) = 1$. Is an N-ergodic process necessarily ergodic? Is an ergodic process necessarily N-ergodic? If m is N-ergodic, is the mixture

$$\overline{m}(F) = \frac{1}{N}\sum_{i=0}^{N-1} m(T^{-i}F)$$

ergodic? Suppose that a process is N-ergodic and N-stationary and has the ergodic property with respect to f. Show that

$$E\langle f\rangle = \frac{1}{N}\sum_{i=0}^{N-1} E(fT^i).$$

4. If a process is stationary and ergodic and T is invertible, show that $m(F) > 0$ implies that

$$m(\bigcup_{i=-\infty}^{\infty} T^i F) = 1.$$

Show also that if $m(F) > 0$ and $m(G) > 0$, then for some i $m(F \cap T^i G) > 0$. Show that, in fact, $m(F \cap T^i G) > 0$ infinitely often.

5. Suppose that $(\Omega, \boldsymbol{B}, m)$ is a directly given stationary and ergodic source with shift T and that $f: \Omega \to \Lambda$ is a measurable mapping of sequences in Ω into another sequence space Λ with shift S. Show that the induced process (with distribution mf^{-1}) is also stationary and ergodic.

6. Two measures m and p are said to be *equivalent* if $m \ll p$ and $p \ll m$. Thus equivalent measures have the same probability zero sets. Show that if μ and λ are both stationary and if $\mu(F) = \lambda(F)$ for all invariant events F, then they are equivalent. Show that if μ is also ergodic, then $\mu(F) = \lambda(F)$ for all events F.

7. Show that if m and p have ergodic properties and are equivalent in the sense of the previous problem, than they have the same sample averages with probability one under both measures.

8. Show that if m is ergodic, then m and mT^{-1} are equivalent in the sense of problem 6.

9. Show that if m has ergodic properties with respect to all indicator functions and it is AMS with stationary mean $\overline{m}$, then

$$\lim_{n\to\infty} \frac{1}{n} \sum_{i=0}^{n-1} \overline{m}(T^{-i}F \cap F) = \lim_{n\to\infty} \frac{1}{n^2} \sum_{i=0}^{n-1} \sum_{j=0}^{n-1} \overline{m}(T^{-i}F \cap T^{-j}F) .$$

10. Show that a process m is stationary and ergodic if and only if for all $f, g \in L^2(m)$ we have

$$\lim_{n\to\infty} \frac{1}{n} \sum_{i=0}^{n-1} E(f\, gT^i) = (Ef)(Eg) .$$

Show that the process is AMS with stationary mean $\overline{m}$ and ergodic if and only if for all bounded measurements f and g

$$\lim_{n\to\infty} \frac{1}{n} \sum_{i=0}^{n-1} E_m(f\, gT^i) = (E_{\overline{m}} f)(E_m g) .$$

11. Suppose that m is a process and $f \in L^2(m)$. Assume that the following hold for all integers k and j:

$$E_m f T^k = E_m f$$

$$E_m\left[(f T^k)(f T^j)\right] = E_m\left[(f)(f T^{|k-j|})\right]$$

$$\lim_{k\to\infty} E_m(f f T^{|k|}) = (E_m f)^2 .$$

Prove that $<f>_n \to E_m f$ in $L^2(m)$.

12. Prove that if a stationary process m is weakly mixing, then there is a subsequence $J = \{j_1 < j_2 < \cdots\}$ of the positive integers that has density 0 in the sense that

$$\lim_{n\to\infty} \frac{1}{n} \sum_{i=0}^{n-1} 1_J(i) = 0$$

and that has the property that

$$\lim_{n\to\infty,\ n\notin J} | m(T^{-n} F \cap G) - m(F)m(G) | = 0$$

for all appropriate events F and G.

13. Are mixtures of mixing processes mixing?
14. Consider the measures m and $N^{-1}\sum_{i=0}^{N-1} mT^{-i}$. Does ergodicity of either imply that of the other?
15. Suppose that m is N-stationary and N-ergodic. Show that the measures mT^{-n} have the same properties.
16. Suppose an AMS process is not stationary but it is ergodic. Is it recurrent? Infinitely recurrent?
17. Is a mixture of recurrent processes also recurrent?

REFERENCES

1. Y. N. Dowker, "Finite and sigma-finite invariant measures," *Annals of Mathematics*, vol. 54, pp. 595-608, November 1951.
2. Y. N. Dowker, "On measurable transformations in finite measure spaces," *Annals of Mathematics*, vol. 62, pp. 504-516, November 1955.
3. R. M. Gray and J. C. Kieffer, "Asymptotically mean stationary measures," *Annals of Probability*, vol. 8, pp. 962-973, Oct. 1980.
4. P. R. Halmos, "Invariant measures," *Ann. of Math.*, vol. 48, pp. 735-754, 1947.
5. U. Krengel, *Ergodic Theorems,* De Gruyter Series in Mathematics, De Gruyter, New York, 1985.
6. K. Petersen, *Ergodic Theory,* Cambridge University Press, Cambridge, 1983.
7. H. Poincaré, *Les méthodes nouvelles de la mécanique céleste,* I, II, III, Gauthiers-Villars, Paris, 1892,1893,1899. Also Dover, New York, 1957.
8. O. W. Rechard, "Invariant measures for many-one transformations," *Duke J. Math.*, vol. 23, pp. 477-488, 1956.
9. F. B. Wright, "The recurrence theorem," *Amer. Math. Monthly*, vol. 68, pp. 247-248, 1961.

7

ERGODIC THEOREMS

7.1 INTRODUCTION

At the heart of ergodic theory are the ergodic theorems: results providing sufficient conditions for dynamical systems or random processes to possess ergodic properties, that is, for sample averages of the form

$$<f>_n = \frac{1}{n} \sum_{i=0}^{n-1} f T^i$$

to converge to an invariant limit. Traditional developments of the pointwise ergodic theorem focus on stationary systems and use a subsidiary result known as the *maximal ergodic lemma* (or *theorem*) to prove the ergodic theorem. The general result for AMS systems then follows since an AMS source inherits ergodic properties from its stationary mean; that is, since the set $\{x: <f>_n(x) \text{ converges }\}$ is invariant and since a system and its stationary mean place equal probability on all invariant sets, one will possess almost everywhere ergodic properties with respect to a class of measurements if and only if the other one does and the limiting sample averages will be the same.

The maximal ergodic lemma is due to Hoph [5], and its simple and elegant proof due to Garsia [2] is almost universally used in published developments of the ergodic theorem. Unfortunately, however, the maximal ergodic lemma is not very intuitive and the proof, although simple and elegant, is somewhat tricky and adds no insight into the ergodic theorem itself. Furthermore, we shall also wish to develop limit theorems for sequences of measurements that are not additive, but are instead subadditive or superadditive. In this case the maximal lemma is insufficient to obtain the desired results without additional techniques.

During the early 1980s several new and simpler proofs of the ergodic theorem for stationary systems were developed and subsequently published by Ornstein and Weiss [11], Jones [6], Katznelson and Weiss [7], and Shields [14]. These developments dealt directly with the behavior of long run sample averages, and the proofs involved no mathematical trickery. We here largely follow the approach of Katznelson and Weiss.

7.2 THE POINTWISE ERGODIC THEOREM

This section will be devoted to proving the following result, known as the *pointwise* or *almost everywhere ergodic theorem* or as *Birkhoff's ergodic theorem* after George Birkhoff, who first proved the result for a stationary transformations. [1]

Theorem 7.2.1. A sufficient condition for a dynamical system $(\Omega,\boldsymbol{B},m,T)$ to possess ergodic properties with respect to all bounded measurements f is that it be AMS. In addition, if the stationary mean of m is $\overline{m}$, then the system possesses ergodic properties with respect to all $\overline{m}$-integrable measurements, and if $f \in L^1(\overline{m})$, then with probability one under m and $\overline{m}$

$$\lim_{n\to\infty} \frac{1}{n} \sum_{i=0}^{n-1} fT^i = E_{\overline{m}}(f \mid \mathrm{I}),$$

where **I** is the sub-σ-field of invariant events.

Observe that the final statement follows immediately from the existence of the ergodic property from Corollary 6.6.1. In this section we will assume that we are dealing with processes to aid interpretations; that is, we will consider points to be sequences and T to be the shift. This is simply for intuition, however, and the mathematics remains valid for the more general dynamical systems. We can assume without loss of generality that the function f is nonnegative (since if it is not it can be decomposed into $f = f^+ - f^-$ and the result for the nonnegative f^+ and f^- and linearity then yield the result for f). Define

$$\underline{f} = \liminf_{n\to\infty} \frac{1}{n} \sum_{i=0}^{n-1} fT^i$$

and

$$\overline{f} = \limsup_{n\to\infty} \frac{1}{n} \sum_{i=0}^{n-1} fT^i .$$

To prove the first part of the theorem we must prove that $\underline{f} = \overline{f}$ with probability one under m. Since both $\overline{f}$ and $\underline{f}$ are invariant, the event $F = \{x: \overline{f}(x) = \underline{f}(x)\} = \{x: \lim_{n\to\infty} < f >_n \text{ exists}\}$ is also invariant and hence has the same probability under m and $\overline{m}$. Thus we must show that $\overline{m}(\overline{f} = \underline{f}) = 1$. Thus we need only prove the theorem for stationary measures and the result will follow for AMS measures. As a result, we can henceforth assume that m is itself stationary.

Since $\underline{f} \le \overline{f}$ everywhere by definition, the desired result will follow if we can show that

$$\int \overline{f}(x)\, dm(x) \le \int \underline{f}(x)\, dm(x) ;$$

that is, if $\overline{f} - \underline{f} \ge 0$ and $\int (\overline{f} - \underline{f}) dm \le 0$, then we must have $\overline{f} - \underline{f} = 0$, m-a.e. We accomplish this by proving the following:

$$\int \overline{f}(x)\, dm(x) \le \int f(x)\, dm(x) \le \int \underline{f}(x)\, dm(x) . \qquad (7.2.1)$$

By assumption $f \in L^1(m)$ since in the theorem statement either f is bounded or it is integrable with respect to $\overline{m}$, which is m since we have now assumed m to be stationary. Thus (7.2.1) implies not only that $\lim < f >_n$ exists, but that it is finite.

We simplify things at first by temporarily assuming that $\overline{f}$ is bounded: Suppose for the moment that the supremum of $\overline{f}$ is $M < \infty$. Fix $\varepsilon > 0$. Since $\overline{f}$ is the limit supremum of $< f >_n$, for each x there must exist a finite n for which

$$\frac{1}{n} \sum_{i=0}^{n-1} f(T^i x) \ge \overline{f}(x) - \varepsilon . \qquad (7.2.2)$$

Define $n(x)$ to be the smallest integer for which (7.2.2) is true. Since $\overline{f}$ is an invariant function, we can rewrite (7.2.2) using the definition of $n(x)$ as

$$\sum_{i=0}^{n(x)-1} \overline{f}(T^i x) \leq \sum_{i=0}^{n(x)-1} f(T^i x) + n(x)\varepsilon . \qquad (7.2.3)$$

Since $n(x)$ is finite for all x (although it is not bounded), there must be an N for which

$$m(\{x: n(x) > N\}) = \sum_{k=N+1}^{\infty} m(\{x: n(x) = k\}) \leq \frac{\varepsilon}{M}$$

since the sum converges to 0 as $N \to \infty$.. Choose such an N and define $B = \{x: n(x) > N\}$. B has small probability and can be thought of as the "bad" sequences, where the sample average does not get close to the limit supremum within a reasonable time. Observe that if $x \in B^c$, then also $T^i x \in B^c$ for $i = 1,2, \cdots, n(x) - 1$; that is, if x is not bad, then the next $n(x) - 1$ shifts of $n(x)$ are also not bad. To see this assume the contrary, that is, that there exists a $p < n(x) - 1$ such that $T^i x \in B^c$ for $i = 1,2, \cdots, p-1$, but $T^p x \in B$. The fact that $p < n(x)$ implies that

$$\sum_{i=0}^{p-1} f(T^i x) < p(\overline{f}(x) - \varepsilon)$$

and the fact that $T^p x \in B$ implies that

$$\sum_{i=p}^{n(x)-1} f(T^i x) < (n(x) - p)(\overline{f}(x) - \varepsilon) .$$

These two facts together, however, imply that

$$\sum_{i=0}^{n(x)-1} f(T^i x) < n(x)(\overline{f}(x) - \varepsilon) ,$$

contradicting the definition of $n(x)$.

We now modify $f(x)$ and $n(x)$ for bad sequences so as to avoid the problem caused by the fact that $n(x)$ is not bounded. Define

$$\tilde{f}(x) = \begin{cases} f(x), & x \notin B \\ M, & x \in B \end{cases}$$

and

$$\tilde{n}(x) = \begin{cases} n(x), & x \notin B \\ 1, & x \in B \end{cases} .$$

Analogous to (7.2.3) the modified functions satisfy

$$\sum_{i=0}^{\tilde{n}(x)-1} \overline{\tilde{f}}(T^i x) \le \sum_{i=0}^{\tilde{n}(x)-1} \tilde{f}(T^i x) + \tilde{n}(x)\varepsilon \tag{7.2.4}$$

and $\tilde{n}(x)$ is bound above by N. To see that (7.2.4) is valid, note that it is trivial if $x \in B$. If $x \notin B$, as argued previously $T^i x \notin B$ for $i = 1,2, \cdots, n(x)-1$ and hence $\tilde{f}(T^i x) = f(T^i x)$ and (7.2.4) follows from (7.2.3).

Observe for later reference that

$$\int \tilde{f}(x)\, dm(x) = \int_{B^c} \tilde{f}(x)\, dm(x) + \int_B \tilde{f}(x)\, dm(x) \le \int_{B^c} f(x)\, dm(x) + \int_B M\, dm(x)$$

$$\le \int f(x)\, dm(x) + Mm(B) \le \int f(x)\, dm(x) + \varepsilon\ , \tag{7.2.5}$$

where we have used the fact that f was assumed positive in the second inequality.

We now break up the sequence into nonoverlapping blocks such that within each block the sample average of $\tilde{f}$ is close to the limit supremum. Since the overall sample average is a weighted sum of the sample averages for each block, this will give a long term sample average near the limit supremum. To be precise, choose an L so large that $NM/L < \varepsilon$ and inductively define $n_k(x)$ by $n_0(x) = 0$ and

$$n_k(x) = n_{k-1}(x) + \tilde{n}(T^{n_{k-1}(x)}x)\ .$$

Let $k(x)$ be the largest k for which $n_k(x) \le L-1$, that is, the number of the blocks in the length L sequence. Since

$$\sum_{i=0}^{L-1} \overline{\tilde{f}}(T^i x) = \sum_{k=1}^{k(x)} \sum_{i=n_{k-1}(x)}^{n_k(x)-1} \overline{\tilde{f}}(T^i x) + \sum_{i=n_{k(x)}(x)}^{L-1} \overline{\tilde{f}}(T^i x)\ ,$$

we can apply the bound of (7.2.4) to each of the $k(x)$ blocks, that is, each of the previous $k(x)$ inner sums, to obtain

$$\sum_{i=0}^{L-1} \overline{\tilde{f}}(T^i x) \le \sum_{i=0}^{L-1} \tilde{f}(T^i x) + L\varepsilon + (N-1)M\ ,$$

where the final term overbounds the values in the final block by M and where the nonnegativity of $\tilde{f}$ allows the sum to be extended to $L - 1$. We can now integrate both sides of the preceding equation, divide by L, use the stationarity of m, and apply (7.2.5) to obtain

$$\int \overline{f}\, dm \leq \int \tilde{f}\, dm + \varepsilon + \frac{(N-1)M}{L} \leq \int f\, dm + 3\varepsilon\,, \tag{7.2.6}$$

which proves the left-hand inequality of (7.2.1) for the case where f is nonnegative and $\overline{f}$ is bounded since ε is arbitrary.

Next suppose that $\overline{f}$ is not bounded. Define for any M the "clipped" function $\overline{f}_M(x) = \min(\overline{f}(x), M)$ and replace $\overline{f}$ by $\overline{f}_M$ from (7.2.2) in the preceding argument. Since $\overline{f}_M$ is also invariant, the identical steps then lead to

$$\int \overline{f}_M\, dm \leq \int f\, dm\,.$$

Taking the limit as $M \to \infty$ then yields the desired left-hand inequality of (7.2.1) from the monotone convergence theorem.

To prove the right-hand inequality of (7.2.1) we proceed in a similar manner. Fix $\varepsilon > 0$ and define $n(x)$ to be the least positive integer for which

$$\frac{1}{n} \sum_{i=0}^{n-1} f(T^i x) \leq \underline{f}(x) + \varepsilon$$

and define $B = \{x\colon n(x) > N\}$, where N is chosen large enough to ensure that

$$\int_B f(x)\, dm(x) < \varepsilon\,.$$

This time define

$$\tilde{f}(x) = \begin{cases} f(x), & x \notin B \\ 0, & x \in B \end{cases}$$

and

$$\tilde{n}(x) = \begin{cases} n(x), & x \notin B \\ 1, & x \in B \end{cases}$$

and proceed as before.

This proves (7.2.1) and thereby shows that the sample averages of all m-integrable functions converge. From Corollary 6.6.1, the limit must be the conditional expectations. (The fact that the limits must be conditional expectations can also be deduced from the preceding arguments by restricting the integrations to invariant sets.) □

Combining the theorem with the results of previous chapter immediately yields the following corollaries.

Corollary 7.2.1. If a dynamical system $(\Omega,\boldsymbol{B},m,T)$ is AMS with stationary mean $\overline{m}$ and if the sequence $n^{-1}\sum_{i=0}^{n-1} fT^i$ is uniformly integrable with respect to m, then the following limit is true m-a.e., $\overline{m}$-a.e., and in $L^1(m)$:

$$\lim_{n\to\infty} \frac{1}{n} \sum_{i=0}^{n-1} fT^i = \langle f \rangle = E_{\overline{m}}(f \mid \mathbf{I}),$$

where $\mathbf{I}$ is the sub-σ-field of invariant events, and

$$E_m \langle f \rangle = E_{\overline{m}} \langle f \rangle = E_{\overline{m}} f.$$

If in addition the system is ergodic, then

$$\lim_{n\to\infty} \frac{1}{n} \sum_{i=0}^{n-1} fT^i = E_{\overline{m}}(f)$$

in both senses.

Corollary 7.2.2. A necessary and sufficient condition for a dynamical system to possess ergodic properties with respect to all bounded measurements is that it be AMS.

Exercises

1. Suppose that m is stationary and that g is a measurement that has the form $g = f - fT$ for some measurement $f \in L^2(m)$. Show that $\langle g \rangle_n$ has an $L^2(m)$ limit as $n\to\infty$. Show that any such g is orthogonal to $\mathbf{I}^2$, the subspace of all square-integrable invariant measurements; that is, $E(gh) = 0$ for all $h \in \mathbf{I}^2$. Suppose that g_n is a sequence of functions of this form that converges to a measurement r in $L^2(m)$. Show that $\langle r \rangle_n$ converges in $L^2(m)$.

2. Suppose that $\langle f \rangle_n$ converges in $L^2(m)$ for a stationary measure m. Show that the sample averages look increasingly like invariant functions in the sense that

$$\| \langle f \rangle_n - \langle f \rangle_n T \| \underset{n\to\infty}{\to} 0.$$

3. Show that if p and m are two stationary and ergodic measures, then they are either identical or singular in the sense that there is an event F for which $p(F) = 1$ and $m(F) = 0$.

4. Show that if p and m are distinct stationary and ergodic processes and $\lambda \in (0,1)$, then the mixture process $\lambda p + (1 - \lambda)m$ is not ergodic.

7.3 BLOCK AMS PROCESSES

Given a dynamical system $(\Omega,\boldsymbol{B},m,T)$ or a corresponding source $\{X_n\}$, one may be interested in the ergodic properties of successive blocks or vectors $X_{nN}^N = (X_{nN},X_{nN+1}, \cdots ,X_{(n+1)N-1})$ or, equivalently, in the N-shift T^N. For example, in communication systems one may use block codes that parse the input source into N-tuples and then map these successive nonoverlapping blocks into codewords for transmission. A natural question is whether or not a source being AMS with respect to T (T-AMS) implies that it is also T^N-AMS or vice versa. This section is devoted to answering this question.

First note that if a source is T-stationary, then it is clearly N-stationary (stationary with respect to T^N) since for any event F $m(T^{-N}F) = m(T^{-N-1}F) = \cdots = m(F)$. The converse is not true, however, since, for example, a process that produces a single sequence that is periodic with period N is N-stationary but not stationary. The following lemma, which is a corollary to the ergodic theorem, provides a needed step for treating AMS systems.

Lemma 7.3.1. If μ and η are probability measures on $(\Omega,\boldsymbol{B})$ and if η is T-stationary and asymptotically dominates μ (with respect to T), then μ is T-AMS.

Proof. If η is stationary, then from the ergodic theorem it possesses ergodic properties with respect to all bounded measurements. Since it asymptotically dominates μ, μ must also possesses ergodic properties with respect to all bounded measurements from Corollary 6.3.5. From Theorem 6.2.1, μ must therefore be AMS. □

The principal result of this section is the following:

Theorem 7.3.1. If a system $(\Omega,\boldsymbol{B},m,T)$ is T^N-AMS for any positive integer N, then it is T^N-AMS for all integers N.

Proof. Suppose first that the system is T^N-AMS and let $\overline{m}_N$ denote its N-stationary mean. Then for any event F

$$\lim_{n\to\infty}\frac{1}{n}\sum_{i=0}^{n-1} m(T^{-iN}T^{-j}F) = \overline{m}_N(T^{-j}F)\ ;\ j = 0,1,\cdots, N-1\ . \tag{7.3.1}$$

We then have

$$\frac{1}{N}\sum_{j=0}^{N-1}\lim_{n\to\infty}\frac{1}{n}\sum_{i=0}^{n-1} m(T^{-iN}T^{-j}F) = \lim_{n\to\infty}\frac{1}{nN}\sum_{j=0}^{N-1}\sum_{i=0}^{n-1} m(T^{-iN}T^{-j}F)$$

$$= \lim_{n\to\infty}\frac{1}{n}\sum_{i=0}^{n-1} m(T^{-i}F)\ , \tag{7.3.2}$$

where, in particular, the last limit must exist for all F and hence m must be AMS

Next assume only that m is AMS with stationary mean $\overline{m}$. This implies from Corollary 6.3.2 that $\overline{m}$ T-asymptotically dominates m, that is, if $\overline{m}(F) = 0$, then also $\limsup_{n\to\infty} m(T^{-n}F) = 0$. But this also means that for any integer K $\limsup_{n\to\infty} m(T^{-nK}F) = 0$ and hence that $\overline{m}$ also T^K-asymptotically dominates m. But this means from Lemma 7.3.1 applied to T^K that m is T^K-AMS. □

Corollary 7.3.1. Suppose that a process μ is AMS and has a stationary mean

$$\overline{m}(F) = \lim_{n\to\infty}\frac{1}{n}\sum_{i=0}^{n-1} m(T^{-i}F)$$

and an N-stationary mean

$$\overline{m}_N(F) = \lim_{n\to\infty}\frac{1}{n}\sum_{i=0}^{n-1} m(T^{-iN}F)\ .$$

Then the two means are related by

$$\frac{1}{N}\sum_{i=0}^{N-1}\overline{m}_N(T^{-i}F) = \overline{m}(F)\ .$$

Proof. The proof follows immediately from (7.3.1)-(7.3.2). □

As a final observation, if an event is invariant with respect to T, then it is also N-invariant, that is, invariant with respect to T^N. Thus if a system is N-ergodic or ergodic with respect to T^N, then all N-invariant events must have probability of 0 or 1 and hence so must all invariant events. Thus the system must also be ergodic. The converse, however, is not true: An ergodic system need not be N-ergodic. A system that is N-ergodic for all N is said to be *totally ergodic*. A strongly mixing system can easily be seen to be totally ergodic.

Exercises

1. Prove that a strongly mixing source is totally ergodic.
2. Suppose that m is N-ergodic and N-stationary. Is the mixture

$$\overline{m} = \frac{1}{N} \sum_{i=0}^{N-1} mT^{-i}$$

stationary and ergodic? Are the limits of the form

$$\frac{1}{n} \sum_{i=0}^{n-1} fT^i$$

the same under both m and $\overline{m}$? What about limits of the form

$$\frac{1}{n} \sum_{i=0}^{n-1} fT^{iN} ?$$

Suppose that you observe a sequence from the process mT^{-i} where you know m but not i. Under what circumstances could you determine i correctly with probability one by observing the infinite output sequence?

7.4 THE ERGODIC DECOMPOSITION

In this section we find yet another characterization of the limiting sample averages of the ergodic theorem. We use the ergodic properties to show that the stationary mean of an AMS measure that describes the limiting sample averages can be considered as a mixture of ergodic stationary measures. Roughly speaking, if we view the outputs of an AMS system, then we are viewing the outputs of an ergodic system, but we do not know *a priori* which one. If we view the limiting sample frequencies, however, we can determine the ergodic source being seen.

We have seen that if a dynamical system $(\Omega,\boldsymbol{B},m,T)$ is AMS with stationary mean $\overline{m}$ and hence has ergodic properties with respect to all indicator functions, then the limiting relative frequencies (the sample averages of the indicator functions) are conditional probabilities, $< 1_F > = \overline{m}(F \mid \mathbf{I})$, where $\mathbf{I}$ is the sigma-field of invariant events. In addition, if the alphabet is standard, one can select a regular version of the conditional probability and hence using iterated expectation one can consider the stationary mean distribution $\overline{m}$ as a mixture of measures $\overline{m}(\,.\mid \mathbf{I})(x);\ x \in \Omega$.

In this section we show that almost all of these measures are both stationary and ergodic, and hence the stationary mean of an AMS system, which describes the sample average behavior of the system, can be written as a mixture of stationary and ergodic distributions. Intuitively, observing the asymptotic sample averages of an AMS system is equivalent to observing the same measurements for a stationary and ergodic system that is an ergodic component of the stationary mean, but we may not know in advance which ergodic component is being observed. This result is known as the *ergodic decomposition* and it is due in various forms to Von Neumann [15], Rohlin [13], Kryloff and Bogoliouboff [9], and Oxtoby [12]. The development here most closely parallels that of Oxtoby. (See also Gray and Davisson [3] for the simple special case of discrete alphabet processes.)

Although we could begin with Theorem 6.6.1, which characterizes the limiting relative frequencies as conditional probabilities, and then use Theorem 5.7.1 to choose a regular version of the conditional probability, it is more instructive to proceed in a direct fashion by assuming ergodic properties. In addition, this yields a stronger result wherein the decomposition is determined by the properties of the points themselves and not the underlying measure; that is, the same decomposition works for all AMS measures.

Let $(\Omega,\boldsymbol{B},m,T)$ be a dynamical system and assume that $(\Omega,\boldsymbol{B})$ is a standard measurable space with a countable generating set $\boldsymbol{F}$.

For each event F define the invariant set $G(F) = \{x: \lim_{n\to\infty} < 1_F >_n(x) \text{ exists}\}$ and denote the invariant limit by $< 1_F >$. Define the set

$$G(\boldsymbol{F}) = \bigcap_{F \in \boldsymbol{F}} G(F),$$

the set of points x for which the limiting relative frequencies of all generating events exist. We shall call $G(\boldsymbol{F})$ the set of *generic points.*

Thus for any $x \in G(\boldsymbol{F})$

$$\lim_{n\to\infty} \sum_{i=0}^{n-1} 1_F(T^i x) = < 1_F >(x); \text{ all } F \in \boldsymbol{F} .$$

Observe that membership in $G(\boldsymbol{F})$ is determined strictly by properties of the points and not by an underlying measure. If m has ergodic properties with respect to the class of all indicator functions, i.e., if it is AMS, then clearly $m(G(F)) = 1$ for any event F and hence we also have for the countable intersection that $m(G(\boldsymbol{F})) = 1$. If $x \in G(\boldsymbol{F})$, then the set function P_x defined by

$$P_x(F) = < 1_F >(x); \; F \in \boldsymbol{F}$$

is nonnegative, normalized, and finitely additive on $\boldsymbol{F}$ from the properties of sample averages. Since the space is standard, this set function extends to a probability measure, which we shall also denote by P_x, on $(\Omega,\boldsymbol{B})$. Thus we easily obtain that every generic point induces a probability measure P_x and hence also a dynamical system $(\Omega,\boldsymbol{B},P_x,T)$ via its convergent relative frequencies. For x not in $G(\boldsymbol{F})$ we can simply fix P_x as some fixed arbitrary stationary and ergodic measure p^*.

We now confine interest to AMS systems so that the set of generic sequences will have probability one. Then $P_x(F) = < 1_F >(x)$ almost everywhere and $P_x(F) = \overline{m}(F \mid \mathbf{I})(x)$ for all x in a set of m measure one. Since this is true for any event F, we can define the countable intersection

$$H = \{x: P_x(F) = \overline{m}(F \mid \mathbf{I}) \text{ all } F \in \boldsymbol{F}\}$$

and infer that $m(H) = 1$. Thus with probability one under m the two measures P_x and $\overline{m}(\,.\mid \mathbf{I})(x)$ agree on $\boldsymbol{F}$ and hence are the same. Thus from the properties of conditional probability and conditional expectation

$$\overline{m}(F \cap D) = \int_D P_x(F)\,d\overline{m}, \text{ all } D \in \mathbf{I}, \tag{7.4.1}$$

and if f is in $L^1(\overline{m})$, then

$$\int_D f\, d\overline{m} = \int_D \left(\int f(y)dP_x(y)\right) d\overline{m}, \text{ all } D \in \mathbf{I}. \tag{7.4.2}$$

Thus choosing $D = \Omega$ we can interpret (7.4.1) as stating that the stationary mean $\overline{m}$ is a mixture of the probability measures P_x and (7.4.2) as stating that the expectation of measurements with respect to the stationary mean can be computed (using iterated expectation) in two steps: First find the expectations with respect to the components P_x for all x and then take the average. Such mixture and average representations will be particularly useful because, as we shall show, the component measures are both stationary and ergodic with probability one. For focus and motivation, we next state the principal result of this section–the ergodic decomposition theorem–which makes this notion rigorous in a useful form. The remainder of the section is then devoted to the proof of this theorem via a sequence of lemmas that are of some interest in their own right.

Theorem 7.4.1. (The Ergodic Decomposition)

Given a standard measurable space $(\Omega,\boldsymbol{B})$, let $\boldsymbol{F}$ denote a countable generating field. Given a transformation $T:\Omega\rightarrow\Omega$ define the set of generic sequences

$$G(\boldsymbol{F}) = \{x: \lim_{n\rightarrow\infty} < 1_F >_n(x) \text{ exists all } F\in\boldsymbol{F}\}.$$

For each $x \in G(\boldsymbol{F})$ let P_x denote the measure induced by the limiting relative frequencies. Define the set $\hat{E}\subset G(\boldsymbol{F})$ by $\hat{E} = \{x: P_x$ is stationary and ergodic$\}$. Define the stationary and ergodic measures p_x by $p_x = P_x$ for $x \in \hat{E}$ and otherwise $p_x = p^*$, some fixed stationary and ergodic measure. (Note that so far all definitions involve properties of the points only and not of any underlying measure.) We shall refer to the collection of measures $\{p_x; x\in \Omega\}$ as the *ergodic decomposition*. We have then that

(a) $\hat{E}$ is an invariant measurable set,

(b) $p_x(F) = p_{Tx}(F)$ for all points $x \in \Omega$ and events $F \in \boldsymbol{B}$,

(c) If m is an AMS measure with stationary mean $\overline{m}$, then

$$m(\hat{E}) = \overline{m}(\hat{E}) = 1\ , \tag{7.4.3}$$

for any $F \in \boldsymbol{B}$

$$\overline{m}(F) = \int p_x(F)d\overline{m}(x) = \int p_x(F)dm(x)\ , \tag{7.4.4}$$

and if $f \in L^1(\overline{m})$, then so is $\int f dp_x$ and

$$\int f\, d\overline{m} = \int (\int f dp_x)d\overline{m}(x). \tag{7.4.5}$$

In addition, given any $f \in L^1(\overline{m})$, then

$$\lim_{n\to\infty} n^{-1} \sum_{i=0}^{n-1} f(T^i x) = E_{p_x} f \text{ , } m\text{–a.e. and } \overline{m}\text{–a.e.} \tag{7.4.6}$$

Comment. The final result is extremely important for both mathematics and intuition: Say that a system source is AMS and that f is integrable with respect to the stationary mean; e.g., f is measurable and bounded and hence integrable with respect to any measure. Then with probability one a point will be produced such that the limiting sample average exists and equals the expectation of the measurement with respect to the stationary and ergodic measure induced by the point. Thus if a system is AMS, then the limiting sample averages will behave as if they were in fact produced by a stationary and ergodic system. These ergodic components are determined by the points themselves and not the underlying measure; that is, the same collection works for all AMS measures.

We begin the proof by observing that the definition of P_x implies that

$$P_{Tx}(F) = \langle 1_F \rangle(Tx) = \langle 1_F \rangle(x) = P_x(F) \text{ ; } \quad \text{all } F \in \boldsymbol{F} \text{ ; all } x \in G(\boldsymbol{F}) \tag{7.4.7}$$

since limiting relative frequencies are invariant. Since P_{Tx} and P_x agree on generating field events, however, they must be identical. Since we will select the p_x to be the P_x on a subset of $G(\boldsymbol{F})$, this proves (b) for $x \in \hat{E}$.

Next observe that for any $x \in G(\boldsymbol{F})$ and $F \in \boldsymbol{F}$

$$P_x(F) = \langle 1_F \rangle(x) = \lim_{n\to\infty} \sum_{i=0}^{n-1} 1_F(T^i x)$$

$$= \lim_{n\to\infty} \sum_{i=0}^{n-1} 1_{T^{-1}F}(T^i x) = \langle 1_{T^{-1}F} \rangle(x) = P_x(T^{-1}F) \text{ .}$$

Thus for $x \in G(\boldsymbol{F})$ the measure P_x looks stationary, at least on generating events. The following lemma shows that this is enough to infer that the P_x are indeed stationary.

Lemma 7.4.1. A measure m is stationary on $(\Omega,\sigma(\boldsymbol{F}))$ if and only if $m(T^{-1}F) = m(F)$ for all $F \in \boldsymbol{F}$.

Proof. Both m and mT^{-1} are measures on the same space, and hence they are equal if and only if they agree on a generating field. □

Note in passing that no similar countable description of an AMS measure is known, that is, if $\lim_{n\to\infty} n^{-1}\sum_{i=0}^{n-1} m(T^{-i}F)$ converges for all F in a generating field, it does not follow that the measure is AMS.

As before let $\hat{E}$ denote the subset of $G(\boldsymbol{F})$ consisting of all of the points x for which P_x is ergodic. From Lemma 6.7.4, the definition of P_x, and the stationarity of the P_x we can write $\hat{E}$ as

$$\hat{E} = \{x\colon x \in G(\boldsymbol{F})\ ;\ \lim_{n\to\infty} n^{-1} \sum_{i=0}^{n-1} P_x(T^{-i}F \cap F) = P_x(F)^2\ ,\ \text{all } F \in \boldsymbol{F}\}\ .$$

Since the probabilities P_x are measurable functions, the set $\hat{E}$ where the given limit exists and has the given value is also measurable. Since $P_x(F) = P_{Tx}(F)$, the set is also invariant. This proves (a) of the theorem. We have already proved (b) for $x \in \hat{E}$. For $x \notin \hat{E}$ we have that $p_x = p^*$. Since $\hat{E}$ is invariant, we also know that $Tx \notin \hat{E}$ and hence also $p_{Tx} = p^*$ so that $p_{Tx} = p_x$.

We now consider an AMS measure m. Although Lemma 6.7.4 provided a countable description of ergodicity that was useful to prove that the collection of all points inducing ergodic measures is an event, the following lemma provides another countable description of ergodicity that is useful for computing the probability of the set $\hat{E}$.

Lemma 7.4.2. Let $(\Omega,\sigma(\boldsymbol{F}),m,T)$ be an AMS dynamical system with stationary mean $\overline{m}$. Define the sets

$$G(F,m) = \{x\colon \lim_{n\to\infty} \frac{1}{n} \sum_{i=0}^{n-1} 1_F(T^i x) = \overline{m}(F)\}\ ;\ F \in \boldsymbol{F}\ .$$

Thus $G(F,m)$ is the set of sequences for which the limiting relative frequency exists for all events F in the generating field and equals the probability of the event under the stationary mean. Define also

$$G(\boldsymbol{F},m) = \bigcap_{F \in \boldsymbol{F}} G(F,m)\ .$$

A necessary and sufficient condition for m to be ergodic is that $m(G(\boldsymbol{F},m)) = 1$.

Comment. The lemma states that an AMS measure m is ergodic if and only if it places all of its probability on the set of generic points that induce $\overline{m}$ through their relative frequencies.

Proof. If m is ergodic then $< 1_F > = \overline{m}(F)$ m-a.e. for all F and the result follows immediately. Conversely, if $m(G(\boldsymbol{F},m)) = 1$ then also $\overline{m}(G(\boldsymbol{F},m)) = 1$ since the set is invariant. Thus integrating to the limit using the dominated convergence theorem yields

$$\lim_{n\to\infty} n^{-1}\sum_{i=0}^{n-1} \overline{m}(T^{-i}F \cap F) = \int_{G(\boldsymbol{F},m)} 1_F (\lim_{n\to\infty} n^{-1}\sum_{i=0}^{n-1} 1_F T^i) d\overline{m}$$

$$= \overline{m}(F)^2 , \text{ all } F \in \boldsymbol{F} ,$$

which with Lemma 6.7.4 proves that $\overline{m}$ is ergodic, and hence so is m from Lemma 6.7.2. □

Proof of the Theorem. If we can show that $m(\hat{E}) = 1$, then the remainder of the theorem will follow from (7.4.1) and (7.4.2) and the fact that $p_x = \overline{m}(\,.\,|\,\mathbf{I}\,)(x)$ $\overline{m}$*–a.e.* From the previous lemma, this will be the case if with p_x probability one

$$< 1_F > = p_x(F) , \text{ all } F \in \boldsymbol{F},$$

or, equivalently, if for all $F \in \boldsymbol{F}$

$$< 1_F > = p_x(F) \quad p_x\text{–}a.e.$$

This last equation will hold if for each $F \in \boldsymbol{F}$

$$\int dp_x(y)\, | < 1_F >(y) - p_x(F)\, |^2 = 0 .$$

This relation will hold on a set of x of m probability one if

$$\int dm(x) \int dp_x(y)\, | < 1_F >(y) - p_x(F)\, |^2 = 0 .$$

Thus proving the preceding equality will complete the proof of the ergodic decomposition theorem. Expanding the square and using the fact that $G(\boldsymbol{F})$ has m probability one yield

$$\int_{G(\boldsymbol{F})} dm(x) \int_{G(\boldsymbol{F})} dp_x(y) < 1_F >(y)^2 -$$

$$- 2\int_{G(\boldsymbol{F})} dm(x) p_x(F) \int_{G(\boldsymbol{F})} dp_x(y) < 1_F >(y) + \int_{G(\boldsymbol{F})} dm(x) p_x(F)^2 .$$

Since $< 1_F >(y) = p_y(F)$, the first and third terms on the right are equal from (7.4.2). (Set $f(x) = < 1_F(x) > = p_x(F)$ in (7.4.2).) Since p_x is stationary, $\int dp_x < 1_F > = p_x(F)$. Thus the middle term cancels the remaining terms and the preceding expression is zero, as was to be shown. This completes the proof of the ergodic decomposition theorem. □

Exercises

1. Suppose that a measure m is stationary and ergodic and N is a fixed positive integer. Show that there is a measure η that is N-stationary and N-ergodic that has the property that

$$m = \frac{1}{N} \sum_{i=0}^{N-1} \eta T^{-i} .$$

This shows that although an ergodic source may not be N-ergodic, it can always be decomposed into a mixture of an N-ergodic source and its shifts. This result is due to Nedoma [10].

7.5 THE SUBADDITIVE ERGODIC THEOREM

Sample averages are not the only measurements of systems that depend on time and converge. In this section we consider another class of measurements that have an ergodic theorem: subadditive measurements. An example of such a measurement is the Levenshtein distance between sequences (see Example 2.5.4).

The subadditive ergodic theorem was first developed by Kingman [8], but the proof again follows that of Katznelson and Weiss [7]. A sequence of functions $\{f_n;\ n = 1,2, \cdots \}$ satisfying the relation

$$f_{n+m}(x) \leq f_n(x) + f_m(T^n x) \ , \text{ all } n,m = 1,2, \cdots$$

for all x is said to be *subadditive*. Note that the unnormalized sample averages $f_n = n < f >_n$ of a measurement f are additive and satisfy the relation with equality and hence are a special case of subadditive functions.

The following result is for stationary systems. Generalizations to AMS systems are considered later.

Theorem 7.5.1. Let $(\Omega,\boldsymbol{B},m,T)$ be a stationary dynamical system. Suppose that $\{f_n;\ n = 1,2,\cdots\}$ is a subadditive sequence of m-integrable measurements. Then there is an invariant function $\phi(x)$ such that m-a.e.

$$\lim_{n\to\infty} \frac{1}{n} f_n(x) = \phi(x) .$$

The function $\phi(x)$ is given by

$$\phi(x) = \inf_n \frac{1}{n} E_m(f_n \mid \mathbf{I}) = \lim_{n\to\infty} \frac{1}{n} E_m(f_n \mid \mathbf{I}) .$$

Most of the remainder of this section is devoted to the proof of the theorem. On the way some lemmas giving properties of subadditive functions are developed. Observe that since the conditional expectations with respect to the σ-field of invariant events are themselves invariant functions, $\phi(x)$ is invariant as claimed.

Since the functions are subadditive,

$$E_m(f_{n+m} \mid \mathbf{I}) \leq E_m(f_n \mid \mathbf{I}) + E_m(f_m T^n \mid \mathbf{I}) .$$

From the ergodic decomposition theorem, however, the third term is $E_{P_x}(f_m T^n) = E_{P_x}(f_m)$ since P_x is a stationary measure with probability one. Thus we have with probability one that

$$E_m(f_{n+m} \mid \mathbf{I})(x) \leq E_m(f_n \mid \mathbf{I})(x) + E_m(f_m \mid \mathbf{I})(x) .$$

For a fixed x, the preceding is just a sequence of numbers, say a_k, with the property that

$$a_{n+k} \leq a_n + a_k .$$

A sequence satisfying such an inequality is said to be *subadditive.* The following lemma (due to Hammersley and Welch [4]) provides a key property of subadditive sequences.

Lemma 7.5.1. If $\{a_n;\ n = 1,2,\cdots\}$ is a subadditive sequence, then

$$\lim_{n\to\infty} \frac{a_n}{n} = \inf_{n\geq 1} \frac{a_n}{n} ,$$

where if the right-hand side is $-\infty$, the preceding means that a_n/n diverges to $-\infty$.

Proof. First observe that if a_n is subadditive, for any k,n

$$a_{kn} \le a_n + a_{(k-1)n} \le a_n + a_n + a_{(k-2)n} \le \cdots \le ka_n . \tag{7.5.1}$$

Call the infimum γ and first consider the case where it is finite. Fix $\varepsilon > 0$ and choose a positive integer N such that $N^{-1}a_N < \gamma + \varepsilon$. Define $b = \max_{1 \le k \le 3N-1} a_k$. For each $n \ge 3N$ there is a unique integer $k(n)$ for which $(k(n)+2)N \le n < (k(n) + 3)N$. Since the sequence is subadditive, $a_n \le a_{k(n)N} + a_{(n-k(n)N)}$. Since $n-k(n)N < 3N$ by construction, $a_{n-k(n)N} \le b$ and therefore

$$\gamma \le \frac{a_n}{n} \le \frac{a_{k(n)N} + b}{n} .$$

Using (7.5.1) this yields

$$\gamma \le \frac{a_n}{n} \le \frac{k(n)a_N}{n} + \frac{b}{n}$$

$$\le \frac{k(n)N}{n}\frac{a_N}{N} + \frac{b}{n} \le \frac{k(n)N}{n}(\gamma + \varepsilon) + \frac{b}{n} .$$

By construction $k(n)N \le n-2N \le n$ and hence

$$\gamma \le \frac{a_n}{n} \le \gamma + \varepsilon + \frac{b}{n} .$$

Thus given an ε, choosing n_0 large enough yields $\gamma \le n^{-1}a_n \le \gamma + 2\varepsilon$ for all $n > n_0$, proving the lemma. If γ is $-\infty$, then for any positive M we can choose n so large that $n^{-1}a_n \le -M$. Proceeding as before for any ε there is an n_0 such that if $n \ge n_0$, then $n^{-1}a_n \le -M + \varepsilon$. Since M is arbitrarily big and ε arbitrarily small, $n^{-1}a_n$ goes to $-\infty$.

Returning to the proof of the theorem, the preceding lemma implies that with probability one

$$\phi(x) = \inf_{n \ge 1} \frac{1}{n} E_m(f_n \mid \mathrm{I})(x) = \lim_{n\to\infty} \frac{1}{n} E_m(f_n \mid \mathrm{I})(x) .$$

As in the proof of the ergodic theorem, define

$$\bar{f}(x) = \limsup_{n\to\infty} \frac{1}{n} f_n(x)$$

and

$$\underline{f}(x) = \liminf_{n\to\infty} \frac{1}{n} f_n(x) .$$

Lemma 7.5.2. Given a subadditive function $f_n(x)$; $n = 1,2, \cdots$, $\overline{f}$ and $\underline{f}$ as defined previously are invariant functions m-a.e.

Proof. From the definition of subadditive functions

$$f_{1+n}(x) \le f_1(x) + f_n(Tx)$$

and hence

$$\frac{f_n(Tx)}{n} \ge \frac{1+n}{n}\frac{f_{1+n}(x)}{1+n} - \frac{f_1(x)}{n} .$$

Taking the limit supremum yields $\overline{f}(Tx) \ge \overline{f}(x)$. Since m is stationary, $E_m \overline{f}T = E_m \overline{f}$ and hence $\overline{f}(x) = \overline{f}(Tx)$ m-a.e. The result for $\underline{f}$ follows similarly. □

We now tackle the proof of the theorem. It is performed in two steps. The first uses the usual ergodic theorem and subadditivity to prove that $\overline{f} \le \phi$ a.e. Next a construction similar to that of the proof of the ergodic theorem is used to prove that $\underline{f} = \phi$ a.e., which will complete the proof.

As an introduction to the first step, observe that since the measurements are subadditive

$$\frac{1}{n} f_n(x) \le \frac{1}{n}\sum_{i=0}^{n-1} f_1(T^i x) .$$

Since $f_1 \in L^1(m)$ for the stationary measure m, we know immediately from the ergodic theorem that

$$\overline{f}(x) \le \langle f_1 \rangle(x) = E_m(f_1 \mid \mathrm{I}) . \tag{7.5.2}$$

We can obtain a similar bound for any of the f_n by breaking it up into fixed size blocks of length, say N and applying the ergodic theorem to the sum of the f_N. Fix N and choose $n \ge N$. For each $i = 0,1,2, \cdots, N-1$ there are integers j,k such that $n = i + jN + k$ and $1 \le k < N$. Hence using subadditivity

$$f_n(x) \leq f_i(x) + \sum_{l=0}^{j-1} f_N(T^{lN+i}x) + f_k(T^{jN+i}x) .$$

Summing over i yields

$$Nf_n(x) \leq \sum_{i=0}^{N-1} f_i(x) + \sum_{l=0}^{jN-1} f_N(T^l x) + \sum_{i=0}^{N-1} f_{n-i-jN}(T^{jN+i}x) ,$$

and hence

$$\frac{1}{n} f_n(x) \leq \frac{1}{nN} \sum_{l=0}^{jN-1} f_N(T^l x) + \frac{1}{nN} \sum_{i=0}^{N-1} f_i(x) + \frac{1}{nN} \sum_{i=0}^{N-1} f_{n-i-jN}(T^{jN+i}x) .$$

As $n \to \infty$ the middle term on the right goes to 0. Since f_N is assumed integrable with respect to the stationary measure m, the ergodic theorem implies that the leftmost term on the right tends to $N^{-1} < f_N >$. Intuitively the rightmost term should also go to zero, but the proof is a bit more involved. The rightmost term can be bound above by

$$\frac{1}{nN} \sum_{i=0}^{N-1} | f_{n-i-jN}(T^{jN+i}x) | \leq \frac{1}{nN} \sum_{i=0}^{N-1} \sum_{l=1}^{N} | f_l (T^{jN+i}x)|$$

where j depends on n (the last term includes all possible combinations of k in f_k and shifts). Define this bound to be w_n/n. We shall prove that w_n/n tends to 0 by using Corollary 4.6.1, that is, by proving that

$$\sum_{n=0}^{\infty} m(\frac{w_n}{n} > \varepsilon) < \infty . \tag{7.5.3}$$

First note that since m is stationary,

$$w_n = \frac{1}{N} \sum_{i=0}^{N-1} \sum_{l=1}^{N} | f_l (T^{jN+i}x)|$$

has the same distribution as

$$w = \frac{1}{N} \sum_{i=0}^{N-1} \sum_{l=1}^{N} | f_l (T^i x)|$$

Thus we will have proved (7.5.3) if we can show that

$$\sum_{n=0}^{\infty} m(w > n\varepsilon) < \infty . \tag{7.5.4}$$

To accomplish this we need a side result, which we present as a lemma.

Lemma 7.5.3. Given a nonnegative random variable X defined on a probability space $(\Omega,\boldsymbol{B},m)$,

$$\sum_{n=1}^{\infty} m(X \geq n) \leq EX .$$

Proof.

$$EX = \sum_{n=1}^{\infty} E(X \mid X \in [n-1,n))m(X \in [n-1,n)) \geq$$

$$\sum_{n=1}^{\infty} (n-1)m(X \in [n-1,n)) = \sum_{n=1}^{\infty} nm(X \in [n-1,n)) - 1 .$$

By rearranging terms the last sum can be written as

$$\sum_{n=0}^{\infty} m(X \geq n) ,$$

which proves the lemma. □

Applying the lemma to the left side of (7.5.4) yields

$$\sum_{n=0}^{\infty} m(\frac{w}{\varepsilon} > n) \leq \sum_{n=0}^{\infty} m(\frac{w}{\varepsilon} \geq n) \leq \frac{1}{\varepsilon} E_m(w) < \infty ,$$

since m-integrability of the f_n implies m-integrability of their absolute magnitudes and hence of a finite sum of absolute magnitudes. This proves (7.5.4) and hence (7.5.3). Thus we have shown that

$$\overline{f}(x) \leq \frac{1}{N} < f_N>(x) = \frac{1}{N} E_m(f_N \mid \mathbf{I})$$

for all N and hence that

$$\overline{f}(x) \leq \phi(x) , \ m\text{–a.e.} . \tag{7.5.5}$$

This completes the first step of the proof.

If $\phi(x) = -\infty$, then (7.5.5) completes the proof since the desired limit then exists and is $-\infty$. Hence we can focus on the invariant set $G = \{x:\phi(x) > -\infty\}$. Define the event $G_M = \{x: \phi(x) \geq -M\}$. We shall show that for all M

$$\int_{G_M} \underline{f} dm \geq \int_{G_M} \phi dm , \tag{7.5.6}$$

which with (7.5.5) proves that $\underline{f} = \phi$ on G_M. We have, however, that

$$G = \bigcup_{M=1}^{\infty} G_M .$$

Thus proving that $\underline{f} = \phi$ on all G_M also will prove that $\underline{f} = \phi$ on G and complete the proof of the theorem.

We now proceed to prove (7.5.6) and hence focus on x in G_M and hence only x for which $\phi(x) \geq -M$. We proceed as in the proof of the ergodic theorem. Fix an $\varepsilon > 0$ and define $\underline{f}_M = \max(\underline{f}, -M-1)$. Observe that this implies that

$$\phi(x) \geq \underline{f}_M(x) , \text{ all } x . \tag{7.5.7}$$

Define $n(x)$ as the smallest integer $n \geq 1$ for which $n^{-1} f_n(x) \leq \underline{f}_M(x) + \varepsilon$. Define $B = \{x: n(x) > N\}$ and choose N large enough so that

$$\int_B (\mid f_1(x) \mid + M + 1) \, dm(x) < \varepsilon . \tag{7.5.8}$$

Define the modified functions as before:

$$\tilde{\underline{f}}_M(x) = \begin{cases} \underline{f}_M(x), & x \notin B \\ f_1(x), & x \in B \end{cases}$$

and

$$\tilde{n}(x) = \begin{cases} n(x), & x \notin B \\ 1, & x \in B \end{cases} .$$

From (7.5.8)

$$\int \tilde{\underline{f}}_M \, dm = \int_{B^c} \tilde{\underline{f}}_M \, dm + \int_B \tilde{\underline{f}}_M \, dm$$

$$= \int_{B^c} \underline{f}_M \, dm + \int_B f_1 dm \leq \int_{B^c} \underline{f}_M \, dm + \int_B \mid f_1 \mid dm$$

$$\leq \int_{B^c} \underline{f}_M \, dm + \varepsilon - \int_B (M+1) dm \leq \int_{B^c} \underline{f}_M \, dm + \varepsilon + \int_B \underline{f}_M \, dm$$

$$\leq \int \underline{f}_M \, dm + \varepsilon . \tag{7.5.9}$$

Since $\underline{f}_M$ is invariant,

$$f_{n(x)}(x) \leq \sum_{i=0}^{n(x)-1} \underline{f}_M(T^i x) + n(x)\varepsilon,$$

and hence also

$$f_{\tilde{n}(x)}(x) \leq \sum_{i=0}^{\tilde{n}(x)-1} \tilde{\underline{f}}_M(T^i x) + \tilde{n}(x)\varepsilon$$

Thus for $L > N$

$$f_L(x) \leq \sum_{i=0}^{L-1} \tilde{\underline{f}}_M(T^i x) + L\varepsilon + N(M-1) + \sum_{i=L-N}^{L-1} | f_1(T^i x) | .$$

Integrating and dividing by L results in

$$\int \phi(x)\, dm(x) \leq \int \frac{1}{L} < f_L > (x)\, dm(x) = \int \frac{1}{L} f_L(x)\, dm(x)$$

$$\leq \int \tilde{\underline{f}}_M(x)\, dm(x) + \varepsilon + \frac{N(M+1)}{L} + \frac{N}{L} \int | f_1(x) |\, dm(x) .$$

Letting $L \to \infty$ and using (7.5.9) yields

$$\int \phi(x)\, dm(x) \leq \int \underline{f}_M(x)\, dm(x) .$$

From (7.5.7), however, this can only hold if in fact $\underline{f}_M(x) = \phi(x)$ a.e. This implies, however, that $\underline{f}(x) = \phi(x)$ a.e. since the event $\underline{f}(x) \neq \underline{f}_M(x)$ implies that $\underline{f}_M(x) = -M - 1$ which can only happen with probability 0 from (7.5.7) and the fact that $\underline{f}_M(x) = \phi(x)$ a.e. This completes the proof of the subadditive ergodic theorem for stationary measures. □

Unlike the ergodic theorem, the subadditive ergodic theorem does not easily generalize to AMS measures since if m is not stationary, it does not follow that $\overline{f}$ or $\underline{f}$ are invariant measurements or that the event $\{\overline{f} = \underline{f}\}$ is either an invariant event or a tail event. The following generalization is a slight extension of unpublished work of Wu Chou and is the most general known subadditive ergodic theorem known to the author.

Theorem 7.5.2. Let $(\Omega,\boldsymbol{B},m,T)$ be an AMS dynamical system with stationary mean $\overline{m}$. Suppose that $\{f_n;\ n = 1,2,\cdots\}$ is a subadditive sequence of $\overline{m}$-integrable measurements. There is an invariant function $\phi(x)$ such that m-a.e.

$$\lim_{n\to\infty} \frac{1}{n} f_n(x) = \phi(x)$$

if and only if $m \ll \overline{m}$, in which case the function $\phi(x)$ is given by

$$\phi(x) = \inf_n \frac{1}{n} E_{\overline{m}}(f_n \mid \mathrm{I}) = \lim_{n\to\infty} \frac{1}{n} E_{\overline{m}}(f_n \mid \mathrm{I}) \,.$$

Comment: Recall from Theorem 6.4.3 that the condition that the AMS measure is dominated by its stationary mean is equivalent to the condition that the system be recurrent. Note that the theorem immediately implies that the Kingman ergodic theorem holds for two-sided AMS processes. What is new here is the result for one-sided AMS processes.

Proof. If $m \ll \overline{m}$, then the conclusions follow immediately from the ergodic theorem for stationary sources since m inherits the probability 1 and 0 sets from $\overline{m}$. Suppose that m is not dominated by $\overline{m}$. Then from Theorem 6.4.3, the system is not conservative and hence there exists a wandering set W with nonzero probability, that is, the sets $T^{-n}W$ are pairwise disjoint and $m(W) > 0$. We now construct a counterexample to the subadditive ergodic theorem, that is, a subadditive function that does not converge. Define

$$f_n(x) = \begin{cases} -2^{k_n} & \text{if } x \in W \text{ and } 2^{k_n} \le n \le 2^{k_n+1},\ k_n = 0,1,2,\cdots \\ f_n(T^j x) - 2^n & \text{if } x \in T^{-j} W \\ 0 & \text{otherwise} \end{cases}$$

We first prove that f_n is subadditive, that is, that $f_{n+m}(x) \le f_n(x) + f_m(T^n x)$ for all $x \in \Omega$. To do this we consider three cases: (1) $x \in W$, (2) $x \in W^* = \bigcup_{i=1}^{\infty} T^{-i} W$, and (3) $x \notin W \cup W^*$. In case (1),

$$f_n(x) = -2^{k_n}$$

and

$$f_{n+m}(x) = -2^{k_{n+m}} \,.$$

To evaluate $f_m(T^n x)$ observe that $x \in W$ implies that $T^n x \notin W \cup W^*$ since otherwise

$$x \in T^{-n} \bigcup_{k=0}^{\infty} T^{-k} W = \bigcup_{k=n}^{\infty} T^{-k} W ,$$

which is impossible because $x \in W$ and W is a wandering set. Thus $f_n(T^m x) = 0$. Thus for $x \in W$

$$f_n(x) + f_m(T^n x) - f_{n+m}(x) = -2^{k_n} + 2^{k_{n+m}} .$$

Since by construction $k_{n+m} \geq k_n$, this is nonnegative and f is subadditive in case (1). For case (3) the result is trivial: If $x \notin W \cup W^*$, then no shift of x is ever in W, and all the terms are 0. This leaves case (2).

Suppose now that $x \in T^{-j} W$ for some k. Then

$$f_{n+m}(x) = f_{n+m}(T^j x) - 2^{n+m} = -2^{k_{n+m}} - 2^{n+m} ,$$

and

$$f_n(x) = f_n(T^j x) - 2^n = -2^{k_n} - 2^n .$$

For the final term there are two cases. If $n > j$, then analogously to case (1) $x \in T^{-j} W$ implies that $T^n x \notin W \cup W^*$ since otherwise

$$x \in T^{-n} \bigcup_{k=0}^{\infty} T^{-k} W = \bigcup_{k=n}^{\infty} T^{-k} W ,$$

which is impossible since W is a wandering set and hence x cannot be in both $T^{-j} W$ and in $\bigcup_{k=n}^{\infty} T^{-k} W$ since $n > j$. If $n \leq j$, then

$$f_m(T^n x) = f_m(T^j x) - 2^m = -2^{k_m} - 2^m .$$

Therefore if $n > j$,

$$f_n(x) + f_m(T^n x) - f_{n+m}(x) = -2^{k_n} - 2^n + 0 + 2^{k_{n+m}} + 2^{n+m} \geq 0 .$$

If $n \leq j$ then

$$f_n(x) + f_m(T^n x) - f_{n+m}(x) = -2^{k_n} - 2^n - 2^{k_m} - 2^m + 2^{k_{n+m}} + 2^{n+m} .$$

Suppose that $n \geq m \geq 1$, then

$$2^{n+m} - 2^n - 2^m = 2^n(2^m - 1 - 2^{-|n-m|}) \geq 0 .$$

Similarly, since then $k_n \geq k_m$, then

$$2^{k_{n+m}} - 2^{k_n} - 2^{k_m} = 2^{k_n}(2^{k_{n+m}-k_n} - 1 - 2^{-|k_n - k_m|} \geq 0 .$$

Similar inequalities follow when $n \leq m$, proving that f_n is indeed subadditive on Ω.

We now complete the proof. The event W has nonzero probability, but if $x \in W$, then

$$\lim_{n \to \infty} \frac{1}{2^n} f_{2^n}(x) = \lim_{n \to \infty} \frac{-2^n}{2^n} = -1$$

whereas

$$\lim_{n \to \infty} \frac{1}{2^n - 1} f_{2^n - 1}(x) = \lim_{n \to \infty} \frac{-2^{n-1}}{2^n - 1} = -\frac{1}{2}$$

and hence $n^{-1} f_n$ does not have a limit on this set of nonzero probability, proving that the subadditive ergodic theorem does not hold in this case.

REFERENCES

1. G. D. Birkhoff, ''Proof of the ergodic theorem,'' *Proc. Nat. Acad. Sci.*, vol. 17, pp. 656-660, 1931.
2. A. Garsia, ''A simple proof of E. Hoph's maximal ergodic theorem,'' *J. Math. Mech.*, vol. 14, pp. 381-2, 1965.
3. R. M. Gray and L. D. Davisson, ''The ergodic decomposition of discrete stationary random processes,'' *IEEE Trans. on Info. Theory*, vol. IT-20, pp. 625-636, September 1974.
4. J. M. Hammersley and D. J. A. Welsh, ''First-passage, percolation, subadditive processes, stochastic networks, and generalized renewal theory,'' *Bernoulli-Bayes-Laplace Anniversary Volume*, Springer, Berlin, 1965.
5. E. Hoph, *Ergodentheorie*, Springer-Verlag, Berlin, 1937.
6. R. Jones, ''New proof for the maximal ergodic theorem and the Hardy-Littlewood maximal inequality,'' *Proc. AMS*, vol. 87, pp. 681-684, 1983.
7. I. Katznelson and B. Weiss, ''A simple proof of some ergodic theorems,'' *Israel Journal of Mathematics*, vol. 42, pp. 291-296, 1982.

8. J. F. C. Kingman, "The ergodic theory of subadditive stochastic processes," *Ann. Probab.*, vol. 1, pp. 883-909, 1973.

9. N. Kryloff and N. Bogoliouboff, "La théorie générale de la mesure dans son application à l'étude des systèmes de la mecanique non linéaire," *Ann. of Math.*, vol. 38, pp. 65-113, 1937.

10. J. Nedoma, "On the ergodicity and r-ergodicity of stationary probability measures," *Zeitschrift Wahrscheinlichkeitstheorie*, vol. 2, pp. 90-97, 1963.

11. D. Ornstein and B. Weiss, "The Shannon-McMillan-Breiman theorem for a class of amenable groups," *Israel J. of Math*, vol. 44, pp. 53-60, 1983.

12. J. C. Oxtoby, "Ergodic Sets," *Bull. Amer. Math. Soc.*, vol. Volume 58, pp. 116-136, 1952.

13. V. A. Rohlin, "Selected topics from the metric theory of dynamical systems," *Uspechi Mat. Nauk.*, vol. 4, pp. 57-120, 1949. AMS Trans. (2) 49

14. P. C. Shields, "The ergodic and entropy theorems revisited," *IEEE Transactions on Information Theory*, vol. IT-33, pp. 263-266, March 1987.

15. J. von Neumann, "Zur operatorenmethode in der klassischen mechanik," *Ann. of Math.*, vol. 33, pp. 587-642, 1932.

8

PROCESS METRICS AND THE ERGODIC DECOMPOSITION

8.1 INTRODUCTION

Given two probability measures, say p and m, on a common probability space, how different or distant from each other are they? Similarly, given two random processes with distributions p and m, how distant are the processes from each other and what impact does such a distance have on their respective ergodic properties? The goal of this final chapter is to develop two quite distinct notions of the distance $d(p,m)$ between measures or processes and to use these ideas to delve further into the ergodic properties of processes and the ergodic decomposition. One metric, the distributional distance, measures how well the probabilities of certain important events match up for the two probability measures, and hence this metric need not have any relation to any underlying metric on the original sample space. In other words, the metric makes sense even when we are not putting probability measures on metric spaces. The second metric, the $\bar{\rho}$-distance (rho-bar distance) depends very strongly on a metric on the output space of the process and measures distance not by how different probabilities are, but by how well one process can be made to approximate another. The second metric is primarily useful in applications in information theory and statistics. Although these applications are beyond the scope of this book, the metric is presented both for comparison and because of the additional insight into ergodic properties the metric provides.

Such process metrics are of preliminary interest because they permit the quantitative assessment of how different two random processes are and how their ergodic properties differ. Perhaps more importantly, however, putting a metric on a space of random processes provides a topology for the space and hence a collection of Borel sets and a measurable space to which we can assign probability measures. This assignment of a probability measure to a space of processes provides a general definition of a mixture process and provides a means of delving more deeply into the ergodic decomposition of stationary measures developed in the previous chapter.

We have seen from the ergodic decomposition that all AMS measures have a stationary mean that can be considered as a mixture of stationary and ergodic components. Thus, for example, any stationary measure m can be considered as a mixture of the form

$$m = \int p_x \, dm(x) \, , \tag{8.1.1}$$

which is an abbreviation for

$$m(F) = \int p_x(F) \, dm(x) \, , \text{ all } F \in \boldsymbol{B} \, .$$

In addition, any m-integrable f can be integrated in two steps as

$$\int f(x) \, dm(x) = \int \left(\int dp_x(y) \, f(y)\right) dm(x). \tag{8.1.2}$$

Thus stationary measures can be considered to be mixtures of stationary and ergodic measures; that is, a stationary and ergodic measure is randomly selected from some collection and then used. In other words, we effectively have a probability measure on the space of all stationary and ergodic probability measures and the first measure is used to select one of the measures of the given class. Alternatively, we have a probability measure on the space of all probability measures on the given measurable space, but this super probability measure assigns probability one to the collection of ergodic and stationary measures. The preceding relations show that both probabilities and expectations of measurements over the resulting mixture can be computed as integrals of the probabilities and expectations over the component measures.

In applications of the theory of random processes such as information and communication theory, one is often concerned not only with ordinary functions or measurements, but also with functionals of probability measures, that is, mappings of the space of probability measures into the real line. Examples include the entropy rate of a dynamical system, the distortion-rate function of a random process with respect to a fidelity criterion, the information capacity of a noisy channel, the rate required to code a given process with nearly zero reproduction error, the rate required to code a given process for transmission with a specified fidelity, and the maximum rate of coded information that can be transmitted over a noisy channel with nearly zero error probability. (See, for example, the papers and commentary in Gray and Davisson [5].) All of these quantities are functionals of measures: a given process distribution or measure yields a real number as a value of the functional.

Given such a functional of measures, say $D(m)$, it is natural to inquire under what conditions the analog of (8.1.2) for functions might hold, that is, conditions under which

$$D(m) = \int D(p_x)\, dm(x) . \tag{8.1.3}$$

This is in general a much more difficult issue than before since unlike an ordinary measurement f, the functional depends on the underlying probability measure. One goal of this chapter to develop conditions under which (8.1.3) holds.

8.2 A METRIC SPACE OF MEASURES

We now focus on spaces of probability measures and on the structure of such spaces. We shall show how the ergodic decomposition provides an example of a probability measure on such a space. In Section 8.4 we shall study certain functionals defined on this space.

Let $(\Omega,\boldsymbol{B})$ be a measurable space such that $\boldsymbol{B}$ is countably generated (e.g., it is standard). We do not assume any metric structure for Ω. Define $\boldsymbol{P}((\Omega,\boldsymbol{B}))$ as the class of all probability measures on $(\Omega,\boldsymbol{B})$. It is easy to see that $\boldsymbol{P}((\Omega,\boldsymbol{B}))$ is a convex subset of the class of all finite measures on $(\Omega,\boldsymbol{B})$; that is, if m_1 and m_2 are in $\boldsymbol{P}((\Omega,\boldsymbol{B}))$ and $\lambda \in (0,1)$, then $\lambda m_1 + (1 - \lambda)m_2$ is also in $\boldsymbol{P}((\Omega,\boldsymbol{B}))$.

We can put a variety of metrics on $\boldsymbol{P}((\Omega,\boldsymbol{B}))$ and thereby make it a metric space. This will provide a notion of closeness of probability measures on the class. It will also provide a Borel field of subsets of $\boldsymbol{P}((\Omega,\boldsymbol{B}))$ on which we can put measures and thereby construct mixtures of measures in the class.

In this section we consider a type of metric that is not very strong, but will suffice for developing the ergodic decomposition of affine functionals. In the next section another metric will be considered, but it is tied to the assumption of Ω being itself a metric space and hence applies less generally.

Given any class $\boldsymbol{G} = \{F_i \; ; i = 1,2,, \cdots \}$ consisting of a countable collection of events we can define for any measures $p,m \in \boldsymbol{P}((\Omega,\boldsymbol{B}))$ the function

$$d_{\boldsymbol{G}}(p,m) = \sum_{i=1}^{\infty} 2^{-i} \mid p(F_i) - m(F_i) \mid . \tag{8.2.1}$$

If $\boldsymbol{G}$ contains a generating field, then $d_{\boldsymbol{G}}$ is a metric on $\boldsymbol{P}((\Omega,\boldsymbol{B}))$ since from Lemma 1.5.5 two measures defined on a common σ-field generated by a field are identical if and only if they agree on the generating field. We shall always assume that $\boldsymbol{G}$ contains such a field. We shall call such a distance a *distributional distance*. Usually we will require that $\boldsymbol{G}$ be a standard field and hence that the underlying measurable space $(\Omega,\boldsymbol{B})$ is standard. Occasionally, however, we will wish a larger class $\boldsymbol{G}$ but will not require that it be standard. The different requirements will be required for different results. When the class $\boldsymbol{G}$ is understood from context, the subscript will be dropped.

Let $(\boldsymbol{P}(\Omega,\boldsymbol{B}),d_{\boldsymbol{G}})$ denote the metric space of $\boldsymbol{P}((\Omega,\boldsymbol{B}))$ with the metric $d_{\boldsymbol{G}}$. A key property of spaces of probability measures on a fixed measurable space is given in the following lemma.

Lemma 8.2.1. The metric space $(\boldsymbol{P}((\Omega,\boldsymbol{B})),d_{\boldsymbol{G}})$ of all probability measures on a measurable space $(\Omega,\boldsymbol{B})$ with a countably generated sigma-field is separable if $\boldsymbol{G}$ contains a countable generating field. If also $\boldsymbol{G}$ is standard (as is possible if the underlying measurable space $(\Omega,\boldsymbol{B})$ is standard), then also $(\boldsymbol{P},d_{\boldsymbol{G}})$ is complete and hence Polish.

Proof. For each n let $\boldsymbol{A}_n$ denote the set of nonempty intersection sets or atoms of $\{F_1, \cdots ,F_n\}$, the first n sets in $\boldsymbol{G}$. (These are Ω sets, events in the original space.) For each set $G \in \boldsymbol{A}_n$ choose an arbitrary point x_G such that $x_G \in G$. We will show that the class of all measures of the form

$$r(F) = \sum_{G \in \boldsymbol{A}_n} p_G 1_F(x_G) ,$$

where the p_G are nonnegative and rational and satisfy

$$\sum_{G \in \boldsymbol{A}_n} p_G = 1 ,$$

forms a dense set in $\boldsymbol{P}((\Omega,\boldsymbol{B}))$. Since this class is countable, $\boldsymbol{P}((\Omega,\boldsymbol{B}))$ is separable. Observe that we are approximating all measures by finite sums of point masses. Fix a measure $m \in (\boldsymbol{P}((\Omega,\boldsymbol{B})),d_{\boldsymbol{G}})$ and an $\varepsilon > 0$. Choose n so large that $2^{-n} < \varepsilon/2$. Thus to match up two measures in $d = d_{\boldsymbol{G}}$, (8.2.1) implies that we must match up the probabilities of the first n sets in $\boldsymbol{G}$ since the contribution of the remaining terms is less that 2^{-n}. Define

$$r_n(F) = \sum_{G \in \boldsymbol{A}_n} m(G)1_F(x_G)$$

and note that

$$m(F) = \sum_{G \in \boldsymbol{A}_n} m(G)m(F|G) ,$$

where $m(F \mid G) = m(F \cap G)/m(G)$ is the elementary conditional probability of F given G if $m(G) > 0$ and is arbitrary otherwise. For convenience we now consider the preceding sums to be confined to those G for which $m(G) > 0$.

Since the G are the atoms of the first n sets $\{F_i\}$, for any of these F_i either $G \subset F_i$ and hence $G \cap F_i = G$ or $G \cap F_i = \emptyset$. In the first case $1_{F_i}(x_G) = m(F_i \mid G) = 1$, and in the second case $1_{F_i}(x_G) = m(F_i \mid G) = 0$, and hence in both cases

$$r_n(F_i) = m(F_i) ; i = 1,2,, \cdots ,n.$$

This implies that

$$d(r_n,m) \le \sum_{i=n+1}^{\infty} 2^{-i} = 2^{-n} \le \frac{\varepsilon}{2} .$$

Enumerate the atoms of $\boldsymbol{A}_n$ as $\{G_l; l = 1,2, \cdots ,L$, where $L \le 2^n$. For all l but the last $(l = L)$ pick a rational number p_{G_l} such that

$$m(G_l) - 2^{-l}\frac{\varepsilon}{4} \le p_{G_l} \le m(G_l) ;$$

that is, we choose a rational approximation to $m(G_l)$ that is slightly less than $m(G_l)$. We define p_{G_L} to force the rational approximations to sum to

1:

$$p_{G_L} = 1 - \sum_{l=1}^{L-1} p_{G_l} .$$

Clearly p_{G_L} is also rational and

$$| p_{G_L} - m(G_L) | = | \sum_{l=1}^{L-1} p_{G_l} - \sum_{l=1}^{L-1} m(G_l) | \le \sum_{l=1}^{L-1} | p_{G_l} - m(G_l) |$$

$$\le \frac{\varepsilon}{4} \sum_{l=1}^{L-1} 2^{-l} \le \frac{\varepsilon}{4} .$$

Thus

$$\sum_{G \in \boldsymbol{A}_n} | p_G - m(G) | = \sum_{l=1}^{L-1} | p_{G_l} - m(G_l) | + | p_{G_L} - m(G_L) | \le \frac{\varepsilon}{2} .$$

Define now the measure t_n by

$$t_n(F) = \sum_{G \in \boldsymbol{A}_n} p_G 1_F(x_G)$$

and observe that

$$d(t_n, r_n) = \sum_{i=1}^{n} 2^{-i} | t_n(F_i) - r_n(F_i) |$$

$$\sum_{i=1}^{n} 2^{-i} | \sum_{G \in \boldsymbol{A}_n} 1_{F_i}(p_G - m(G)) | \le \sum_{i=1}^{n} 2^{-i} \sum_{G \in \boldsymbol{A}_n} | p_G - m(G) | \le \frac{\varepsilon}{4} .$$

We now know that

$$d(t_n, m) \le d(t_n, r_n) + d(r_n, m) \le \frac{\varepsilon}{2} + \frac{\varepsilon}{2} = \varepsilon ,$$

which completes the proof of separability.

Next assume that m_n is a Cauchy sequence of measures. From the definition of d this can only be if also $m_n(F_i)$ is a Cauchy sequence of real numbers for each i. Since the real line is complete, this means that for each i $m_n(F_i)$ converges to something, say $\alpha(F_i)$. The set function α is defined on the class $\boldsymbol{G}$ and is clearly nonnegative, normalized, and finitely additive. Hence if $\boldsymbol{G}$ is also standard, then α extends to a probability measure on $(\Omega, \boldsymbol{B})$; that is, $\alpha \in \boldsymbol{P}((\Omega, \boldsymbol{B}))$. By construction $d(m_n, \alpha) \to 0$ as $n \to \infty$, and hence in this case $\boldsymbol{P}((\Omega, \boldsymbol{B}))$ is complete. □

Lemma 8.2.2. Assume as in the previous lemma that $(\Omega,\mathbf{B})$ is standard and $\mathbf{G}$ is a countable generating field. Then $(\mathbf{P}((\Omega,\mathbf{B})),d_{\mathbf{G}})$ is sequentially compact; that is, if $\{\mu_n\}$ is a sequence of probability measures in $\mathbf{P}(\Omega,\mathbf{B})$, then it has a subsequence μ_{n_k} that converges.

The proof of the lemma is based on a technical result that we state and prove separately.

Lemma 8.2.3. The space $[0,1]$ with the Euclidean metric $|x-y|$ is sequentially compact. The space $[0,1]^{\mathbf{I}}$ with $\mathbf{I}$ countable and the metric of Example 2.5.11 is sequentially compact.

Proof. The second statement follows from the first and Corollary 2.5.1. To prove the first statement, let S_i^n, $n = 1,2,\cdots,n$ denote the closed spheres $[(i-1)/n,i/n]$ having diameter $1/n$ and observe that for each $n = 1,2,\cdots$

$$[0,1] \subset \bigcup_{i=1}^{n} S_i^n .$$

Let $\{x_i\}$ be an infinite sequence of distinct points in $[0,1]$. For $n = 2$ there is an integer k_1 such that $K_1 = S_{k_1}^2$ contains infinitely many x_i. Since $S_{k_1}^n \subset \bigcup_i S_i^3$ we can find an integer k_2 such that $K_2 = S_{k_1}^2 \cap S_{k_2}^3$ contains infinitely many x_i. Continuing in this manner we construct a sequence of sets K_n such that $K_n \subset K_{n-1}$ and the diameter of K_n is less than 1/n. Since $[0,1]$ is complete, it follows from the finite intersection property of Lemma 3.2.2 that $\bigcap_n K_n$ consists of a single point $x \in [0,1]$. Since every neighborhood of x contains infinitely many of the x_i, x must be a limit point of the x_i. □

Proof of Lemma 8.2.2. Since $\mathbf{G}$ is a countable generating field, a probability measure μ in $\mathbf{P}$ is specified via the Carathéodory extension theorem by the infinite dimensional vector $\mu = \{\mu(F);F \in \mathbf{G}\} \in [0,1]^{\mathbf{Z}_+}$. Suppose that μ_n is a sequence of probability measures and let μ_n denote the corresponding sequence of vectors. From Lemma 8.2.3 $[0,1]^{\mathbf{Z}_+}$ is sequentially compact with respect to the product space metric, and hence there must exist a $\bar{\mu} \in [0,1]^{\mathbf{Z}_+}$ and a subsequence μ_{n_k} such that $\mu_{n_k} \to \bar{\mu}$. The vector $\bar{\mu}$ provides a set function $\bar{\mu}$ assigning a value $\bar{\mu}(F)$ for all sets F in the generating field $\mathbf{G}$. This function is nonnegative, normalized, and finitely additive. In particular, if $F = \bigcup_i F_i$ is a union of a finite number of disjoint events in $\mathbf{G}$, then

$$\bar{\mu}(F) = \lim_{k\to\infty} \mu_{n_k}(F) = \lim_{k\to\infty} \sum_i \mu_{n_k}(F_i)$$

$$= \sum_i \lim_{k\to\infty} \mu_{n_k}(F_i) = \sum_i \bar{\mu}(F_i) .$$

Since the alphabet is standard, $\bar{\mu}$ must extend to a probability measure on $(\Omega,\boldsymbol{B})$. By construction $\mu_{n_k}(F) \to \bar{\mu}(F)$ for all $F \in \boldsymbol{G}$ and hence $d_{\boldsymbol{G}}(\mu_{n_k},\bar{\mu}) \to 0$, completing the proof.

An Example

As an example of the previous construction, suppose that Ω is itself a Polish space with some metric d and that $\boldsymbol{G}$ is the field generated by the countable collection of spheres V_Ω of Lemma 3.2.1. Thus if $m_n \to \mu$ under $d_{\boldsymbol{G}}$, $m_n(F) \to \mu(F)$ for all spheres in the collection and all finite intersections and unions of such spheres. From (3.2.2) any open set G can be written as a union of elements of V_Ω, say $\bigcup_i V_i$. Given ε, choose an M such that

$$m(\bigcup_{i=1}^{M} V_i) > m(G) - \varepsilon .$$

Then

$$m(G) - \varepsilon < m(\bigcup_{i=1}^{M} V_i) = \lim_{n\to\infty} m_n(\bigcup_{i=1}^{M} V_i) \leq \liminf_{n\to\infty} m_n(G) .$$

Thus we have proved that convergence of measures under $d_{\boldsymbol{G}}$ implies that

$$\liminf_{n\to\infty} m_n(G) \geq m(G) \text{ ; all open } G .$$

Since the complement of a closed set is open, this also implies that

$$\limsup_{n\to\infty} m_n(G) \leq m(G) \text{ ; all closed } G .$$

Along with the preceding limiting properties for open and closed sets, convergence under $d_{\boldsymbol{G}}$ has strong implications for continuous functions. Recall from Section 3.1 that a function $f: \Omega \to R$ is *continuous* if given $\omega \in \Omega$ and $\varepsilon > 0$ there is a δ such that $d(\omega,\omega') < \delta$ ensures that $|f(\omega) - f(\omega')| < \varepsilon$. We consider two such results, one in the form we will later need and one to relate our notion of convergence to that of weak convergence of measures.

Lemma 8.2.4. If f is a nonnegative continuous function and $(\Omega,\boldsymbol{B})$ and $d_{\boldsymbol{G}}$ are as previously, then $d_{\boldsymbol{G}}(m_n,m) \to 0$ implies that

$$\limsup_{n\to\infty} E_{m_n} f \le E_m f .$$

Proof. For any n we can divide up $[0,\infty)$ into the countable collection of intervals $[k/n,(k+1)/n)$, $k = 0,1,2,\cdots$ that partition the nonnegative real line. Define the sets $G_k(n) = \{r\colon k/n \le f(r) < (k+1)/n\}$ and define the closed sets $F_k(n) = \{r\colon k/n \le f(r)\}$. Observe that $G_k(n) = F_k(n) - F_{k+1}(n)$. Since f is nonnegativefor any n

$$\sum_{k=0}^{\infty} \frac{k}{n} m(G_k(n)) \le E_m f < \sum_{k=0}^{\infty} \frac{k+1}{n} m(G_k(n)) = \sum_{k=0}^{\infty} \frac{k}{n} m(G_k(n)) + \frac{1}{n} .$$

The sum can be written as

$$\frac{1}{n}\sum_{k=0}^{\infty} k\left[m(F_k(n)) - m(F_{k+1}(n)\right] = \frac{1}{n}\sum_{k=0}^{\infty} m(F_k(n)) ,$$

and therefore

$$\frac{1}{n}\sum_{k=0}^{\infty} m(F_k(n)) \le E_m f .$$

By a similar construction

$$E_{m_n} f < \frac{1}{n}\sum_{k=0}^{\infty} m_n(F_k(n)) + \frac{1}{n}$$

and hence from the property for closed sets

$$\limsup_{n\to\infty} E_{m_n} f \le \frac{1}{n}\sum_{k=0}^{\infty} \limsup_{n\to\infty} m_n(F_k(n)) + \frac{1}{n}$$

$$\le \frac{1}{n}\sum_{k=0}^{\infty} m(F_k(n)) + \frac{1}{n} \le E_m f + \frac{1}{n} .$$

Since this is true for all n, the lemma is proved. □

Corollary 8.2.1. Given $(\Omega,\boldsymbol{B})$ and $d_{\boldsymbol{G}}$ as in the lemma, suppose that f is a bounded, continuous function. Then $d_{\boldsymbol{G}}(m_n,m) \to 0$ implies that

$$\lim_{n\to\infty} E_{m_n} f = E_m f .$$

Proof. If f is bounded we can find a finite constant c such that $g = f + c \geq 0$. g is clearly continuous so we can apply the lemma to g to obtain

$$\limsup_{n\to\infty} E_{m_n} g \leq E_m g = c + E_m f .$$

Since the left-hand side is $c + \limsup_{n\to\infty} E_{m_n} f$,

$$\limsup_{n\to\infty} E_{m_n} f \leq E_m f ,$$

as we found with positive functions. Now, however, we can apply the same result to $-f$ (which is also bounded) to obtain

$$\liminf_{n\to\infty} E_{m_n} f \geq E_m f ,$$

which together with the limit supremum result proves the corollary. □

Given a space of measures on a common metric space $(\Omega,\boldsymbol{B})$, a sequence of measures m_n is said to *converge weakly to* m if

$$\lim_{n\to\infty} E_{m_n} = E_m$$

for all bounded continuous functions f. We have thus proved the following:

Corollary 8.2.2. Given the assumptions of Lemma 8.2.4, if $d_{\boldsymbol{G}}(m_n,m) \to 0$, then m_n converges weakly to m.

We will not make much use of the notion of weak convergence, but the preceding result is included to relate the convergence that we use to weak convergence for the special case of Polish alphabets. For a more detailed study of weak convergence of measures, the reader is referred to Billingsley [1, 2].

8.3 THE RHO-BAR DISTANCE

In the previous section we considered a distance measure on probability measures. In this section we introduce a very different distance measure, which can be viewed as a metric on random variables and random processes. Unlike the distributional distances that force probabilities to be close, this distance will force the average distance between the random variables in the sense of an underlying metric to be close. Suppose now that we have two random variables X and Y with distributions P_X and P_Y with a common standard alphabet A. Let ρ be a pseudo-metric or a metric on A. Define the $\overline{\rho}$-*distance* between the random variables X and Y by

$$\overline{\rho}(P_X,P_Y) = \inf_{p \in \boldsymbol{P}} E_p \, \rho(X,Y) \, ,$$

Where $\boldsymbol{P} = \boldsymbol{P}(P_X,P_Y)$ is the collection of all measures on $(A \times A, \boldsymbol{B}_A \times \boldsymbol{B}_A)$ with P_X and P_Y as marginals; that is,

$$p(A \times F) = P_Y(F) \; ; \; F \in \boldsymbol{B}_A \, ,$$

and

$$p(G \times A) = P_X(G) \; ; \; G \in \boldsymbol{B}_A \, .$$

Note that $\boldsymbol{P}$ is not empty since, for example, it contains the product measure $P_X \times P_Y$.

Levenshtein [9] and Vasershtein [16] studied this quantity for the special case where A is the real line and ρ is the Euclidean distance. Vallender [15] provided several specific evaluations of the distance in this case. Ornstein [11] developed the distance and many of its properties for the special case where A is a discrete space and ρ is the Hamming distance. In this case the $\overline{\rho}$-distance is called the $\overline{d}$-distance. R. Dobrushin [3] has suggested that because of the common suffix in the names of its originators this distance between distributions should be called the *shtein* or *stein distance*.

The $\overline{\rho}$-distance can be extended to vectors and processes in a natural way. Suppose now that $\{X_n\}$ is a process with process distribution m_X and that $\{Y_n\}$ is a process with process distribution m_Y. Let P_{X^n} and P_{Y^n} denote the induced finite dimensional distributions. If ρ_1 is a metric on the coordinate alphabet A, then let ρ_n be the induced metric on the product space A^n defined by

$$\rho_n(x^n, y^n) = \sum_{i=0}^{n-1} \rho_1(x_i, y_i) .$$

Let $\overline{\rho}_n$ denote the corresponding $\overline{\rho}$ distance between the n dimensional distributions describing the random vectors. We define the $\overline{\rho}$-distance between the processes by

$$\overline{\rho}(m_X, m_Y) = \sup_n \frac{1}{n} \overline{\rho}_n(P_{X^n}, P_{Y^n}) ,$$

that is, the $\overline{\rho}$-distance between two processes is the maximum of the $\overline{\rho}$-distances per symbol between n-tuples drawn from the process. The $\overline{\rho}$-distance for processes was introduced by Ornstein for the special case of the Hamming distance [12, 11] and was extended to general distances by Gray, Neuhoff, and Shields [6]. The results of this section largely follow these two references.

The following theorem contains the fundamental properties of this process metric.

Theorem 8.3.1. Suppose that m_X and m_Y are stationary process distributions with a common standard alphabet A and that ρ_1 is a pseudo-metric on A and that ρ_n is defined on A^n in an additive fashion as before. Then

(a) $\lim_{n \to \infty} n^{-1} \overline{\rho}_n(P_{X^n}, P_{Y^n})$ exists and equals $\sup_n n^{-1} \overline{\rho}_n(P_{X^n}, P_{Y^n})$.

(b) $\overline{\rho}_n$ and $\overline{\rho}$ are pseudo-metrics. If ρ_1 is a metric, then $\overline{\rho}_n$ and $\overline{\rho}$ are metrics.

(c) If m_X and m_Y are both IID, then $\overline{\rho}(m_X, m_Y) = \overline{\rho}_1(P_{X_0}, P_{Y_0})$.

(d) Let $\boldsymbol{P}_s = \boldsymbol{P}_s(m_X, m_Y)$ denote the collection of all stationary distributions p_{XY} having m_X and m_Y as marginals, that is, distributions on $\{X_n, Y_n\}$ with coordinate processes $\{X_n\}$ and $\{Y_n\}$ having the given distributions. Define the process distance measure $\overline{\rho}'$

$$\overline{\rho}'(m_X, m_Y) = \inf_{p_{XY} \in \boldsymbol{P}_s} E_{p_{XY}} \rho(X_0, Y_0) .$$

Then

$$\overline{\rho}(m_X, m_Y) = \overline{\rho}'(m_X, m_Y) ;$$

that is, the limit of the finite dimensional minimizations is given by a minimization over stationary processes.

(e) Suppose that m_X and m_Y are both stationary and ergodic. Define $\boldsymbol{P}_e = \boldsymbol{P}_e(m_X,m_Y)$ as the subset of $\boldsymbol{P}_s$ containing only ergodic processes, then

$$\overline{\rho}(m_X,m_Y) = \inf_{p_{XY} \in \boldsymbol{P}_e} E_{p_{XY}} \rho(X_0,Y_0) ,$$

(f) Suppose that m_X and m_Y are both stationary and ergodic. Let G_X denote a collection of generic sequences for m_X in the sense of Section 7.4. Recall that by measuring relative frequencies on generic sequences one can deduce the underlying stationary and ergodic measure that produced the sequence. Similarly let G_Y denote a set of generic sequences for m_Y. Define the process distance measure

$$\overline{\rho}''(m_X,m_Y) = \inf_{x \in G_X, y \in G_Y} \limsup_{n\to\infty} \frac{1}{n} \sum_{i=0}^{n-1} \rho_1(x_0,y_0) .$$

Then

$$\overline{\rho}(m_X,m_Y) = \overline{\rho}''(m_X,m_Y) .$$

that is, the $\overline{\rho}$ distance gives the minimum long term time average distance obtainable between generic sequences from the two processes.

(g) The infima defining $\overline{\rho}_n$ and $\overline{\rho}'$ are actually minima.

Proof.

(a) Suppose that p^N is a joint distribution on $(A^N \times A^N, \boldsymbol{B}_A{}^N \times \boldsymbol{B}_A{}^N)$ describing (X^N, Y^N) that approximately achieves $\overline{\rho}_N$, e.g., p^N has P_{X^N} and P_{Y^N} as marginals and for $\varepsilon > 0$

$$E_{p^N} \rho(X^N, Y^N) \le \overline{\rho}_N(P_{X^N},P_{Y^N}) + \varepsilon .$$

For any $n < N$ let p^n be the induced distribution for (X^n, Y^n) and p_n^{N-n} that for (X_n^{N-n}, Y_n^{N-n}). Then since the processes are stationary $p^n \in \boldsymbol{P}_n$ and $p_n^{N-n} \in \boldsymbol{P}_{N-n}$ and hence

$$E_{p^N} \rho(X^N, Y^N) = E_{p^n} \rho(X^n, Y^n) + E_{p_n^{N-n}} \rho(X_n^{N-n}, Y_n^{N-n})$$

$$\ge \overline{\rho}_n(P_{X^n},P_{Y^n}) + \overline{\rho}_{N-n}(P_{X^{N-n}},P_{Y^{N-n}}) .$$

Since ε is arbitrary we have shown that if $a_n = \bar{\rho}_n(P_{X^n}, P_{Y^n})$, then for all $N > n$

$$a_N \geq a_n + a_{N-n} ;$$

that is, the sequence a_n is *superadditive.* It then follows from Lemma 7.5.2 (since the negative of a superadditive sequence is a subadditive sequence) that

$$\lim_{n\to\infty} \frac{a_n}{n} = \sup_{n\geq 1} \frac{a_n}{n} ,$$

which proves (a).

(b) Since ρ_1 is a pseudo-metric, ρ_n is also a pseudo-metric and hence $\bar{\rho}_n$ is nonnegative and symmetric in its arguments. Thus $\bar{\rho}$ is also nonnegative and symmetric in its arguments. Thus we need only prove that the triangle inequality holds. Suppose that m_X, m_Y, and m_Z are three process distributions on the same sequence space. Suppose that p approximately yields $\bar{\rho}_n(P_{X^n}, P_{Z^n})$ and that r approximately yields $\bar{\rho}_n(P_{Z^n}, P_{Y^n})$. These two distributions can be used to construct a joint distribution q for the triple (X^n, Z^n, Y^n) such that $X^n \to Z^n \to Y^n$ is a Markov chain and such that the coordinate vectors have the correct distribution. In particular, r and m_Z^n together imply a conditional distribution $p_{Y^n | Z^n}$ and p and P_{X^n} together imply a conditional distribution $p_{Z^n | X^n}$. Since the spaces are standard these can be chosen to be regular conditional probabilities and hence the distribution specified by its values on rectangles as a cascade of finite dimensional channels: If we define $\nu_x(F) = P_{Z^n | X^n}(F \mid x^n)$, then

$$q(F\times G\times D) = \int_D dP_{X^n}(x^n) \int_G d\nu_x(z^n) P_{Y^n | Z^n}(F \mid z^n) .$$

For this distribution, since ρ_n is a pseudo-metric,

$$\bar{\rho}_n(P_{X^n}, P_{Y^n}) \leq E\rho_n(X^n, Y^n) \leq E\rho_n(X^n, Z^n) + E\rho_n(Z^n, Y^n)$$

$$\leq \bar{\rho}_n(P_{X^n}, P_{Z^n}) + \bar{\rho}_n(P_{Z^n}, P_{Y^n}) + 2\varepsilon ,$$

which proves the triangle inequality for $\bar{\rho}_n$. Taking the limit as $n \to \infty$ proves the result for $\bar{\rho}$. If ρ_1 is a metric, then so is ρ_n. Suppose now that $\bar{\rho} = 0$ and hence also $\bar{\rho}_n = 0$ for all n. This implies that there exists a $p^n \in \boldsymbol{P}_n$ such that $p^n(x^n, y^n : x^n \neq y^n) = 0$ since $\rho_n(x^n, y^n)$ is 0 only if $x^n = y^n$. This is only possible, however, if $P_{X^n}(F) = P_{Y^n}(F)$ for all events F, that is, the two marginals are identical. Since this is true for all N, the two processes must be the same.

(c) Suppose that p^1 approximately yields $\overline{\rho}_1$, that is, has the correct marginals and

$$E_{p^1}\,\rho_1(X_0,Y_0) \le \overline{\rho}_1(P_{X_0},P_{Y_0}) + \varepsilon\,.$$

Then for any n the product measure p^n defined by

$$p^n = \mathop{\times}_{i=0}^{n-1} p^1$$

has the correct vector marginals for X^n and Y^n since they are IID, and hence $p^n \in \boldsymbol{P}_n$, and therefore

$$\overline{\rho}_n(P_{X^n},P_{Y^n}) \le E_{p^n}\,\rho_n(X^n,Y^n)$$
$$= nE_{p^1}\,\rho(X_0,Y_0) \le n(\overline{\rho}_1(P_{X_0},P_{Y_0}) + \varepsilon)\,.$$

Since by definition $\overline{\rho}$ is the supremum of $n^{-1}\overline{\rho}_n$, this proves (c).

(d) Given $\varepsilon > 0$ let $p \in \boldsymbol{P}_s$ be such that $E_p\,\rho_1(X_0,Y_0) \le \overline{\rho}'(m_X,m_Y) + \varepsilon$. The induced distribution on $\{X^n,Y^n\}$ is then contained in $\boldsymbol{P}_n$, and hence using the stationarity of the processes

$$\overline{\rho}_n(P_{X^n},P_{Y^n}) \le E\rho_n(X^n,Y^n) = nE\rho_1(X_0,Y_0) \le n\overline{\rho}'(m_X,m_Y)\,,$$

and therefore $\overline{\rho}' \ge \overline{\rho}$ since ε is arbitrary.

Let $p^n \in \boldsymbol{P}_n$, $n=1,2,\cdots$ be a sequence of measures such that

$$E_{p^n}[\rho_n(X^n,Y^n)] \le \overline{\rho}_n + \varepsilon_n$$

where $\varepsilon_n \to 0$ as $n\to\infty$. Let q_n denote the product measure on the sequence space induced by the p_n, that is, for any N and N-dimensional rectangle $F = \times_{i \in \boldsymbol{I}} F_i$ with all but a finite number N of the F_i being $A \times B$,

$$q(F) = \prod_{j \in \boldsymbol{I}} p^n(\mathop{\times}_{i=jn}^{(j+1)n-1} F_i)\,;$$

that is, q_n is the pair process distribution obtained by gluing together independent copies of p^n. This measure is obviously N-stationary and we form a stationary process distribution p_n by defining

$$p_n(F) = \frac{1}{n}\sum_{i=0}^{n-1} q_n(T^{-i}F)$$

for all events F. If we now consider the metric space of all process distributions on $(A^I \times A^I, B_A{}^I \times B_A{}^I)$ of Section 8.2 with a metric d_G based on a countable generating field of rectangles (which exists since the spaces are standard), then from Lemma 8.2.2 the space is sequentially compact and hence there is a subsequence p_{n_k} that converges in the sense that there is a measure p and $p_{n_k}(F) \to p(F)$ for all F in the generating field. Assuming the generating field to contain all shifts of field members, the stationarity of the p_n implies that $p(T^{-1}F) = p(F)$ for all rectangles F in the generating field and hence p is stationary. A straightforward computation shows that for all n p induces marginal distributions on X^n and Y^n of P_{X^n} and P_{Y^n} and hence $p \in \boldsymbol{P}_s$. In addition,

$$E_p \rho_1(X_0,Y_0) = \lim_{k\to\infty} E_{p_{n_k}} \rho(X_0,Y_0)$$

$$= \lim_{k\to\infty} n_k^{-1} \sum_{i=0}^{n_k-1} E_{q_{n_k}} \rho(X_i,Y_i) = \lim_{k\to\infty} (\bar{\rho}_{n_k} + \varepsilon_{n_k}) = \bar{\rho} ,$$

proving that $\bar{\rho}' \le \bar{\rho}$ and hence that they are equal.

(e) If $p \in \boldsymbol{P}_s$ nearly achieves $\bar{\rho}'$, then from the ergodic decomposition it can be expressed as a mixture of stationary and ergodic processes p_λ and that

$$E_p \rho_1 = \int dW(\lambda) E_{p_\lambda} \rho_1 .$$

All of the p_λ must have m_X and m_Y as marginal process distributions (since they are ergodic) and at least one value of λ must yield an $E_{p_\lambda} \rho_1$ no greater than the preceding average. Thus the infimum over stationary and ergodic pair processes can be no worse than that over stationary processes. (It can obviously be no better.)

(f) Suppose that $x \in G_X$ and $y \in G_Y$. Fix a dimension n and define for each N and each $F \in \boldsymbol{B}_A^N$, $G \in \boldsymbol{B}_A^N$

$$\mu_N(F \times G) = \frac{1}{N} \sum_{i=0}^{N-1} 1_{F\times G}(T^i(x,y)) .$$

μ_N extends to a measure on $(A^N \times A^N, \boldsymbol{B}_A^N \times \boldsymbol{B}_A^N)$. Since the sequences are assumed to be generic,

$$\mu_N(F \times A^I) \underset{N\to\infty}{\to} P_{X^n}(F) ,$$

$$\mu_N(A^I \times G) \underset{N\to\infty}{\to} P_{Y^n}(G) ,$$

for all events F and G and generating fields for $\boldsymbol{B}_A^N$ and $\boldsymbol{B}_A^N$, respectively. From Lemma 8.2.2 (as used previously) there is a convergent subsequence of the μ_N. If μ is such a limit point, then for generating F, G

$$\mu(F \times A^I) = P_{X^n}(F) ,$$

$$\mu(A^I \times G) = P_{Y^n}(G) ,$$

which implies that the same relation must hold for all F and G (since the measures agree on generating events). Thus $\mu \in \boldsymbol{P}_n$. Since μ_N is just a sample distribution,

$$E_{\mu_N}[\rho_n(X^n, Y^n)] = \frac{1}{N}\sum_{j=0}^{N-1} \rho_n(x_j^n, y_j^n) = \frac{1}{N}\sum_{j=0}^{N-1} [\sum_{i=0}^{n-1} \rho_1(x_{i+j}, y_{i+j})] .$$

Thus

$$\frac{1}{n}E_\mu[\rho_n(X^n, Y^n)] \le \limsup_{N\to\infty} \frac{1}{N}\sum_{i=0}^{N-1} \rho_1(x_i, y_i)$$

and hence $\overline{\rho} \le \overline{\rho}''$.

Choose for $\varepsilon > 0$ a stationary measure $p \in \boldsymbol{P}_s$ such that

$$E_p\, \rho_1(X_0, Y_0) = \overline{\rho}' + \varepsilon .$$

Since p is stationary,

$$\rho_\infty(x,y) = \lim_{n\to\infty} \frac{1}{n}\sum_{i=0}^{n-1} \rho_1(x_i, y_i)$$

exists with probability one. Furthermore, since the coordinate processes are stationary and ergodic, with probability one x and y must be generic. Thus

$$\rho_\infty(x,y) \ge \overline{\rho}'' .$$

Taking the expectation with respect to p and invoking the ergodic theorem,

$$\overline{\rho}'' \le E_p\, \rho_\infty = E_p\, \rho_1(X_0, Y_0) \le \overline{\rho}' + \varepsilon .$$

Hence $\bar{\rho}'' \leq \bar{\rho}' = \bar{\rho}$ and hence $\bar{\rho}'' = \bar{\rho}$.

(g) Consider $\bar{\rho}'$. Choose a sequence $p_n \in \boldsymbol{P}_s$ such that

$$E_{p_n} \rho_1(X_0,Y_0) \leq \bar{\rho}'(m_X,m_Y) + 1/n .$$

From Lemma 8.2.2 p_n must have a convergent subsequence with limit, say, p. p must be stationary and have the correct marginal processes (since all of the p_n are) and from the preceding equation

$$E_p \rho_1(X_0,Y_0) \leq \bar{\rho}'(m_X,m_Y) .$$

A similar argument works for $\bar{\rho}_n$. □

The Prohorov Distance

Another distance measure on random variables that is often encountered in the literature is the *Prohorov distance*. This distance measure resembles the $\bar{\rho}$-distance for random variables and has many applications in probability theory, ergodic theory, and statistics. (See, for example, Dudley [4], Billingsley [1,2], Moser, *et al.* [10], Strassen [14], and Hample [7].) The Prohorov distance unfortunately does not extend in a useful way to provide a distance measure between processes and it does not ensure the generic sequence closeness properties that the $\bar{\rho}$ distortion possesses. We present a brief description, however, for completeness and to provide the relation between the two distance measures on random vectors. The development follows [13].

Analogous to the definition of the nth order $\bar{\rho}$ distance, we define for positive integer n the nth order Prohorov distance $\Pi_n = \Pi_n(P_{X^n},P_{Y^n})$ between two distributions for random vectors X^n and Y^n with a common standard alphabet A^n and with a pseudo-metric or metric ρ_n defined on A^n by

$$\Pi_n = \inf_{p \in \boldsymbol{P}} \inf \{\gamma : p(x^n,y^n : \frac{1}{n}\rho_n(x^n,y^n) > \gamma) \leq \gamma\}$$

where $\boldsymbol{P}$ is as in the definition of $\bar{\rho}_n$, that is, it is the collection of all joint distributions P_{X^n,Y^n} having the given marginal distributions P_{X^n} and P_{Y^n}. It is known from Strassen [14] and Dudley [4] that the infimum is actually a minimum. It can be proved in a manner similar to that for the corresponding $\bar{\rho}_n$ result.

Lemma 8.3.1. The $\bar{\rho}$ and Prohorov distances satisfy the following inequalities:

$$\Pi_n(P_{X^n}, P_{Y^n})^2 \leq \frac{1}{n}\bar{\rho}_n(P_{X^n}, P_{Y^n}) \leq \bar{\rho}(m_X, m_Y) . \tag{8.3.1}$$

If ρ_1 is bounded, say by $\rho_{\max}$, then

$$\frac{1}{n}\bar{\rho}_n(P_{X^n}, P_{Y^n}) \leq \Pi_n(P_{X^n}, P_{Y^n})(1 + \rho_{\max}) . \tag{8.3.2}$$

If there exists an a^* such that

$$\begin{aligned} E_{P_{X_0}}(\rho_1(X_0, a^*)^2) &\leq \rho^* < \infty , \\ E_{P_{Y_0}}(\rho_1(Y_0, a^*)^2) &\leq \rho^* < \infty , \end{aligned} \tag{8.3.3}$$

then

$$\frac{1}{n}\bar{\rho}_n(P_{X^n}, P_{Y^n}) \leq \Pi_n(P_{X^n}, P_{Y^n}) + 2(\rho^* \Pi_n(P_{X^n}, P_{Y^n}))^{1/2} . \tag{8.3.4}$$

Comment. The Lemma shows that the $\bar{\rho}$ distance is stronger than the Prohorov in the sense that making it small also makes the Prohorov distance small. If the underlying metric either is bounded or has a reference letter in the sense of (8.3.3), then the two distances provide the same topology on nth order distributions.

Proof. From Theorem 8.3.1 the infimum defining the $\bar{\rho}_n$ distance is actually a minimum. Suppose that p achieves this minimum; that is,

$$E_p \, \rho_n(X^n, Y^n) = \bar{\rho}_n(P_{X^n}, P_{Y^n}) .$$

For simplicity we shall drop the arguments of the distance measures and just refer to Π_n and $\bar{\rho}_n$. From the Markov inequality

$$p(x^n, y^n : \frac{1}{n}\rho_n(x^n, y^n) > \varepsilon) \leq \frac{E_p \frac{1}{n}\rho_n(X^n, Y^n)}{\varepsilon} = \frac{\frac{1}{n}\bar{\rho}_n}{\varepsilon} .$$

Setting $\varepsilon = (n^{-1}\bar{\rho}_n)^{1/2}$ then yields

$$p(x^n, y^n : \frac{1}{n}\rho_n(x^n, y^n) > (n^{-1}\bar{\rho}_n)^{1/2}) \leq (n^{-1}\bar{\rho}_n)^{1/2} ,$$

proving the first part of the first inequality. The fact that $\bar{\rho}$ is the supremum of $\bar{\rho}_n$ completes the first inequality.

If ρ_1 is bounded, let p yield Π_n and write

$$\frac{1}{n}\bar{\rho}_n \leq E_p \frac{1}{n}\rho_n(X^n, Y^n)$$

$$\leq \Pi_n + p(x^n, y^n : \frac{1}{n}\rho_n(x^n, y^n) > \Pi_n)\rho_{\max} = \Pi_n(1 + \rho_{\max}) \, .$$

Next consider the case where (8.3.3) holds. Again let p yield Π_n, and write, analogously to the previous equations,

$$\frac{1}{n}\bar{\rho}_n \leq \Pi_n + \int_G dp(x^n, y^n)\frac{1}{n}\rho_n(x^n, y^n)$$

where

$$G = \{x^n, y^n : n^{-1}\rho_n(x^n, y^n) > \Pi_n\} \, .$$

Let 1_G be the indicator of G and let E denote expectation with respect to p. Since ρ_n is a pseudo-metric, the triangle inequality and the Cauchy-Schwartz inequality yield

$$\frac{1}{n}\bar{\rho}_n \leq \frac{1}{n}E\rho_n(X^n, Y^n) \leq \Pi_n + \frac{1}{n}E(\rho_n(X^n, a^{*n})1_G) + E(\frac{1}{n}\rho_n(Y^n, a^{*n})1_G)$$

$$\leq \Pi_n + [E(\frac{1}{n}\rho_n(X^n, a^{*n}))^2]^{1/2}[E(1_G^2)]^{1/2} + [E(\frac{1}{n}\rho_n(Y^n, a^{*n}))^2]^{1/2}[E(1_G^2)]^{1/2} \, .$$

Applying the Cauchy-Schwartz inequality for sums and invoking (8.3.3) yield

$$E(\frac{1}{n}\rho_n(X^n, a^{*n}))^2 = E\left[(\frac{1}{n}\sum_{i=0}^{n-1}\rho_1(X_i, a^*))^2\right]$$

$$\leq E(\frac{1}{n}\sum_{i=0}^{n-1}\rho_1(X_i, a^*)^2) \leq \rho^* \, .$$

Thus

$$\frac{1}{n}\bar{\rho}_n \leq 2(\rho^*)^{1/2}(p(G))^{1/2} = \Pi_n + 2(\rho^*)^{1/2}\Pi_n^{1/2} \, ,$$

completing the proof.

8.4 MEASURES ON MEASURES

We next consider placing measures on the space of measures. To accomplish this we focus on a distributional distance for a particular $\boldsymbol{G}$ and hence on a particular metric for generating the Borel field of measures. We require that $(\Omega,\boldsymbol{B})$ be standard and take $\boldsymbol{G}$ to be a standard generating field. We will denote the resulting metric simply by d. Thus $\boldsymbol{P}((\Omega,\boldsymbol{B}))$ will itself be Polish from Lemma 8.2.1. We will usually wish to assign nonzero probability only to a nice subset of this space, e.g., the subset of all stationary measures. Henceforth let $\hat{\boldsymbol{P}}$ be a closed and convex subset of $\boldsymbol{P}((\Omega,\boldsymbol{B}))$. Since $\hat{\boldsymbol{P}}$ is a closed subset of a Polish space, it is also Polish under the same metric. Let $\hat{\boldsymbol{B}} = Borel(\hat{\boldsymbol{P}},d)$ denote the Borel field of subsets of $\hat{\boldsymbol{P}}$ with respect to the metric d. Since $\hat{\boldsymbol{P}}$ is Polish, this measurable space $(\hat{\boldsymbol{P}},\hat{\boldsymbol{B}})$ is standard.

Now consider a probability measure W on the standard space $(\hat{\boldsymbol{P}},\hat{\boldsymbol{B}})$. This probability measure thus is a member of the class $\boldsymbol{P}((\hat{\boldsymbol{P}},\hat{\boldsymbol{B}}))$ of all probability measures on the standard measurable space $(\hat{\boldsymbol{P}},\hat{\boldsymbol{B}})$, where $\hat{\boldsymbol{P}}$ is a closed convex subset of the collection of all probability measures on the original standard measurable space $(\Omega,\boldsymbol{B})$. (The author promises at this point not to put measures on the space $\boldsymbol{P}((\hat{\boldsymbol{P}},\hat{\boldsymbol{B}}))$. Spaces of measures on spaces of measures are complicated enough!)

We now argue that any such W induces a new measure $m \in \hat{\boldsymbol{P}}$ via the mixture relation

$$m = \int \alpha \, dW(\alpha) \, . \tag{8.4.1}$$

To be precise and to make sense of the preceding formula, consider for any fixed $F \in \boldsymbol{G}$ and any measure $\alpha \in \hat{\boldsymbol{P}}$ the expectation $E_\alpha 1_F$. We can consider this as a function (or functional) mapping probability measures in $\hat{\boldsymbol{P}}$ into $\mathbf{R}$, the real line. This function is continuous, however, since given $\varepsilon > 0$, if $F = F_i$ we can choose $d(\alpha,m) < \varepsilon 2^{-i}$ and find that

$$| E_\alpha 1_F - E_m 1_F | = | \alpha(F) - m(F) | \le 2^i \sum_{j=1}^{\infty} 2^{-j} | \alpha(F_j) - m(F_j) | \le \varepsilon \, .$$

Since the mapping is continuous, it is measurable from Lemma 3.1.6. Since it is also bounded, the following integrals all exist:

$$m(F) = \int E_\alpha 1_F \, dW(\alpha) \; ; \; F \in \boldsymbol{G} \, . \tag{8.4.2}$$

This defines a set function on $\boldsymbol{G}$ that is nonnegative, normalized, and finitely additive from the usual properties of integration. Since the space

$(\Omega,\boldsymbol{B})$ is standard, this set function extends to a unique probability measure on $(\Omega,\boldsymbol{B})$, which we also denote by m. This provides a mapping from any measure $W \in \boldsymbol{P}(\hat{\boldsymbol{P}},\hat{\boldsymbol{B}})$ to a unique measure $m \in \hat{\boldsymbol{P}} \subset \boldsymbol{P}((\Omega,\boldsymbol{B}))$. This mapping is denoted symbolically by (8.4.1) and is defined by (8.4.2). We will use the notation $W \rightarrow m$ to denote it. The induced measure m is called the *barycenter* of W.

Before developing the properties of barycenters, we return to the ergodic decomposition to show how it fits into this framework.

8.5 THE ERGODIC DECOMPOSITION REVISITED

Say we have an stationary dynamical system $(\Omega,\boldsymbol{B},m,T)$ with a standard measurable space. The ergodic decomposition then implies a mapping ψ: $\Omega,\rightarrow\boldsymbol{P}((\Omega,\boldsymbol{B}))$ defined by $x \rightarrow p_x$ as given in the ergodic decomposition theorem. In fact, this mapping is into the subset of ergodic and stationary measures. In this section we show that this mapping is measurable, relate integrals in the two spaces, and determine the barycenter of the induced measure on measures. In the process we obtain an alternative statement of the ergodic decomposition.

Lemma 8.5.1. Given a stationary system $(\Omega,\boldsymbol{B},m,T)$ with a standard measurable space, let $\boldsymbol{P} = \boldsymbol{P}(\Omega,\boldsymbol{B})$ be the space of all probability measures on $(\Omega,\boldsymbol{B})$ with the topology of the previous section. Then the mapping ψ: $\Omega\rightarrow\boldsymbol{P}$ defined by the mapping $x \rightarrow p_x$ of the ergodic decomposition is a measurable mapping. (All generic points x are mapped into the corresponding ergodic component; all remaining points are mapped into a common arbitrary stationary ergodic process.)

Proof. To prove ψ measurable, it suffices to show that any set of the form $V = \{\beta\colon d(\alpha,\beta) < r\}$ has an inverse image $\psi^{-1}(V) \in \boldsymbol{B}$. This inverse image is given by

$$\{x : \sum_{i=1}^{\infty} 2^{-i} \mid p_x(F_i) - \alpha(F_i) \mid < r\} .$$

Defining the function γ: $\Omega\rightarrow\mathbf{R}$ by

$$\gamma(x) = \sum_{i=1}^{\infty} 2^{-i} \mid p_x(F_i) - \alpha(F_i) \mid ,$$

then the preceding set can also be expressed as $\gamma^{-1}((-\infty,r))$. The function γ, however, is a sum of measurable real-valued functions and hence is itself measurable, and hence this set is in $\boldsymbol{B}$. □

Given the ergodic decomposition $\{p_x\}$ and a stationary measure m, define the measure $W = m\psi^{-1}$ on $(\hat{\boldsymbol{P}},\hat{\boldsymbol{B}})$. Given any measurable function $f: \hat{\boldsymbol{P}} \to \mathbf{R}$, from the measurability of ψ and the chain rule of integration

$$\int f(\alpha)\, dW(\alpha) = \int f(\alpha) dm\psi^{-1}(\alpha) =$$

$$\int f(\psi x) dm(x) = \int f(p_x)\, dm(x) .$$

Thus the rather abstract looking integrals on the space of measures equal the desired form suggested by the ergodic decomposition, that is, one integral exists if and only if the other does, in which case they are equal. This demonstrates that the existence of certain integrals in the more abstract space implies their existence in the more concrete space. The more abstract space, however, is more convenient for developing the desired results because of its topological structure. In particular, the ergodic decomposition does not provide all possible probability measures on the original probability space, only the subset of ergodic and stationary measures. The more general space is required by some of the approximation arguments. For example, one cannot use the previous separability lemma to approximate arbitrary ergodic processes by point masses on ergodic processes since a mixture of distinct ergodic processes is not ergodic!

Letting $f(\alpha) = E_\alpha 1_F = \alpha(F)$ for $F \in \boldsymbol{G}$ we see immediately that

$$\int E_\alpha 1_F \; dm\psi^{-1}(\alpha) = \int p_x(F)\, dm(x) ,$$

and hence since the right side is

$$m(F) = \int E_\alpha 1_F \, dm\psi^{-1}(\alpha) ;$$

that is, m is the barycenter of the measure induced on the space of all measures by the original stationary measure m and the mapping defined by the ergodic decomposition.

Since the ergodic decomposition puts probability one on stationary and ergodic measures, we should be able to cut down the measure W to these sets and hence confine interest to functionals of stationary measures. In order to do this we need only show that these sets are measurable, that is, in the Borel field of all probability measures on $(\Omega,\boldsymbol{B})$. This is accomplished next.

Lemma 8.5.2. Given the assumptions of the previous lemma, assume also that $\boldsymbol{G}$ is chosen to include a countable generating field and that $\boldsymbol{G}$ is chosen so that $F \in \boldsymbol{G}$ implies $T^{-1}F \in \boldsymbol{G}$. (If this is not true, simply expand $\boldsymbol{G}$ to $\bigcup_i T^{-i}\boldsymbol{G}$. For example, $\boldsymbol{G}$ could be a countable generating family of cylinders.) Let $\boldsymbol{P}_s$ and $\boldsymbol{P}_e$ denote the subsets of $\hat{\boldsymbol{P}}$ consisting respectively of all stationary measures and all stationary and ergodic measures. Then $\boldsymbol{P}_s$ is closed (and hence measurable) and convex. $\boldsymbol{P}_e$ is measurable.

Proof. Let $\alpha_n \in \boldsymbol{P}_s$ be a sequence of stationary measures converging to a measure α and hence for every $F \in \boldsymbol{G}$ $\alpha_n(F) \to \alpha(F)$. Since $T^{-1}F \in \boldsymbol{G}$ this means that $\alpha_n(T^{-1}F) \to \alpha(T^{-1}F)$. Since the α_n are stationary, however, this sequence is the same as $\alpha_n(F)$ and hence converges to $\alpha(F)$. Thus $\alpha(T^{-1}F) = \alpha(F)$ on a generating field and hence α is stationary. Thus $\boldsymbol{P}_s$ is a closed subset under d and hence must be in the Borel field.

From Lemma 6.7.4, an $\alpha \in \boldsymbol{P}_s$ is also in $\boldsymbol{P}_e$ if and only if

$$\alpha \in \bigcap_{F \in \boldsymbol{G}} G(F),$$

where here

$$G(F) = \{m: \lim_{n\to\infty} n^{-1} \sum_{i=0}^{n-1} m(T^{-i}F \cap F) = m(F)^2\}$$

$$= \{m: \lim_{n\to\infty} n^{-1} \sum_{i=0}^{n-1} E_m 1_{T^{-i}F \cap F} = (E_m 1_F)^2\}.$$

The $E_m 1_F$ terms are all measurable functions of m since the sets are all in the generating field. Hence the collection of m for which the limit exists and has the given value is also measurable. Since the $G(F)$ are all in $\hat{\boldsymbol{B}}$, so is their countable intersection. □

Thus we can take $\hat{\boldsymbol{P}}$ to be $\boldsymbol{P}_s$ and observe that for any AMS measure $W(\boldsymbol{P}_s) = m(x: p_x \in \boldsymbol{P}_s) = 1$ from the ergodic decomposition.

The following theorem uses the previous results to provide an alternative and intuitive statement of the ergodic decomposition theorem.

Theorem 8.5.1. Fix a standard measurable space $(\Omega,\boldsymbol{B})$ and a transformation $T: \Omega \rightarrow \Omega$. Then there are a standard measurable space $(\Lambda,\boldsymbol{\Lambda})$, a family of stationary ergodic measures $\{m_\lambda;\ \lambda \in \Lambda\}$ on $(\Omega,\boldsymbol{B})$, and a mapping $\psi: \Omega \rightarrow \Lambda$ such that

(a) ψ is measurable and invariant, and

(b) if m is a stationary measure on $(\Omega,\boldsymbol{B})$ and W_ψ is the induced distribution; that is, $W_\psi(G) = m(\psi^{-1}(G))$ for $G \in \Lambda$ (which is well defined from (a)), then

$$m(F) = \int dm(x) m_{\psi(x)}(F) = \int dW_\psi(\lambda) m_\lambda(F)\ ,\ \text{all } F \in \boldsymbol{B}\ ,$$

and if $f \in L^1(m)$, then so is $\int f\, dm_\lambda$ $m\psi$-a.e. and

$$E_m f = \int dW_\psi(\lambda) E_{m_\lambda} f\ .$$

Finally, for any event F $m_\psi(F) = m(F \mid \psi)$, that is, given the ergodic decomposition and a stationary measure m, the ergodic component λ is a version of the conditional probability under m given $\psi = \lambda$.

Proof. Let $\boldsymbol{P}$ denote the space of all probability measures described in Lemma 8.2.2 along with the Borel sets generated by using the given distance and a generating field. Then from Lemma 8.2.1 the space is Polish. Let $\Lambda = \boldsymbol{P}_e$ denote the subset of $\boldsymbol{P}$ consisting of all stationary and ergodic measures. From Lemma 8.2.1 this set is measurable and hence is a Borel set. Theorem 3.3.2 then implies that Λ together with its Borel sets is standard. Define the mapping $\psi: \Omega \rightarrow \boldsymbol{P}$ as in Lemma 8.5.1. From Lemma 8.5.1 the mapping is measurable. Since ψ maps points in Ω into stationary and ergodic processes, it can be considered as a mapping $\psi: \Omega \rightarrow \Lambda$ that is also measurable since it is the restriction of a measurable mapping to a measurable set. Simply define $m_\lambda = \lambda$ (since each λ is a stationary ergodic process). The formulas for the probabilities and expectations then follow from the ergodic decomposition theorem and change of variable formulas. The conditional probability $m(F \mid \psi = \lambda)$ is defined by the relation

$$m(F \cap \{\psi \in G\}) = \int_G m(F \mid \psi = \lambda)\, dW_\psi(\lambda)\ ,$$

for all $G \in \boldsymbol{\Lambda}$. Since ψ is invariant, $\sigma(\psi)$ is contained in the σ-field of all invariant events. Combining the ergodic decomposition theorem and the ergodic theorem we have that $E_{p_x} f = E_m(f \mid \mathrm{I})$, where I is the σ-field of invariant events. (This follows since both equal $\lim n^{-1}\sum_i f T^i$.) Thus the left-hand side is

$$m(F \cap \psi^{-1}(G)) = \int_{\psi^{-1}(G)} p_x(F)\, dm(x)$$

$$\int_{\psi^{-1}(G)} m_{\psi(x)}(F)\, dm(x) = \int_G m_\lambda(F) dW_\psi(\lambda) ,$$

where the first equality follows using the conditional expectation of $f = 1_F$, and the second follows from a change of variables. □

The theorem can be viewed as simply a reindexing of the ergodic components and a description of a stationary measure as a weighted average of the ergodic components where the weighting is a measure on the indices and not on the original points. In addition, one can view the function ψ as a special random variable defined on the system or process that tells which ergodic component is in effect and hence view the ergodic component as the conditional probability with respect to the stationary measure given the ergodic component in effect. For this reason we will occasionally refer to the ψ of Theorem 8.5.1 as the *ergodic component function.*

8.6 THE ERGODIC DECOMPOSITION OF MARKOV PROCESSES

The ergodic decomposition has some rather surprising properties for certain special kinds of processes. Suppose we have a process $\{X_n\}$ with standard alphabet and distribution m. For the moment we allow the process to be either one-sided or two-sided. The process is said to be *kth order Markov* or *k-step Markov* if for all n and all $F \in \sigma(X_n, \cdots)$

$$m((X_n,X_{n+1}, \cdots) \in F \mid X_i ; i \le n-1)$$

$$= m((X_n,X_{n+1}, \cdots) \in F \mid X_i ; i = n-k, \cdots ,n-1) ;$$

that is, if the probability of a "future" event given the entire past depends only on the k most recent samples. When k is 0 the process is said to be memoryless. The following provides a test for whether or not a process is Markov.

Lemma 8.6.1. A process $\{X_n\}$ is k-step Markov if and only if the conditional probabilities $m((X_n,X_{n+1},\cdots)\in F \mid X_i\,;\, i\le n-1)$ are measurable with respect to $\sigma(X_{n-k},\cdots,X_{n-1})$, i.e., if for all F $m((X_n,X_{n+1},\cdots)\in F \mid X_i;\ i\le n-1)$ depends on $X_i;\ i\le n-1$ only through $X_{n-k},\cdots,X_{n-i}$.

Proof. By iterated conditional probabilities,

$$E(m((X_n,X_{n+1},\cdots)\in F \mid X_i\,;\, i\le n-1) \mid X_{n-k},\cdots,X_{n-1})$$
$$= m((X_n,X_{n+1},\cdots)\in F \mid X_i\,;\, i=n-k,\cdots,n-1)\,.$$

Since the spaces are standard, we can treat a conditional expectation as an ordinary expectation, that is, an expectation with respect to a regular conditional probability. Thus if the argument $\zeta = m((X_n,X_{n+1},\cdots)\in F \mid X_i\,;\, i\le n-1)$ is measurable with respect to $\sigma(X_{n-k},\cdots,X_{n-1})$, then from Corollary 5.9.2 the conditional expectation of ζ is ζ itself and hence the left-hand side equals the right-hand side, which in turn implies that the process is k-step Markov. This conclusion can also be drawn by using Lemma 5.9.5. □

If a k-step Markov process is stationary, then it has an ergodic decomposition. What do the ergodic components look like? In particular, must they also be Markov? First note that in the case of $k = 0$ the answer is trivial: If the process is memoryless and stationary, then it must be IID, and hence it is strongly mixing and hence it is ergodic. Thus the process puts all of its probability on a single ergodic component, the process itself. In other words, a stationary memoryless process produces with probability one an ergodic memoryless process. Markov processes exhibit a similar behavior: We shall show in this section that a stationary k-step Markov process puts probability one on a collection of ergodic k-step Markov processes having the same transition or conditional probabilities. The following lemma will provide an intuitive explanation.

Lemma 8.6.2. Suppose that $\{X_n\}$ is a two-sided stationary process with distribution m. Let $\{m_\lambda\,;\,\lambda\in\Lambda\}$ denote the ergodic decomposition for a standard space Λ of Theorem 8.5.1 and let ψ be the ergodic component function. Then the mapping ψ of Theorem 8.5.1 can be taken to be measurable with respect to $\sigma(X_{-1},X_{-2},\cdots)$. Furthermore,

$$m((X_0,X_1,\cdots)\in F \mid X_{-1},X_{-2},\cdots)$$
$$= m((X_0,X_1,\cdots)\in F \mid \psi,X_{-1},X_{-2},\cdots)$$

$$= m_{\psi}((X_0,X_1,\cdots) \in F \mid X_{-1},X_{-2},\cdots),$$

m-a.e.

Comment. The lemma states that we can determine the ergodic component of a two-sided process by looking at the past alone. Since the past determines the ergodic component, the conditional probability of the future given the past and the component is the same as the conditional probability given the past alone.

Proof. The ergodic component in effect can be determined by the limiting sample averages of a generating set of events. The mapping of Theorem 8.5.1 produced a stationary measure $\psi(x) = p_x$ specified by the relative frequencies of a countable number of generating sets computed from x. Since the finite-dimensional rectangles generate the complete σ-field and since the probabilities under stationary measures are not affected by shifting, it is enough to determine the probabilities of all events of the form $X_i \in F_i$; $i = 0, \cdots ,k-1$ for all one-dimensional generating sets F_i. Since the process is stationary, the shift T is invertible, and hence for any event F of this type we can determine its probability from the limiting sample average

$$\lim_{n\to\infty} \frac{1}{n} \sum_{i=k}^{n-1} 1_F(T^{-i}x)$$

(we require in the definition of generic sets that both these time averages as well as those with respect to T converge). Since these relative frequencies determine the ergodic component and depend only on $x_{-1},x_{-2},\cdots$, ψ has the properties claimed (and still satisfies the construction that was used to prove Theorem 8.6.1, we have simply constrained the sets considered and taken averages with respect to T^{-1}). Since ψ is measurable with respect to $\sigma(X_{-1},X_{-2},\cdots)$, the conditional probability of a future event given the infinite past is unchanged by also conditioning on ψ. Since conditioning on ψ specifies the ergodic component in effect, these conditional probabilities are the same as those computed using the ergodic component. (The final statement can be verified by plugging into the defining equations for conditional probabilities.)

We are now ready to return to k-step Markov processes. If m is two-sided and stationary and k-step Markov, then from the Markov property and the previous lemma

$$m((X_0, \cdots) \in F \mid X_{-1}, \cdots, X_{-k})$$
$$= m((X_0, \cdots) \in F \mid X_{-1}, X_{-2}, \cdots)$$
$$= m_\psi((X_0, \cdots) \in F \mid X_{-1}, X_{-2}, \cdots) .$$

But this says that $m_\psi((X_0, \cdots) \in F \mid X_{-1}, X_{-2}, \cdots)$ is measurable with respect to $\sigma(X_{-1}, \cdots, X_{-k})$ (since it is equal to a function that is), which implies from Lemma 8.6.1 that it is k-step Markov. We have now proved the following lemma for two-sided processes.

Lemma 8.6.3. If $\{X_n\}$ is a stationary Markov process and $\{m_\lambda; \lambda \in \Lambda\}$ is the ergodic decomposition, and ψ is the ergodic component function, then

$$m_\psi((X_n, \cdots) \in F \mid X_i; i \le n) = m((X_n, \cdots) \in F \mid X_{n-k}, \cdots, X_{n-1})$$

almost everywhere. Thus with probability one the ergodic components are also k-step Markov with the same transition probabilities.

Proof. For two-sided processes the statement follows from the previous argument and the stationarity of the measures. If the process is one-sided, then we can "embed" the process in a two-sided process with the same behavior. Given a one-sided process m, let p be the stationary two-sided process for which $p((X_k, \cdots, X_{k+n-1}) \in F) = m((X_0, \cdots, X_{n-1}) \in F)$ for all integers k and n dimensional sets F. This uniquely specifies the two-sided process. Similarly, given any stationary two-sided process there is a unique one-sided process assigning the same probabilities to finite-dimensional events. It is easily seen that the ergodic decomposition of the one-sided process yields that for the two-sided process. The one-sided process is Markov if and only if the two-sided process is and the transition probabilities are the same. Application of the two-sided result to the induced two-sided process then implies the one-sided result.

8.7 BARYCENTERS

Barycenters have some useful linearity and continuity properties. These are the subject of this section.

To begin recall from Section 8.3 and (8.4.1) that the barycentric mapping is a mapping of a measure $W \in \boldsymbol{P}(\hat{\boldsymbol{P}}, \hat{\boldsymbol{B}})$, the space of probability measures on the space $\hat{\boldsymbol{P}}$, into a measure $m \in \hat{\boldsymbol{P}}$, that is, m is a probability measure on the original space $(\Omega,\boldsymbol{B})$. The mapping is denoted by $W \rightarrow m$ or by $m = \int \alpha dW(\alpha)$.

First observe that if $W_1 \rightarrow m_1$ and $W_2 \rightarrow m_2$ and if $\lambda \in (0,1)$, then

$$\lambda W_1 + (1-\lambda)W_2 \rightarrow \lambda m_1 + (1-\lambda)m_2 ;$$

that is, the mapping $W \rightarrow m$ of a measure into its barycenter is *affine*. The following lemma provides a countable version of this property.

Lemma 8.7.1. If $W_i \rightarrow m_i$, $i = 1,2, \cdots$, and if

$$\sum_{i=1}^{\infty} a_i = 1$$

for a nonnegative sequence $\{a_i\}$, then

$$\sum_{i=1}^{\infty} a_i W_i \rightarrow \sum_{i=1}^{\infty} a_i m_i .$$

Comment. The given countable mixtures of probability measures are easily seen to be themselves probability measures. As an example of the construction, suppose that W is described by a probability mass function, that is, W places a probability a_i on a probability measure $\alpha_i \in \hat{P}$ for each i. We can then take the W_i of the lemma as the point masses placing probability 1 on α_i. Then $W_i \rightarrow \alpha_i$ and the mixture W of these point masses has the corresponding mixture $\sum \alpha_i$ as barycenter.

Proof. Call the weighted average of the W_i W and the weighted average of the m_i m. The assumptions imply that for all i

$$m_i(F) = \int E_\alpha 1_F \, d\, W_i(\alpha) \; ; \; F \in \boldsymbol{G} .$$

(That $E_\alpha 1_F$ is measurable was discussed prior to (8.4.2).) Thus to prove the lemma we must prove that

$$\sum_{i=1}^{\infty} a_i \int E_\alpha 1_F \, d\, W_i(\alpha) = \int E_\alpha 1_F \, dW(\alpha).$$

for all $F \in \boldsymbol{G}$. We have by construction that for any indicator function of an event $D \in \hat{\boldsymbol{B}}$ that

$$\sum_{i=1}^{\infty} a_i \int 1_D(\alpha) \, d\, W_i(\alpha) = \int 1_D dW .$$

Via the usual integration arguments this implies the corresponding equality for simple functions and then for bounded nonnegative measurable functions and hence for the $E_\alpha 1_F$.

The next lemma shows that if a measure W puts probability 1 on a closed ball in $\hat{\boldsymbol{P}}$, then the barycenter must be inside the ball.

Lemma 8.7.2. Given a closed ball $S = \{\alpha: d(\alpha,\beta) \le r\}$ in $\boldsymbol{P}((\Omega,\boldsymbol{B}))$ and a measure W such that $W(S) = 1$, then $W \rightarrow m$ implies that $m \in S$.

Proof. We have

$$d(m,\beta) = \sum_{i=1}^{\infty} 2^{-i} \mid m(F_i) - \beta(F_i) \mid =$$

$$\sum_{i=1}^{\infty} 2^{-i} \left| \int_S \alpha(F_i) dW(\alpha) - \beta(F_i) \right| \le \sum_{i=1}^{\infty} 2^{-i} \int_S \mid \alpha(F_i) - \beta(F_i) \mid d\,W(\alpha)$$

$$= \int_S d(\alpha,\beta) dW(\alpha) \le r ,$$

where the integral is pulled out of the sum via the dominated convergence theorem. This completes the proof. □

The final result of this section provides two continuity properties of the barycentric mapping.

Lemma 8.7.3. (a) Assume as earlier that the original measurable space $(\Omega,\boldsymbol{B})$ is standard. Form a metric space from the subset $\hat{\boldsymbol{P}}$ of all measures on $(\Omega,\boldsymbol{B})$ using a distributional metric $d = d_{\boldsymbol{W}}$ with $\boldsymbol{W}$ a generating field for $\boldsymbol{B}$. If $\hat{\boldsymbol{B}}$ is the resulting Borel field, then the Borel space $(\hat{\boldsymbol{P}},\hat{\boldsymbol{B}})$ is also standard. Given these definitions, the barycentric mapping $W \rightarrow m$ is continuous when considered as a mapping from the metric space $(\boldsymbol{P}(\hat{\boldsymbol{P}},\hat{\boldsymbol{B}}),d_{\boldsymbol{G}})$ to the metric space $(\hat{\boldsymbol{P}},d)$, where $d_{\boldsymbol{G}}$ is the metric based on the countable collection of sets $\boldsymbol{G} = \boldsymbol{F} \cup \boldsymbol{S}$, where $\boldsymbol{F}$ is a countable generating field of the standard sigma-field $\hat{\boldsymbol{B}}$ of subsets of $\hat{\boldsymbol{P}}$ and where $\boldsymbol{S}$ is the collection of all sets of the form

$$\{\alpha: \alpha(F) = E_\alpha 1_F \in [\frac{k}{n},\frac{k+1}{n})\}$$

for all F in the countable field that generates $\boldsymbol{B}$ and all $n = 1,2,\cdots$, and $k = 0,1,\cdots,n-1$. (These sets are all in $\hat{\boldsymbol{B}}$ since the functions defining them are measurable.)

(b) Given the preceding and a nonnegative bounded measurable function $D:\hat{\boldsymbol{P}} \to \mathbf{R}$ mapping probability measures into the real line, define the class $\boldsymbol{H}$ of subsets $\boldsymbol{P}$ as all sets of the form

$$\{\alpha: \frac{D(\alpha)}{D_{MAX}} \in [\frac{k}{n}, \frac{k+1}{n})\}$$

where D_{MAX} is the maximum absolute value of D, $n = 1,2, \cdots$, and $k = 0, \pm 1, \cdots, \pm n - 1$. Let $d_{\boldsymbol{G}}$ denote the metric on $\boldsymbol{P}((\hat{\boldsymbol{P}},\hat{\boldsymbol{B}}))$ induced by the class of sets $\boldsymbol{G} = \boldsymbol{F} \cup \boldsymbol{H}$, with $\boldsymbol{F}$ and $\boldsymbol{H}$ as before. Then $\int D(\alpha)\, d\, W(\alpha)$ is continuous with respect to $d_{\boldsymbol{G}}$; that is, if $d_{\boldsymbol{G}}(W_n,W) \underset{n\to\infty}{\to} 0$, then also

$$\lim_{n\to\infty} \int D(\alpha)dW_n(\alpha) = \int D(\alpha)dW(\alpha) .$$

Comment. We are simply adding enough sets to the metric to make sure that the various integrals have the required limits.

Proof. (a) Define the uniform quantizer

$$q_n(r) = \frac{k}{n} \quad r \in [\frac{k}{n}, \frac{k+1}{n}) .$$

Assume that $d_{\boldsymbol{G}}(W_N,W) \to 0$ and consider the integral

$$\begin{aligned}
| m_N(F) - m(F) | &= | \int E_\alpha 1_F \, d\, W_N(\alpha) - \int E_\alpha 1_F \, d\, W(\alpha) | \\
&\leq | \int E_\alpha 1_F \, d\, W_N(\alpha) - \int q_n(E_\alpha 1_F) \, d\, W_N(\alpha)| \\
&+ | \int q_n(E_\alpha 1_F) \, d\, W_N(\alpha) - \int q_n(E_\alpha 1_F) \, d\, W(\alpha)| \\
&+ | \int q_n(E_\alpha 1_F) \, d\, W(\alpha) - \int E_\alpha 1_F \, d\, W(\alpha) | .
\end{aligned}$$

The uniform quantizer provides a uniformly good approximation to within $1/n$ and hence given ε we can choose n so large that the leftmost and rightmost terms of the bound are each less than, say, $\varepsilon/3$ uniformly in N. The center term is the difference of the integrals of a fixed simple function. Since $\boldsymbol{S}$ was included in $\boldsymbol{G}$, taking $N\to\infty$ forces the probabilities under W_n of these events to match the probabilities under W and hence the finite weighted combinations of these probabilities goes to zero as $N\to\infty$. In particular N can be chosen large enough to ensure that this term is less than $\varepsilon/3$ for the given n. This proves (a). Part (b) follows in essentially the same way. □

8.8 AFFINE FUNCTIONS OF MEASURES

We now are prepared to develop the principal result of this chapter. Let $\hat{P} \subset \boldsymbol{P}((\Omega,\boldsymbol{B}))$ denote as before a convex closed subset of the space of all probability measures on a standard measurable space $(\Omega,\boldsymbol{B})$. Let d be the metric on this space induced by a standard generating field of subsets of Ω. A real-valued function (or functional) $D: \hat{\boldsymbol{P}} \to \mathrm{R}$ is said to be measurable if it is measurable with respect to the Borel field $\hat{\boldsymbol{B}}$, affine if for every $\alpha_1,\alpha_2 \in \hat{\boldsymbol{P}}$ and $\lambda \in (0,1)$ we have

$$D(\lambda\alpha_1 + (1-\lambda)\alpha_2) = \lambda D(\alpha_1) + (1-\lambda)D(\alpha_2) ,$$

and upper semicontinuous if

$$d(\alpha_n,\alpha) \underset{n\to\infty}{\to} 0 \Rightarrow D(\alpha) \geq \limsup_{n\to\infty} D(\alpha_n) .$$

If the inequality is reversed and the limit supremum replaced by a limit infimum, then D is said to be lower semicontinuous. An upper or lower semicontinuous function is measurable. We will usually assume that D is nonnegative, that is, that $D(m) \geq 0$ for all m.

Theorem 8.8.1. If $D: \hat{\boldsymbol{P}} \to \mathrm{R}$ is a nonnegative upper semicontinuous affine functional on a closed and convex set of measures on a standard space $(\Omega,\boldsymbol{B})$, if $D(\alpha)$ is W-integrable, and if m is the barycenter of W, then

$$D(m) = \int D(\alpha)\, d\, W(\alpha) .$$

Comments. The import of the theorem is that if a functional of probability measures is affine, then it can be expressed as an integral. This is a sort of converse to the usual implication that a functional expressible as an integral is affine.

Proof. We begin by considering countable mixtures. Suppose that W is a measure assigning probability to a countable number of measures e.g., W assigns probability a_i to α_i for $i = 1,2,, \cdots$. Since the barycenter of W is

$$m = \int \alpha dW(\alpha) = \sum_{i=1}^{\infty} \alpha_i a_i ,$$

and

$$D(\sum_{i=1}^{\infty} \alpha_i a_i) = D(m)$$

As in Lemma 8.7.1, we consider a slightly more general situation. Suppose that we have a sequence of measures $W_i \in \boldsymbol{P}((\hat{\boldsymbol{P}},\hat{\boldsymbol{B}}))$, the corresponding barycenters m_i, and a nonnegative sequence of real numbers a_i that sum to one. Let $W = \sum a_i W_i$ and $m = \sum a_i m_i$. From Lemma 8.7.1 $W \rightarrow m$. Define for each n

$$b_n = \sum_{i=n+1}^{\infty} a_i$$

and the probability measures

$$m^{(n)} = b_n^{-1} \sum_{i=n+1}^{\infty} a_i m_i ,$$

and hence we have a finite mixture

$$m = \sum_{i=1}^{n} m_i a_i + b_n m^{(n)} .$$

From the affine property

$$D(m) = D(\sum_{i=1}^{n} a_i m_i + b_n m^{(n)})$$

$$= \sum_{i=1}^{n} a_i D(m_i) + b_n D(m^{(n)}). \tag{8.8.1}$$

Since D is nonnegative, this implies that

$$D(m) \geq \sum_{i=1}^{\infty} D(m_i)a_i . \tag{8.8.2}$$

Making the W_i point masses as in the comment following Lemma 8.7.1 yields the fact that

$$D(m) \geq \sum_{i=1}^{\infty} D(\alpha_i)a_i . \tag{8.8.3}$$

Observe that if D were bounded, but not necessarily nonnegative, then (8.8.1) would imply the countable version of the theorem, that is, that

$$D(m) = \sum_{i=1}^{\infty} D(\alpha_i) a_i .$$

First fix M large and define $D_M(\alpha) = \min(D(\alpha),M)$. Note that D_M may not be affine (it is if D is bounded and M is larger than the maximum). D_M is clearly measurable, however, and hence we can apply the metric of part (b) of the previous lemma together with Lemma 8.2.1 to get a sequence of measures W_n consisting of finite sums of point masses (as in Lemma 8.2.1) with the property that if $W_n \to W$ and m_n is the barycenter of W_n, then also $m_n \to m$ (in d) and

$$\int D_M(\alpha)\, d\, W_n(\alpha) \underset{n\to\infty}{\to} \int D_M(\alpha)\, d\, W(\alpha)$$

as promised in the previous lemma. From upper semicontinuity

$$D(m) \geq \limsup_{n\to\infty} D(m_n) ,$$

and from (8.8.3)

$$D(m_n) \geq \int D(\alpha)\, d\, W_n(\alpha) .$$

Since $D(\alpha) \geq D_M(\alpha)$, therefore

$$D(m) \geq \limsup_{n\to\infty} \int D_M(\alpha) dW_n(\alpha) =$$

$$\int D_M(\alpha)\, d\, W(\alpha) .$$

Since $D_M(\alpha)$ is monotonically increasing to $D(\alpha)$, from the monotone convergence theorem

$$D(m) \geq \int D(\alpha) dW(\alpha) ,$$

which half proves the theorem.

For each $n = 1,2, \cdots$ carve up the separable (under d) space $\hat{\boldsymbol{P}}$ into a sequence of closed balls

$$S_{1/n}(\alpha_i) = \{\beta: d(\alpha_i, \beta) \leq 1/n\} \;;\; i = 1,2,, \cdots$$

Call the collection of all such sets $\boldsymbol{G} = \{G_i\,,\, i=1,2, \cdots\}$. We use this collection to construct a sequence of finite partitions $\boldsymbol{V}_n = \{V_i(n), i = 0,1,2, \cdots ,N_n\}$, $n=1,2, \ldots$. Let $\boldsymbol{V}_n$ be the collection of all nonempty intersection sets of the form

$$\bigcap_{i=1}^{n} G_i^* ,$$

where G_i^* is either G_i or G_i^c. For convenience we always take $V_0(n)$ as the set $\bigcap_{i=1}^{n} G_i^c$ of all points in the space not yet included in one of the first n G_i (assuming that the set is not empty). Note that $S_{1/1}(\alpha_i)$ is the whole set since d is bound above by one. Observe also that $\boldsymbol{V}_n$ has no more than 2^n atoms and that each $\boldsymbol{V}_n$ is a refinement of its predecessor. Finally note that for any fixed α and any $\varepsilon > 0$ there will always be for sufficiently large n an atom in $\boldsymbol{V}_n$ that contains α and that has diameter less than ε. If we denote by $V^n(\alpha)$ the atom of $\boldsymbol{V}_n$ that contains α (and one of them must), this means that

$$\lim_{n\to\infty} diam(V^n(\alpha)) = 0 \; ; \text{ all } \alpha \in \hat{\boldsymbol{P}} . \tag{8.8.4}$$

For each n, define for each $V \in \boldsymbol{V}_n$ for which $W(V) > 0$ the elementary conditional probability W^V; that is, $W^V(F) = W(V \cap F)/W(V)$. Break up W using these elementary conditional probabilities as

$$W = \sum_{V \in \boldsymbol{V}_n, W(V) > 0} W(V) W^V .$$

For each such V let m_V denote the barycenter of W^V, that is, the mixture of the measures in V with respect to the measure W^V that gives probability one to V. Since W has barycenter m and since it has the finite mixture form given earlier, Lemma 8.7.1 implies that

$$m = \sum_{V \in \boldsymbol{V}_n,\, W(V) > 0} m_V W(V) \; ;$$

that is, for each n we can express W as a finite mixture of measures W^V for $V \in \boldsymbol{V}_n$ and m will be the corresponding mixture of the barycenters m_V.

Thus since D is affine

$$D(m) = \sum_{V \in \boldsymbol{V}_n, W(V) > 0} W(V) D(m_V) .$$

Define $D_n(\alpha)$ as $D(m_V)$ if $\alpha \in V$ for $V \in \boldsymbol{V}_n$. The preceding equation then becomes

$$D(m) = \int D_n(\alpha) dW(\alpha) . \tag{8.8.5}$$

Recall that $V^n(\alpha)$ is the set in $\boldsymbol{V}_n$ that contains α. Let G denote those α for which $W(V^n(\alpha)) > 0$ for all n. We have $W(G) = 1$ since at each stage we can remove any atoms having zero probability. The union of all such atoms is countable and hence also has zero probability.

For $\alpha \in G$ given ε we have as in (8.8.4) that we can choose N large enough to ensure that for all $n > N$ α is in an atom $V \in \boldsymbol{V}_n$ that is contained in a closed ball of diameter less than ε. Since the measure W^V on V can also be considered as a measure on the closed ball containing V, Lemma 8.7.2 implies that the barycenter of V must be within the closed ball and hence also

$$d(\alpha, m_{V^n(\alpha)}) \leq \varepsilon \; ; \; n \geq N ,$$

and hence the given sequence of barycenters $m_{V^n(\alpha)}$ converges to α. From upper semicontinuity this implies that

$$D(\alpha) \geq \limsup_{n\to\infty} D(m_{V^n(\alpha)}) = \limsup_{n\to\infty} D_n(\alpha).$$

Since the $D_n(\alpha)$ are nonnegative and $D(\alpha)$ is W-integrable, this means that the $D_n(\alpha)$ are uniformly integrable. To see this, observe that the preceding equation implies that given $\varepsilon > 0$ there is an N such that

$$D_n(\alpha) \leq D(\alpha) + \varepsilon \; ; \; n \geq N ,$$

and hence for all n

$$D_n(\alpha) \leq \sum_{i=1}^{N} D_i(\alpha) + D(\alpha) + \varepsilon ,$$

which from Lemma 4.4.4 implies the D_n are uniformly integrable since the right-hand side is integrable by (8.8.5) and the integrability of $D(\alpha)$. Since the sequence is uniformly integrable, we can apply Lemma 4.4.5 and (8.8.5) to write

$$\int D(\alpha)\, dW(\alpha) \geq \int (\limsup_{n\to\infty} D_n(\alpha))\, dW(\alpha)$$

$$\geq \limsup_{n\to\infty} \int D_n(\alpha)\, dW(\alpha) = D(m) ,$$

which completes the proof of the theorem. □

8.9 THE ERGODIC DECOMPOSITION OF AFFINE FUNCTIONALS

Now assume that we have a standard measurable space $(\Omega,\boldsymbol{B})$ and a measurable transformation $T: \Omega \to \Omega$. Define $\hat{P} = P_s$. From Lemma 8.5.2 this is a closed and convex subset of $\boldsymbol{P}((\Omega,\boldsymbol{B}))$, and hence Theorem 8.8.1 applies. Thus we have immediately the following result:

Theorem 8.9.1. Let $(\Omega,\boldsymbol{B},m,T)$ be a dynamical system with a standard measurable space $(\Omega,\boldsymbol{B})$ and an AMS measure m with stationary mean $\overline{m}$. Let $\{p_x \,;\, x \in \Omega,\}$ denote the ergodic decomposition. Let D: $\boldsymbol{P}_s \to \mathrm{R}$ be a nonnegative functional defined for all stationary measures with the following properties:

(a) $D(p_x)$ is $\overline{m}$-integrable.

(b) D is affine.

(c) D is an upper semicontinuous function; that is, if P_n is a sequence of stationary measures converging to a stationary measure P in the sense that $P_n(F) \to P(F)$ for all F in a standard generating field for $\boldsymbol{B}$, then

$$D(P) \geq \limsup_{n\to\infty} D(P_n) .$$

Then

$$D(\overline{m}) = \int D(p_x)\, dm(x) .$$

Comment: The functional is required to have the given properties only for stationary measures since we only require its value in the preceding expression for the stationary mean (the barycenter) and for the stationary and ergodic components. We cannot use the previous techniques to evaluate D of the AMS process m because we have not shown that it has a representation as a mixture of ergodic components. (It is easy to show when the transformation T is invertible and the measure is dominated by its stationary mean.) We can at least partially relate such functionals for the AMS measure and its stationary mean since

$$n^{-1} \sum_{i=0}^{n-1} mT^{-i}(F) \underset{n\to\infty}{\to} \overline{m}(F) ,$$

and hence the affine and upper semicontinuity properties imply that

$$\lim_{n\to\infty} n^{-1} \sum_{i=0}^{n-1} D(mT^{-i}) \leq D(\overline{m}) . \tag{8.9.1}$$

The preceding development is patterned after that of Jacobs [8] for the case of bounded (but not necessarily nonnegative) D except that here the distribution metric is used and there weak convergence is used.

REFERENCES

1. P. Billingsley, *Convergence of Probability Measures,* Wiley Series in Probability and Mathematical Statistics, Wiley, New York, 1968.
2. P. Billingsley, *Weak Convergence of Measures: Applications in Probability,* SIAM, Philadelphia, 1971.
3. R. Dobrushin. private correspondence
4. R. M. Dudley, ''Distances of probability measures and random variables,'' *Ann. Math. Statist.*, vol. 39, pp. 1563-1572, 1968.
5. R. M. Gray and L. D. Davisson, *Ergodic and Information Theory,* Benchmark Papers in Electrical Engineering and Computer Science, 19, Dowden, Hutchninson, & Ross, Stroudsbug PA, 1977.
6. R. M. Gray, D. L. Neuhoff, and P. C. Shields, ''A generalization of Ornstein's d-bar distance with applications to information theory,'' *Annals of Probability*, vol. 3, pp. 315-328, April 1975.
7. F. R. Hample, ''A general qualitative definition of robustness,'' *Ann. Math. Statist.*, vol. 42, pp. 1887-1896, 1971.
8. K. Jacobs, ''The ergodic decomposition of the Kolmogorov-Sinai invariant,'' in *Ergodic Theory*, ed. F. B. Wright, Academic Press, New York, 1963.
9. V. I. Levenshtein, ''Binary codes capable of correcting deletions, insertions, and reversals,'' *Sov. Phys. -Dokl.*, vol. 10, pp. 707-710, 1966.
10. J. Moser, E. Phillips, and S. Varadhan, *Ergodic Theory: A Seminar,* Courant Institute of Math. Sciences, New York, 1975.
11. D. Ornstein, ''An application of ergodic theory to probability theory,'' *Ann. Probability*, vol. 1, pp. 43-58, 1973.
12. D. Ornstein, *Ergodic Theory, Randomness, and Dynamical Systems,* Yale University Press, New Haven, 1975.
13. P. Papantoni-Kazakos and R. M. Gray, ''Robustness of estimators on stationary observations,'' *Annals of Probability*, vol. 7, pp. 989-1002, Dec. 1979.

14. V. Strassen, ''The existence of probability measures with given marginals,'' *Ann. Math. Statist.*, vol. 36, pp. 423-429, 1965.

15. S. S. Vallender, ''Computing the Wasserstein distance between probability distributions on the line,'' *Theor. Probability Appl.*, vol. 18, pp. 824-827, 1973.

16. L. N. Vasershtein, ''Markov processes on countable product space describing large systems of automata,'' *Problemy Peredachi Informatsii*, vol. 5, pp. 64-73, 1969.

BIBLIOGRAPHY

1. R.B. Ash, *Real Analysis and Probability,* Academic Press, New York, 1972.
2. P. Billingsley, *Ergodic Theory and Information,* Wiley, New York, 1965.
3. P. Billingsley, *Convergence of Probability Measures,* Wiley Series in Probability and Mathematical Statistics, New York, 1968.
4. P. Billingsley, *Weak Convergence of Measures: Applications in Probability,* SIAM, Philadelphia, 1971.
5. O.J. Bjornsson, "A note on the characterization of standard borel spaces," *Math. Scand.*, vol. 47, pp. 135-136, 1980.
6. D. Blackwell, "On a class of probability spaces," *Proc. 3rd Berkeley Symposium on Math. Sci. and Prob.*, vol. II, pp. 1-6, Univ. California Press, Berkeley, 1956.
7. N. Bourbaki, *Elements de mathematique, Livere VI, Integration,* Hermann, Paris, 1956-1965.
8. L. Breiman, *Probability,* Addison-Wesley, Menlo Park, Calif., 1968.
9. J.C. Candy, "A use of limit cycle oscillations to obtain robust analog-to-digital converters," *IEEE Transactions on Communications*, vol. COM-22, pp. 298-305, March 1974.
10. J. P. R. Christensen, *Topology and Borel Structure,* Mathematics Studeies 10, North-Holland/American Eslevier, New York, 1974.
11. K.L. Chung, *A Course in Probability Theory,* Academic Press, New York, 1974.
12. D.C. Cohn, *Measure Theory,* Birkhauser, New York, 1980.
13. H. Cramér, *Mathematical Methods of Statistics,* Princeton University Press, Princeton, NJ, 1946.

14. M. Denker, C. Grillenberger, and K. Sigmund, *Ergodic Theory on Compact Spaces,* Lecture Notes in Mathematics, 57, Springer-Verlag, New York, 1970.

15. J.L. Doob, *Stochastic Processes,* Wiley, New York, 1953.

16. Y. N. Dowker, ''Finite and sigma-finite invariant measures,'' *Annals of Mathematics*, vol. 54, pp. 595-608, November 1951.

17. Y. N. Dowker, ''On measurable transformations in finite measure spaces,'' *Annals of Mathematics*, vol. 62, pp. 504-516, November 1955.

18. R.M. Dudley, ''Distances of probability measures and random variables,'' *Ann. Math. Statist.*, vol. 39, pp. 1563-1572, 1968.

19. N.A. Friedman, *Introduction to Ergodic Theory,* Van Nostrand Reinhold Company, New York, 1970.

20. A. Garcia, ''A simple proof of E. Hoph's maximal ergodic theorem,'' *J. Math. Mech.*, vol. 14, pp. 381-2, 1965.

21. R.M. Gray, ''Oversampled Sigma-Delta Modulation,'' *IEEE Transactions on Communications*, vol. COM-35, pp. 481-489, April 1987.

22. R. M. Gray and L.D. Davisson, ''The ergodic decomposition of discrete stationary random processes,'' *IEEE Trans. on Info. Theory*, vol. IT-20, pp. 625-636, September 1974.

23. R.M. Gray and L.D. Davisson, *Ergodic and Information Theory,* Benchmark Papers in Electrical Engineering and Computer Science, 19, Dowden, Hutchninson, & Ross, Stroudsbug PA, 1977.

24. R.M. Gray and L.D. Davisson, *Random Processes: A Mathematical Approach for Engineers,* Prentice Hall, New Jersey, 1986.

25. R.M. Gray and J.C. Kieffer, ''Asymptotically mean stationary measures,'' *Annals of Probability*, vol. 8, pp. 962-973, Oct. 1980.

26. R. M. Gray, D. L. Neuhoff, and P. C. Shields (1975), , ''A generalization of Ornstein's d-bar distance with applications to information theory,'' *Annals of Probability*, vol. 3, , pp. 315-328, April 1975.

27. P.R. Halmos, ''Invariant measures,'' *Ann. of Math.*, vol. 48, pp. 735-754, 1947.

28. P.R. Halmos, *Measure Theory,* Van Nostrand Reinhold, New York, 1950.

29. P.R. Halmos, *Lectures on Ergodic Theory,* Chelsea, New York, 1956.

30. P.R. Halmos, *Introduction to Hilbert Space,* Chelsea, New York, 1957.

31. J.M. Hammersley and D.J.A. Welsh, "First-passage, percolation, subadditive processes, stochastic networks, and generalized renewal theory," *Bernoulli-Bayes-Laplace Anniversary Volume*, Springer, Berlin, 1965.

32. F.R. Hample, "A general qualitative definition of robustness," *Ann. Math. Statist.*, vol. 42, pp. 1887-1896, 1971.

33. E. Hoph, *Ergodentheorie,* Springer-Verlag, Berlin, 1937.

34. H. Inose and Y. Yasuda, "A unity bit coding method by negative feedback," *Proceedings of the IEEE,* vol. 51, pp. 1524-1535., November 1963.

35. K. Jacobs, "The ergodic decomposition of the Kolmogorov-Sinai invariant," in *Ergodic Theory,* ed. F.B. Wright, Academic Press, New York, 1963.

36. R. Jones, "New proof for the maximal ergodic theorem and the Hardy-Littlewood maximal inequality," *Proc. AMS*, vol. 87, pp. 681-684, 1983.

37. I. Katznelson and B. Weiss, "A simple proof of some ergodic theorems," *Israel Journal of Mathematics*, vol. 42, pp. 291-296, 1982.

38. J.F.C. Kingman, "The ergodic theory of subadditive stochastic processes," *Ann. Probab.*, vol. 1, pp. 883-909, 1973.

39. A.N. Kolmogorov, *Foundations of the Theory of Probability,* Chelsea, New York, 1950. (Translated from the German).

40. U. Krengel, *Ergodic Theorems,* De Gruyter Series in Mathematics, De Gruyter, New York, 1985.

41. N. Kryloff and N. Bogoliouboff, "La théorie générale de la mesure dans son application à l'étude des systèmes de la mecanique non linéaire," *Ann. of Math.*, vol. 38, pp. 65-113, 1937.

42. V.I. Levenshtein, "Binary codes capable of correcting deletions, insertaions, and reversals," *Sov. Phys.-Dokl.*, vol. 10, pp. 707-710, 1966.

43. M. Loève, *Probability Theory,* D. Van Nostrand, Princeton, N.J., 1963. Third Edition.

44. D. G. Luenberger, *Optimization by Vector Space Methods,* Wiley, New York, 1969.

45. G. Mackey, ''Borel structures in groups and their duals,'' *Trans. Am. Math. Soc.*, vol. 85, pp. 134-165, 1957.

46. L.E. Maistrov, *Probability Theory: A Historical Sketch,* Academic Press, New York, 1974. Translated by S. Kotz.

47. J. Moser, E. Phillips, and S. Varadhan, *Ergodic Theory: A Seminar,* Courant Institute of Math. Sciences, New York, 1975.

48. J. Nedoma, ''On the ergodicity and r-ergodicity of stationary probability measures,'' *Zeitschrift Wahrscheinlichkeitstheorie*, vol. 2, pp. 90-97, 1963.

49. D. Ornstein, ''An application of ergodic theory to probability theory,'' *Ann. Probability*, vol. 1, pp. 43-58, 1973.

50. D. Ornstein, *Ergodic Theory, Randomness, and Dynamical Systems,* Yale University Press, New Haven, 1975.

51. D. Ornstein and B. Weiss, ''The Shannon-McMillan-Breiman theorem for a class of amenable groups,'' *Israel J. of Math*, vol. 44, pp. 53-60, 1983.

52. J.C. Oxtoby, ''Ergodic Sets,'' *Bull. Amer. Math. Soc.*, vol. Volume 58, pp. 116-136, 1952.

53. P. Papantoni-Kazakos and R.M. Gray, ''Robustness of estimators on stationary observations,'' *Annals of Probability*, vol. 7, pp. 989-1002, Dec. 1979.

54. K.R. Parthasarathy, *Probability Measures on Metric Spaces,* Academic Press, New York, 1967.

55. I. Paz, *Stochastic Automata Theory,* Academic Press, New York, 1971.

56. K. Petersen, ''On a series of cosecants related to a problem in ergodic theory,'' *Compositio Mathematica*, vol. 26, pp. 313-317, 1973.

57. K. Petersen, *Ergodic Theory,* Cambridge University Press, Cambridge, 1983.

58. H. Poincaré, *Les méthodes nouvelles de la méecanique céleste,* I, II, III, Gauthiers-Villars, Paris, 1892,1893,1899. Also Dover, New York, 1957.

59. O.W. Rechard, ''Invariant measures for many-one transformations,'' *Duke J. Math.*, vol. 23, pp. 477-488, 1956.

60. V.A. Rohlin, ''Selected topics from the metric theory of dynamical systems,'' *Uspechi Mat. Nauk.*, vol. 4, pp. 57-120, 1949. AMS Trans. (2) 49

61. W. Rudin, *Principles of Mathematical Analysis,* McGraw-Hill, New York, 1964.

62. L. Schwartz, *Radon Measures on Arbitrary Topological Spaces and Cylindrical Measures,* Oxford University Press, Oxford, 1973.

63. P. Shields, *The Theory of Bernoulli Shifts,* The University of Chicago Press, Chicago, IL, 1973.

64. P. Shields, ''The ergodic and entropy theorems revisited,'' *IEEE Transactions on Information Theory*, 1987.

65. G.F. Simmons, *Introduction to Topology and Modern Analysis,* McGraw-Hill, New York, 1963.

66. Ya. G. Sinai, *Introduction to Ergodic Theory,* Mathematical Notes, Princeton University Press, Princeton, 1976.

67. V. Strassen, ''The existence of probability measures with given marginals,'' *Ann. Math. Statist.*, vol. 36, pp. 423-429, 1965.

68. E. Tanaka and T. Kasai, ''Synchronization and subsititution error correcting codes for the Levenshtein metric,'' *IEEE Transactions on Information Theory*, vol. IT-22, pp. 156-162, 1976.

69. S.S. Vallender, ''Computing the Wasserstein distance between probability distributions on the line,'' *Theor. Probability Appl.*, vol. 18, pp. 824-827, 1973.

70. L.N. Vasershtein, ''Markov processes on countable product space describing large systems of automata,'' *Problemy Peredachi Informatsii*, vol. 5, pp. 64-73, 1969.

71. P. Walters, *Ergodic Theory-Introductory Lectures,* Lecture Notes in Mathematics No. 458, Springer-Verlag, New York, 1975.

72. F.B. Wright, ''The recurrence theorem,'' *Amer. Math. Monthly*, vol. 68, pp. 247-248, 1961.

INDEX